THE MOVING STATUES OF SEVENTEENTH-CENTURY AMSTERDAM

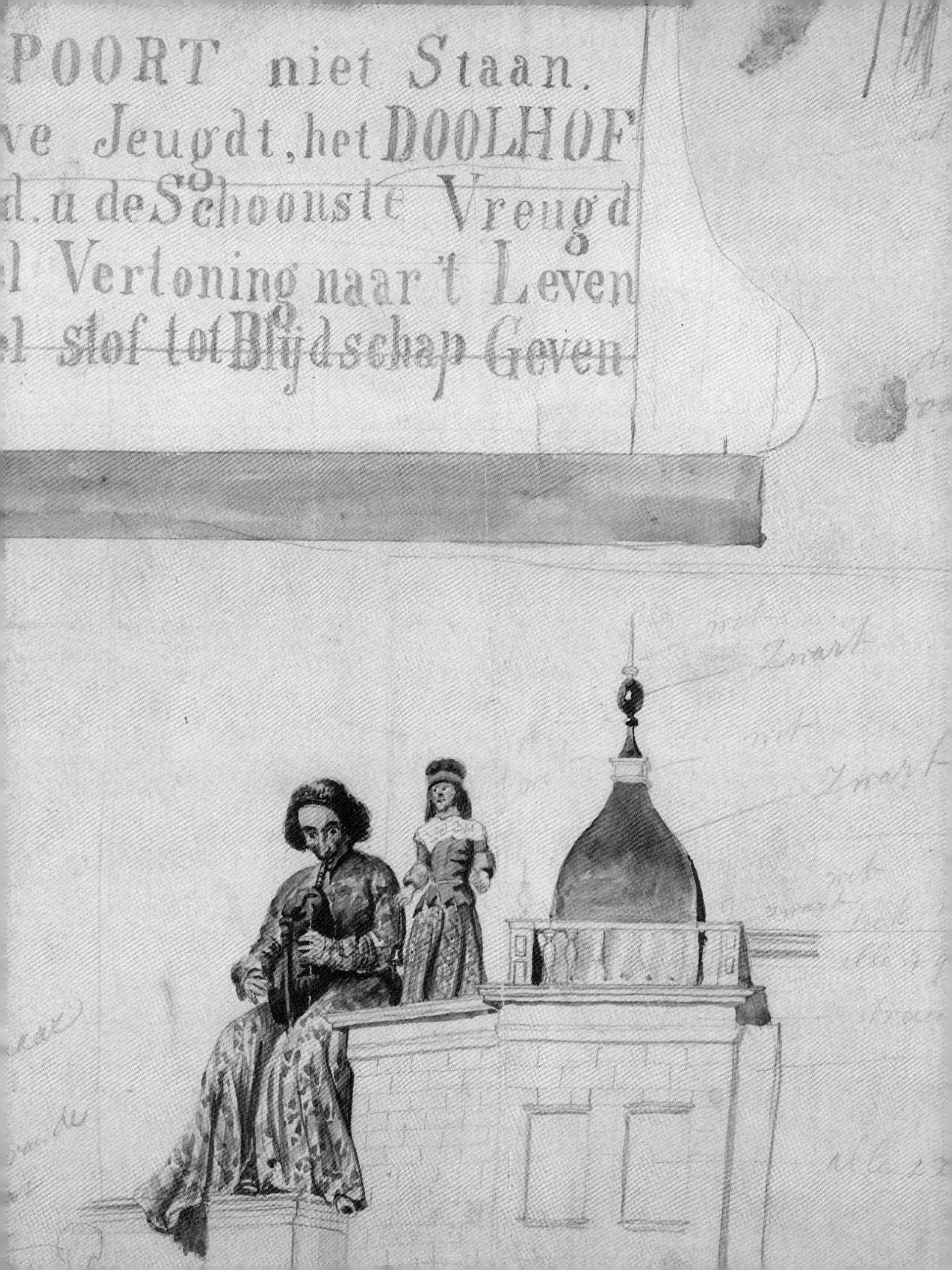

POORT niet Staan.
ve Jeugdt, het DOOLHOF
d. u de Schoonste Vreugd
l Vertoning naar 't Leven
l stof tot Blydschap Geven
wit
Zwart
wit
Zwart

ANGELA VANHAELEN

The Moving Statues of Seventeenth-Century Amsterdam

Automata, Waxworks, Fountains, Labyrinths

The Pennsylvania State University Press | University Park, Pennsylvania

Publication of this book has been aided by funds from the Social Sciences and Humanities Research Council of Canada.

Names: Vanhaelen, Angela, author.
Title: The moving statues of seventeenth-century Amsterdam : automata, waxworks, fountains, labyrinths / Angela Vanhaelen.
Description: University Park, Pennsylvania : The Pennsylvania State University Press, [2022] | Includes bibliographical references and index.
Summary: "Explores the visual and cultural history of Amsterdam in the early modern era, focusing on the doolhoven: winding mazes behind pubs and taverns that featured pleasure gardens, waterworks, wax galleries, and automata"—Provided by publisher.
Identifiers: LCCN 2021027948 | ISBN 9780271091402 (hardback)
Subjects: LCSH: Labyrinths—Netherlands—Amsterdam—History—17th century. | Robots—Netherlands—Amsterdam—History—17th century. | Wax figures—Netherlands—Amsterdam—History—17th century. | Protestantism in art. | Fountains—Netherlands—Amsterdam—History—17th century.
Classification: LCC SB475 .V36 2022 | DDC 629.8/9209492352—dc23
LC record available at https://lccn.loc.gov/2021027948

Printed in Lithuania by BALTO print
Published by The Pennsylvania State University Press, University Park, PA 16802–1003

The Pennsylvania State University Press is a member of the Association of University Presses.

It is the policy of The Pennsylvania State University Press to use acid-free paper. Publications on uncoated stock satisfy the minimum requirements of American National Standard for Information Sciences—Permanence of Paper for Printed Library Material, ANSI Z39.48–1992.

Additional credits: frontispiece, J. G. L. Rieke, *Oude Doolhof, Diverse Sketches*, ca. 1861 (plate 5), detail; 1, Anonymous drawing, *Fountains in the Oude Doolhof*, ca. 1622 (fig. 3), detail; 31, Anonymous, *Map of Amsterdam, 1612*, from *Civitates Orbis Terrarum* (1617–18) (fig. 24), detail; 75, *David and Goliath*, from *Verklaringe van Verscheyden Kunst-Rijcke Wercken en haer Beweginghe* (1657) (fig. 26), detail; 119, *Hessel Gerritsz, Nieuwe Doolhof on Looiersgracht*, ca. 1632 (fig. 44), detail.

La métamorphose est liée au labyrinthe.

—Michel Foucault, *Death and the Labyrinth*, 1963

CONTENTS

ILLUSTRATIONS

Color Plates (*following page 68*)

Figures

ACKNOWLEDGMENTS

The research and writing of this book were supported by a series of research grants from the Social Sciences and Humanities Research Council of Canada. The project also benefited from time spent as a Guest Scholar at the Getty Research Institute in 2018 and as a Clark Professor at the UCLA Center for 17th- & 18th-Century Studies & William Andrews Clark Memorial Library from 2018 to 2019. Sincere thanks for the opportunities generated by these institutions. In Montreal, I appreciate the support of a vibrant academic community, which includes colleagues in the Department of Art History and Communication Studies, team members of the collaborative research initiative Early Modern Conversions, and participants in Le Séminaire des nouveaux modernes. I have benefited from the insights and research assistantship of numerous graduate students, including Tomasz Grusiecki, Braden Scott, Isabelle Masse, Zifeng Zhao, Peter Boudreau, Lindsay Corbett, and Saraphina Masters. Special thanks to Bronwen Wilson for our ongoing research collaborations and conversations, especially the *Making Worlds* initiative. I truly appreciate the guidance offered by Ellie Goodman and Maddie Caso at Penn State University Press, the insightful suggestions of the peer reviewers, and the careful editing of Annika Fisher. Parts of this book have previously been published as "Strange Things for Strangers: Automata and Transcultural Encounters in Early Modern Amsterdam," *Art Bulletin* 103, no. 3 (September 2021), and I thank the editors and readers for their incisive feedback. Above all, thank you to my husband, Ray Geurkink, for his unfailing support and to our daughters, Hannah and Ella, who are a source of inspiration.

INTRODUCTION

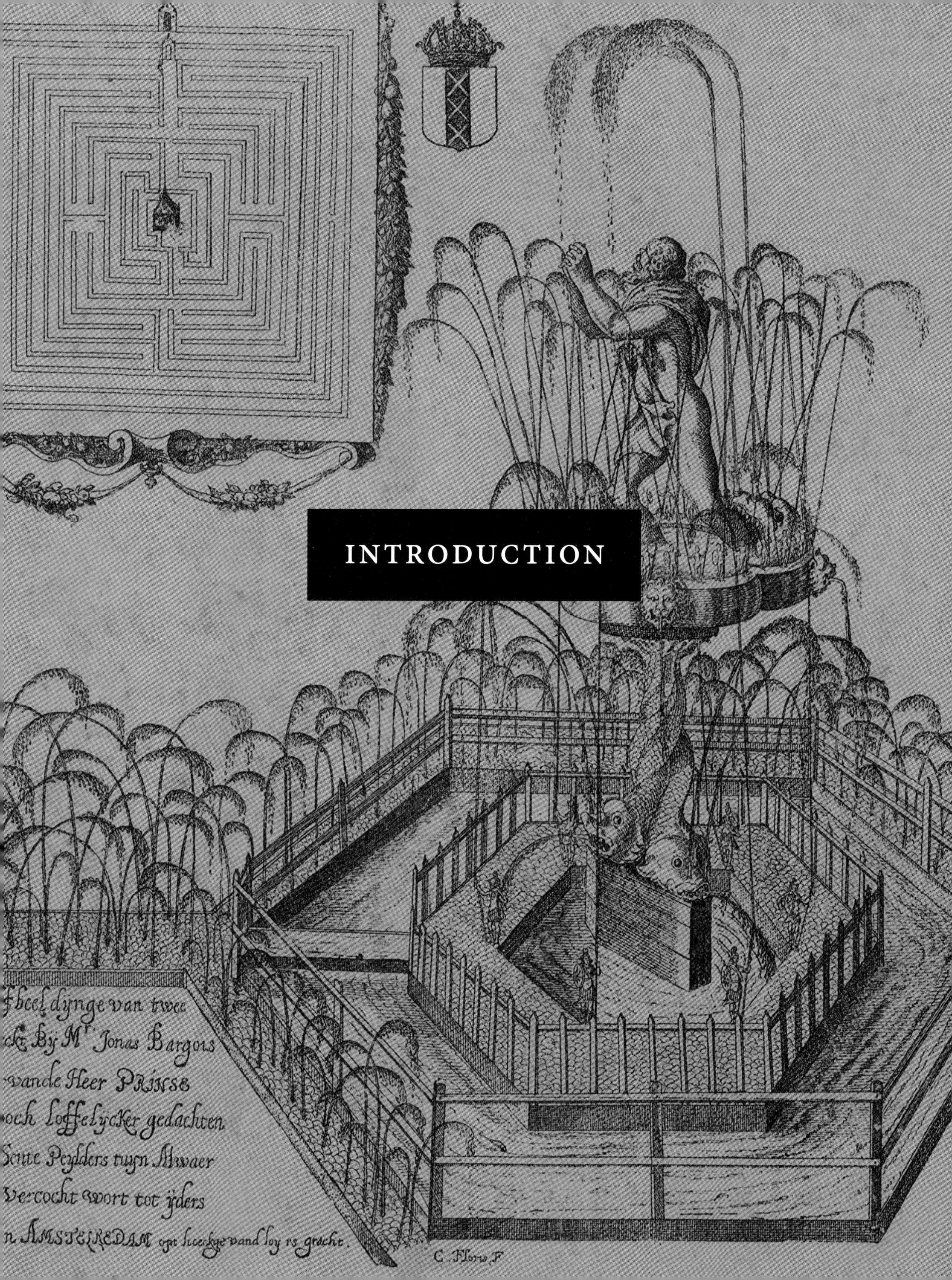

THE CLOSED DOOR

Walking In

Chapter 1

In brief, you find all things suddenly reversed,
when you open up the Silenus.
—Erasmus, *The Praise of Folly*, 1511

The Moving Statues of Seventeenth-Century Amsterdam is a study of idiosyncratic attractions called *Doolhoven*, or labyrinths, which were unique to early modern Amsterdam. These licensed public houses offered guest lodgings and also tendered alcoholic refreshments and the enjoyment of artful recreations in their gardens and galleries. While early modern curiosity cabinets and private collections have received much scholarly attention as precursors to the museum, less is known about how inns and taverns were used as exhibition spaces. Seventeenth-century guidebooks to the Doolhof sites outline a prescribed route that transported visitors into the realm of ritual.[1] From the street, a green door led into a treed tavern yard, where travelers and locals could enjoy a drink as well as the play of a spectacular figural fountain of the triumph of Bacchus; the intoxicating powers of the pagan god flowed both from the libations and the hydraulics. A small admission fee was charged to enter the actual doolhof, an intricate hedge maze with forking and twisting paths designed to confound. Within this mythic structure, pleasures and constrictions were experienced with equal intensity. If solved, the labyrinth delivered peripatetic sightseers into a sculpture gallery housing curious creations billed as both ancient and novel. There, the performances of animated mechanical statues struck spectators with wonder, arousing curiosity about ingenious inventions. Equally astonishing were the life-sized wax portraits of European rulers and notorious persons, which vividly channeled the lively presences of significant historical personages into the here and now, denoting the present moment as a critical age. Visitors also were confronted with intricate clocks and astronomical apparatuses, devices that demanded decisive action in revolutionary times as their moving mechanisms inexorably ushered in the end of the world. Distinctive and renowned in the seventeenth century, these innovative exhibition venues have since fallen into disregard and call for further investigation. Focusing on the peculiar combination of fountains, labyrinths, automata, and waxworks that characterize these sites, *Moving Statues* is structured as a walk through the Doolhof gardens, where drinking, entertainment, and the acquisition of knowledge mingled in emotive encounters with lively displays of animated artifacts.

Rough Exteriors

This introductory chapter begins at the entrance to the site by considering nineteenth-century imagery of the Doolhof portal. These images of the front door serve as an opening to my discussion of the historiography of these amusement parks and the methodologies that they provoke. Our starting point is a rough pencil sketch of a closed door, which is preserved in the Amsterdam City Archives (fig. 1). At first glance, it is an unpromising beginning, as the most remarkable thing about this work appears to be the fact that the artist stopped to make it in the first place. There is nothing arresting about the door. The drawing looks hasty and inept, as if the sketcher disparaged his own efforts. The edges of architectural elements—doorframe, window, pilaster, lintel, and transom—are indicated with wavering and broken strokes of the pencil. There are some particularly maladroit passages: the left side of the window frame is narrow at bottom and bulges out at top, and the delineation of the outer edge of the left pilaster stops midway, leaving a blank area that skews the whole picture. This awkward rendering, together with the long inward-leaning line that runs down the left of the sheet, gives the door and its minimal surrounds a precarious slant. The area of interface between portal and street is filled in with horizontal strokes underlined by a series of parallel lines. These lines might indicate the planes of steps, but the parallel hatching extends too far at the left, confusing the area where we expect to see a post and railing matching the one at right. The drawing is also marred by a trailing line that runs unevenly over its surface: it cuts across the bottom of the left pilaster, over the threshold, overlaps the side and front of the base of the right pilaster, and then drops and cuts across a corner of the footing of the railing post. This meandering stroke looks like a trace of the pencil in motion, as if it had skidded or accidentally was dropped or dragged across the paper. The scrabbling line trails down to the bottom of the sheet and ends by crossing the letter *t* in one of the words scrawled there. The phrase is difficult to decipher, for the artist's handwriting, like his drawing, verges on the illegible. The word crossed by the scribble line might spell *muurtrans* (battlement) or misspell *illustrum* (illustrious). The bottom word is easier to decode: *doolhof* (labyrinth).

This rough sketch does not showcase the fact that its maker, J. G. L. Rieke (1817–1898), was a distinguished antiquarian and skilled draftsman. Rieke's oeuvre includes numerous prints, drawings,

FIG. 1 J. G. L. Rieke, *Oude Doolhof, Entrance Door on Prinsengracht 338*, 1862. Photo: Amsterdam City Archives (010097007130).

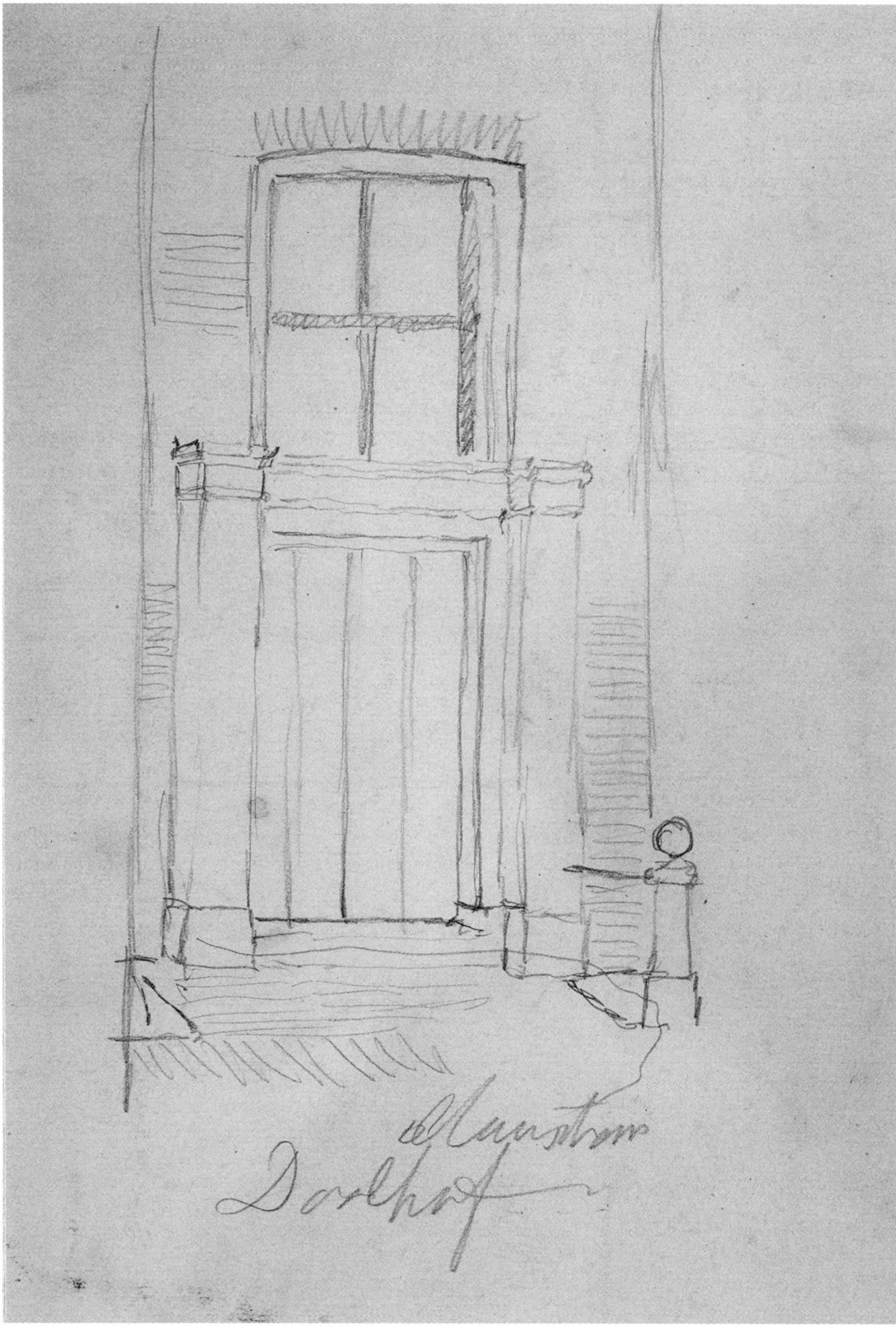

watercolors, and lithographs of notable sites in and around Amsterdam.[2] These works are characterized by a meticulous approach to the precise perspectival rendering of buildings, demonstrating the sure technique of an artist trained in architectural drawing. If Rieke was dissatisfied with his clumsy Doolhof door drawing, he undoubtedly had the ability to better it.

The archive surprisingly contains more sketches of this door. An ink-and-watercolor drawing by Amsterdam schoolmaster Jan ter Gouw (1814–1894) depicts the same unexceptional entrance (color plate 1). Ter Gouw, like Rieke, was an antiquarian, and his book *De Uithangteekens*, coauthored with Jacob van Lennep in the late 1860s, still serves as a rich source of information about old Dutch street signs.[3] Ter Gouw dedicated himself to the study of Amsterdam's doors—he knew their histories and connected them to a wide range of civic and cultural traditions. Nevertheless, his sketch, like Rieke's, seems awkwardly rendered to convey the uninviting ugliness of the Doolhof entryway. Its coarse wooden boards, painted a muddy bluish green, are scratched and worn. These marks of wear and tear register the passage of time and people, indicating centuries of use and the damage done by countless feet and hands. A graffiti stick figure scrawled on the lower part of the left pilaster is another type of mark made by the anonymous person in the street, as is the mild act of defacement registered by the ripped remnants of a printed broadside that has been torn from the right pilaster. All that remains of this notice is what looks to be part of a letter *D*. A similarly sized, intact advertisement appears on the left pilaster, inexpertly glued so that it folds in, out, and around the wide fluting. If the success of a street poster depends on its ability to attract the attention of the passerby, then this one, with its solid block of unreadable text, is not much of an eye-catcher.[4] Serving more as label than enticement, it nevertheless offers a point of access through its only legible words: "Oude Doolhof" (Old Labyrinth).

Pictures often are compared to open windows: the aim of almost any work of art is to invite the viewer to enter. These sketches, by contrast, operate as closed doors. Each employs ungainliness and reticence as if attempting to be as visually uninviting as possible. Awkwardness of style works in tandem with subject matter: a barred, battered, handleless wooden door frustrates and repels visual access. As artistic strategies go, this is a risky one, since the contradictory objective of these drawings seems to be to convince the viewer that they are unworthy of notice. If the game is visual frustration, then Rieke's badly executed pencil sketch is a particularly skilled player: it comes so close to succeeding that it could easily have been discarded. To fail at this game is to win, and to win is to fail.

If we look at the sketches as deliberate artistic exercises, however, then their pictorial strategies may be conveying something specific about this uninviting door. One way to make sense of these unappealing drawings is to look at them as works of rhopography, depictions of insignificant subject matter. Adapting this classical visual convention, Netherlandish art had a long and particularly inventive rhopographic tradition. Playing with paradox, rhopographers often lavished dazzling technical virtuosity on mundane things that do not normally merit close attention, as with luminous still-life paintings of overturned wine glasses, loaves of bread, and dead fish surrounded by scatterings of cracked nuts,[5] or the Dutch specialty of depicting domestic interiors in which beams of sunlight play over commonplace chores like combing lice from children's hair. Another form of rhopography is to employ rough artistic style to depict rough subject matter, as with the streaky brushstrokes and splotchy impasto used to render a dunescape, a peasant's face, or a hut in a bog. For all their ugliness, such works were valued. To give a pertinent example, one of Peter Paul Rubens's most expensive paintings was an unfinished-looking, brushy rendering of an unidealized drunken and corpulent Bacchus.[6]

Matching rough technique with rough motif, Rieke and Ter Gouw are heirs to this well-established visual tradition. If anything, they push it to an extreme, for their labored sketches do not convey effortless effort, the *sprezzatura* virtuosity of a Rubens, a Rembrandt, or a Ruisdael. While their pictorial qualities do not engage, each drawing thwarts complete disregard through the insertion of the written label. "Doolhof" gives the door a hinge, linking mundane portal with mythic place. If this is the entry to a labyrinth, then something intricate, artful, and potentially wondrous or dangerous is coiled inside the uninteresting exterior.

The depictions of an unappealing doorway evoke a variation of rhopography called the Silenus box—something that is unsightly on the outside but contains beautiful and precious things. Rubens was playing on this tradition with his coarse Bacchus painting, and a number of other Netherlandish artists made sophisticated use of the Silenus device in their works.[7] The influential account of such an art object appears in Plato's *Symposium*. Socrates's student Alcibiades describes "those Sileni you find sitting in sculptors' shops . . . which when opened up are found to contain effigies of gods inside."[8] Silenus, the satyr associated with Bacchus, usually is represented as old, fat, flushed, balding, pug-nosed, and drunken. In spite of his repellent appearance, however, Silenus possessed divine wisdom, every virtue, and the gift of prophecy. The term "Silenus box" designates more than just a tricky artifact. A Silenus may look insignificant on the outside, but it opens into sacred mystery. Plato extends the metaphor to Socrates, whose ugly face, foolish behavior, and awkward manner of speech concealed wondrous powers: "His conversations too are just like those Sileni you can open up . . . if you can get through to what's under the surface . . . his arguments abound with divinity and effigies of goodness. They . . . cover absolutely everything which needs to be taken into consideration on the path to true goodness."[9]

It seems a leap to suggest that the Doolhof door contained this kind of promise or that its labyrinth and effigies offered a path to true goodness. However, something of the sort is suggested by yet another picture of the plain portal (fig. 2), this one by Rieke's son, Johan (1851–1899). Notably, Rieke junior records the distinguishing notice above this entry, which both his father and Ter Gouw, the historian of street signs, deliberately left out of their drawings. Between the doorway and transom window is a large curved lintel whose inscription commands the passerby, "Don't just stand in front of this door. Walk in."[10] Fountains, organs, and lively performances are mentioned as specific tantalizers. The incised text promises that walking through the Doolhof door would bring "the purest joy and contentment" to all who entered. The green door is ajar, enhancing anticipation but, disappointingly, not admitting any glimpse of the pleasures inside.

In fact, this door is probably swinging shut rather than open. Although the drawing was made in 1870, the cartouche at the bottom indicates

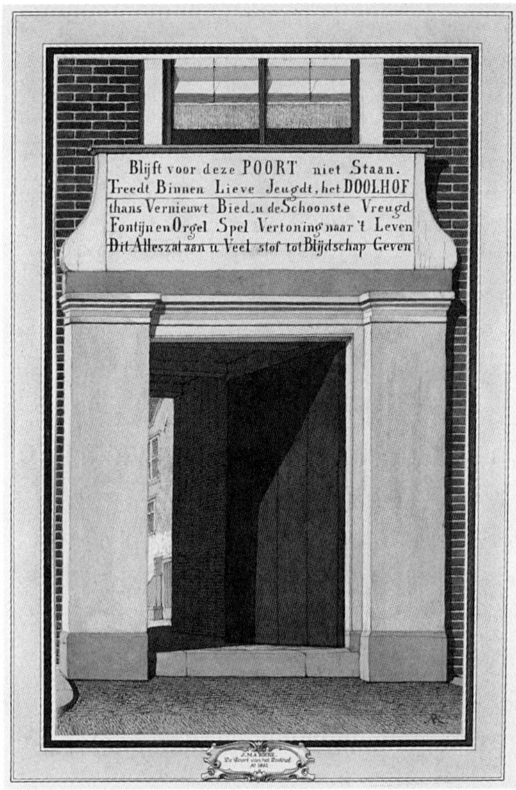

FIG. 2 J. M. A. Rieke, *Door of the Doolhof on Prinsengracht and Looijersgracht*, ca. 1870. Photo: Amsterdam City Archives (010094001704).

1862, a year before the Doolhof closed its door for good. Rieke's picture thus commemorates the closure of a venerable and unique type of civic establishment. No other European city had public urban labyrinth gardens of this sort, whereas in Amsterdam, five exhibition spaces called Doolhoven were established in the first half of the seventeenth century.[11] The Oude Doolhof (Old Labyrinth), pictured by Ter Gouw and the Riekes, was at the corner of Prinsengracht and Looiersgracht. It was the longest-lived of the Doolhoven, remaining in operation for close to two hundred and fifty years, from around 1620 until 1863. In 1626, its proprietor purchased an adjacent garden on Looiersgracht and built a Nieuwe Doolhof (New Labyrinth). In 1648, the Nieuwe Doolhof was sold, and the business was updated and moved to a larger site on the Rozengracht, where it remained until 1717. There was also a Roode Doolhof (Red Labyrinth) near the Regulierspoort from 1630 to 1663, and a Franse Doolhof (French Labyrinth) by the St. Anthoniespoort from 1637 to 1679.[12] While highly celebrated in their day, none of the Doolhof inns and very few of their attractions survive. The dearth of extant artifacts, together with nineteenth-century disparagement of amusements that were no longer innovative, has contributed to their obsolescence. The approach advised by Johan Huizinga accordingly inspires this study: "It is not enough to gaze admiringly on the masterpieces that have been preserved; one must consider what has been lost."[13] Following from this directive, *Moving Statues* is in part an experiment in recovering what can be known about nonextant works and their modes of display.

Ever since the closure of the Oude Doolhof, scholars consistently have derided the Amsterdam Doolhoven as marginal spaces of low-class urban entertainment, quirky forms of popular culture that do not merit close analysis. The sites have been dismissed as somewhat primitive ancestors of the theme park or the wax museum.[14] In the late nineteenth and early twentieth centuries, a number of articles and short pieces were published by historians, antiquarians, and archivists, as if to salvage the memory of a vanished form of Amsterdam street life. These studies summarize specific archival sources and take a folkloric approach, registering strong class, gender, and nationalist biases. Like Rieke and Ter Gouw, the early chroniclers seem to belittle their own efforts, describing and dismissing the Doolhoven as naive amusements for the *volk*, the common people. The salvage paradigm—an effort to preserve the memory of lost cultural practices—is employed together with another customary approach to folk culture: the safety valve model. The theory is that these fun parks gave the volk a bit of rowdy pleasure, allowing them to let off steam, a contained disruption that kept urban populations happy. *Volk* is a capacious category in these early histories of the Doolhoven: it does not denote a specific socioeconomic group so much as a perceived lack of cultural sophistication.[15] The folklorists singled out the types they assumed would most enjoy this kind of low-class entertainment: women, laborers, tradespeople, maidservants, the petty bourgeoisie, farmers, foreigners, youth, and the proletariat. The reiterated conclusion was that these popular urban pleasure parks mainly appealed to mothers, nursemaids, and children. The implication is that—with such an audience—the Doolhoven were places of frivolity rather than serious cultural institutions. One of the earliest articles about the Oude Doolhof, by historian D. C. Meijer Jr., was published in the first issue of the journal *Oud Holland* in 1883. Meijer's conclusions have perpetuated the predominant view: a Doolhof is a *kinderspeltuin*—a children's playground.[16] Amsterdam city archivist and art historian Nicolaas de Roever, cofounder of *Oud Holland*, followed up on Meijer's research with an 1888 article on the Nieuwe Doolhof and an 1890 book chapter about all of the Doolhoven. De Roever sums up the entertainment as "common merrymaking, typically Dutch, perhaps a bit rough, but artless and indecent."[17] In short, this was coarse fare for uncultured folk.

The aim of my study, by contrast, is to approach the Doolhof courtyards as Silenus boxes: as equally rough and artful. As Desiderius Erasmus observes in *The Praise of Folly*: "You find all things suddenly reversed, when you open up the Silenus."[18] We could begin, then, by reversing the conclusions of the folklorists. The obvious starting point is to reconsider the primary sources, which indicate that the Amsterdam Doolhoven were not characterized as rough-and-tumble sites of marginal mass culture in the seventeenth century. To the contrary, they successfully attracted widespread attention in and beyond Amsterdam, as evidenced by numerous prints, posters, drawings, and souvenir pamphlets; by the inclusion of the sites in officially commissioned histories and maps of the city; and by descriptions in the travel accounts of well-educated upper-class international curiosity seekers.[19] While the Doolhoven may have been perceived as childish folk entertainment in the nineteenth century, this clearly was not the case in the seventeenth century.

Among the artists and artisans whose work was exhibited at these sites were Amsterdam's leading architect and sculptor, Hendrik de Keyser (1565–1621); the engraver and publisher Crispijn van de Passe the Younger (1594/95–1670); the city's official sculptor, Albert Vinckenbrinck (1604–1665); the Parisian fountain maker Jonas Bargois (1582–1629); and Frankfurt horologists David Lingelbach (1592–1653) and his son Philips Lingelbach (1622–1673).[20] As advertised by Doolhof publicity materials, these highly accomplished artists and artisans made "things that have never been seen or even thought of before"[21]—works that were inventive and distinctive rather than artless and common. We need only look again at the list: labyrinths, automata, wax portraits, monumental fountains, clocks, astronomical devices, hydraulic installations, mechanical inventions, and large-scale multimedia sculpture groups. These creations represent some

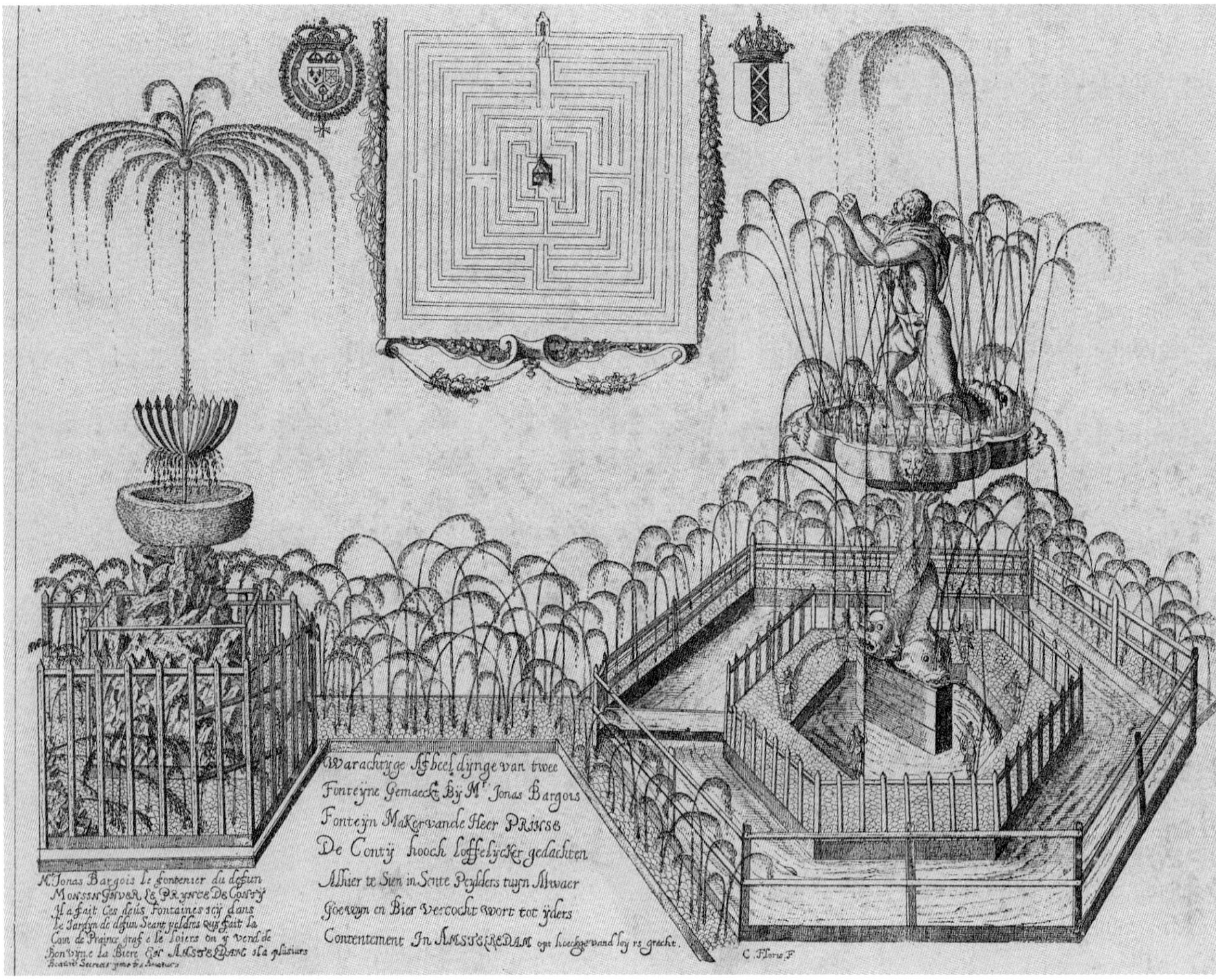

Warachtyge Afbeeldynge van twee
Fonteyne Gemaeckt By Mr. Jonas Bargois
Fonteyn Maker vande Heer PRINSE
De Contij hooch loffelycker gedachten
Alhier te Sien in Sente Peylders tuyn Alwaer
Goewyn en Bier Vercocht wort tot yders
Contentement In AMSTELREDAM opt hoeckge vand ley rs gracht.
M. Jonas Bargois le fontenier du defun
MONSSNGNUER LE PRYNCE DE CONTY
Ha fait Ces deus Fontaines icy dans
Le Jardyn de defun Seant peldres qui fait la
Com de Prainx graf e le loiers on y vend de
Bon vyn e la Biere En AMSTELDAM il a plusiurs

of the most innovative and complex artistic and scientific technologies of the day. In fact, artistry of this type had hitherto been the preserve of elite and courtly gardens and was featured in private collections and cabinets. In other European cities, such inventive arts adorned important public sites like town squares and cathedrals, promoting the cultural accomplishments of centers like Florence, Paris, Rome, Ghent, Venice, and Strasbourg, places with which Amsterdam vied as it strove to establish itself as a cosmopolitan center in the early decades of the seventeenth century.

The earliest surviving Doolhof advertising poster, dated to around 1622, proffers a way into this urban pleasure park (fig. 3). Unlike the unadorned (and probably fictive) poster for the Oude Doolhof depicted in Ter Gouw's image of the closed green door (color plate 1), this seventeenth-century exemplar was designed to intrigue. It pictures two fountains joined by a pebbled path that spouts water jets. Inset above is the detailed plan of a labyrinth. Pendant garlands of fruit hang from the mouths of beasts to frame the diagram, which is flanked by the coats of arms of Amsterdam at right and the Kingdom of France and Navarre at left. There is much to catch the eye here. Unlike Paris, Amsterdam could not boast of any impressive civic fountains in the 1620s. Its low, flat, and boggy terrain made the creation of fountains particularly challenging, as gravity could not easily be harnessed to force water upward into spectacular sprays such as those pictured on the poster. The very unfamiliarity of the depicted fountains poses questions about their hidden workings. For the person in the street, this poster would have presented astounding feats of engineering and artistry. These waterworks actually were "things that had never been seen or thought of before"—at least in Amsterdam, for they represent the city's first monumental public fountains.[22]

FIG. 3 Anonymous drawing, after Cornelis Florisz van Berckenrode, *Fountains in the Oude Doolhof*, ca. 1622. Photo: Amsterdam City Archives (010094004316).

The poster's explanatory text is in Dutch and French, soliciting both local and international interest. The skills of the French émigré Bargois are emphasized, and he is identified as the fountain maker to (in larger text) *Monseigneur the Prince of Conti*. The fountain at right, picturing Jonah emerging from the mouth of the fish, must then be Bargois's signature piece, a play on his name. The labyrinth is not labeled or explained: its plan simply offers a visual enigma. The text gives the location of the garden at the corner of Prinsengracht and Looiersgracht, together with the enticement that good wine and beer can be purchased there. The inn's proprietor is also named: Cente Peijlder, the nickname of Vincent Coster, who held the office of civic wine-assayer and also served as municipal treasurer.[23] The alliance of Bargois and Coster is aggrandized by the coats of arms: the combined forces of France and Amsterdam bring you these marvels. There are no allusions to typically Dutch peasantlike merrymaking or kindergartens here.

The French text curiously includes an extra bit of information, tacked on in smaller print at the end. It is as though the poster whispers to a select group, the readers of French: "Il a plusiurs beaux secreets pour les Amatures" (There are many fine secrets for amateurs). Almost all Doolhof advertisements appeal to these amateurs. Subsequent posters address "liefhebbers vande konst" (lovers of art) and "alle Heeren, Koop-luyden, Burgers en Liefhebbers" (all gentlemen, merchants, citizens, and enthusiasts).[24] The terms *liefhebbers* and *amateurs* designate a self-selecting group of discerning connoisseurs, local and international virtuosi in search of artistry advertised as "new," "rare," "curious," "strange," and "wondrous." Like *volk*, *liefhebber* is a capacious category. In fact, the terms are binary opposites, related concepts with reversed meanings: *liefhebber* distinguishes the cultured art lover from the uncultured folk.

The poster thus calls up an intriguing array of attractions: an intricate labyrinth, the princely creations of a Parisian fountain maker, a well-known Amsterdam proprietor, good wines and beers, and many fine secrets. It addresses every passerby as well as a select and self-selecting group, educated readers of French and lovers of art, and offers them things that are abstruse and difficult, beyond ordinary comprehension and general public knowledge. By simultaneously promising and withholding, this advertisement indicates that a visit would provide further initiation: access to novel technologies and the challenge of finding that hidden place at the heart of the labyrinth. The poster publicizes secrets, and it is this contradictory approach that ensures its success. It tantalizes viewers with the promise of an exhibition space that is both closed and open, princely and popular, cosmopolitan and local, challenging and accessible, clandestine and convivial.

We thus have to revisit the question of the social status and function of these exhibitions. The advertising strategy of making secrets public indicates the complex cultural dynamics of the sites. Street posters, especially ones that include imagery as well as text, address a public composed of any pedestrian who stops to look or read. As commercial establishments, the Doolhoven obviously attempted to attract as many paying customers as possible—they were accessible to all who could pay the admission fee. Drawing customers was just savvy business practice: without a doubt, making a profit was a priority. While the publicity throws the Doolhof open to anyone, however, it simultaneously singles out a select group—the art lovers. As innovative and elite artistic forms were made available to a wide audience, divisions were drawn between those with skillful understanding and those perceived to be in need of such training.

In his field-defining studies of early modern popular culture, Roger Chartier has observed that it is impossible to establish exclusive relationships between specific cultural forms and discrete social groups, since the fluid circulation of objects and practices blurs such distinctions. The intermingling of diverse people and things within shared culture—as at the Doolhoven—frequently is countered by assertions about proper and improper uses and understandings, which are a means to inculcate normative responses and reinforce social boundaries.[25] The Oude Doolhof poster, which targets a preferred group within the broad audience, represents such an attempt to make distinctions between learned and unlearned visitors, separating out the initiated from the uninitiated, an issue I return to below.

The evidence that we have of the Doolhof displays indicates that the general public was given access to a variety of artifacts and ideas derived from classical antiquity. Labyrinths, fountains, automata, and waxworks were all billed as ancient art forms that had been revived. The Oude Doolhof included large-scale statues of Theseus battling the Minotaur and a fountain of the abandoned Ariadne discovered by Bacchus. Besides sources like Ovid and Catullus, these exhibits specifically reference the vernacular classicism that was being developed by leading literary figures like Daniel Heinsius and Pieter Cornelisz Hooft. Heinsius's controversial Dutch-language poem *Ode to Bacchus* (1614) probably inspired the Bacchus and Ariadne fountain, which I take up in the next chapter. In one of the few substantive recent studies of the Doolhoven, literary scholar Marijke Spies shows that the theatrical works of Hooft were especially relevant. His well-known play *Theseus and Ariadne* (1614) was performed regularly in Amsterdam in the early decades of the seventeenth century, and its themes were central to these labyrinth sculpture parks, as explored in chapter 3.[26] Another of Hooft's popular plays, the pastoral *Granida* (1605), was featured at the Oude Doolhof, where scenes were acted out by automata in a moving mechanical picture show displayed in the sculpture gallery; chapter 7 considers this peculiar use of machinery to animate imagery of a technology-free golden age.

The strong connections with the literary works of Hooft offer a way to understand the civic functions of the Doolhoven. Not only were his plays referenced in many of the exhibits, but Hooft also had personal links to these sites. Antonia Cloeck, proprietor of the Franse Doolhof, was Hooft's niece, the daughter of his sister Jannetje Cornelisdr Hooft. Cloeck's grandparents were the Amsterdam burgomaster Cornelis Pieters Hooft and Anna Jacobsdr Blaeu, of the renowned publishing and map-making family. Coster, the first proprietor of the Oude Doolhof, also hailed from this exclusive social circle: he held civic office, and his uncle was a burgomaster.[27] The Hoofts, Cloecks, Costers, and Blaeus were among Amsterdam's leading families. They were of the regent class, the oligarchs who served in municipal government. Members of these families held positions on the governing boards of various civic institutions, like the prisons, orphanages, and old people's homes, as well as the powerful and profitable Dutch East and West India trading companies. As civic leaders, the regent families deliberately shaped the cultural life of the city. The Doolhof concept thus seems to have been the brainchild of Amsterdam's political, cultural, and intellectual elite. These innovative exhibition sites evidently expressed patrician as well as proletarian interests.

To situate the populist aims of the establishments, it is useful to consider a pioneering effort made by P. C. Hooft to disseminate colloquial classicism to the urban populace. Together with fellow playwrights Dr. Samuel Coster and G. A. Bredero, Hooft was one of the founders of the *Nederduysch Academie*, or Dutch Academy. Built on the Keizersgracht in 1617, the academy was Amsterdam's first permanent theater building. More than just a performance venue, it was conceived as a cultural and academic center, with the aim of offering vernacular public instruction in the arts and sciences. The Dutch Academy deliberately distinguished itself from Leiden University, where teaching was in Latin. Two professors, both Mennonites, were appointed to teach at the academy: Sybrandt Hansz Cardinael specialized in arithmetic, and Jan Theunisz taught Hebrew. The academy also differed from Leiden University because it had no official connection to the Reformed church; its Dutch-language instruction focused on classical rather than theological learning. As a contemporary source notes, the academy was "meant for the edification and amusement of everyone."[28]

The ambitious aims of this civic cultural center were short-lived, however. There were repeated condemnations by Reformed church leaders, who especially complained that its classical curriculum veered rather too far in the direction of paganism.

Moreover, they disapproved of the hiring of non-Calvinist professors. The early decades of the seventeenth century were a volatile period in the religious life of the Dutch Republic, when theological disputes between opposing Calvinist political factions almost erupted into civil war. Although the Amsterdam magistrates rarely took an ultraorthodox stance, they could not afford to take risks, and in 1619 they decided to appease the hardline Calvinists and keep the peace by curtailing the academy's activities. Decades later, the academy was reconfigured—the Amsterdam Schouwburg was established in 1637 for public theater performances, whereas higher learning became the domain of the Athenaeum Illustre (precursor to the University of Amsterdam), inaugurated in 1632.

At about the same time that the Dutch Academy fell under censure, the first Doolhof appeared on Prinsengracht, just a canal away from the academy building on Keizersgracht. Besides their proximity, there are other parallels between the academy and the Oude Doolhof, where the vernacular classicism of experimental works of art, theater, science, technology, and astronomy was presented to the general public. Another connection is Theunisz, the academy's Hebrew instructor, who was the proprietor of a tavern and music house called D'Os in de Bruyloft. Theunisz's inn displayed ingenious mechanical and hydraulic works, which I will explore in more detail in chapters 4 and 5. With the restrictions on civic cultural institutions in these decades, it seems that Amsterdam's inns emerged as sites that advanced various forms of community instruction, including the demonstration of artworks, artifacts, curiosities, and new technologies. Other well-known Amsterdam inns included De Blauw Jan and De Witte Oliphant, whose courtyards and outbuildings housed and displayed exotic birds, animals, and even people from around the world.[29] Like the Doolhoven, these public houses attracted international visitors and dignitaries and were frequented by a wide range of local people, which may have comprised servants, laborers, housewives, children, and farmers but also civic magistrates and notable scholars. What is of particular interest about the use of inns and taverns as exhibition sites is their potential to generate new social formations through broad access to shared culture. The public, urban displays of moving statues thus prompt an approach that shifts emphasis from artists and patrons to consideration of the effects of lively images on the diverse audiences who interacted with them.

The Moving Image

In what follows, I contend that the Doolhoven shared some of the higher ambitions of the short-lived Dutch Academy: their distinctive brand of colloquial classicism was paired with artistic and scientific experimentation in a manner devised to offer education and entertainment to Amsterdam residents as well as visitors to the city. Given the diversity of the audience, we can assume that not every guest would have been equipped with the same background knowledge, and the evidence indicates that concerted efforts were made to control the reception and understanding of the exhibits. The various artworks were staged as "shows" (*vertooningen*) and explained to audiences by a presenter, likely one of the proprietors. Theatrical modes of display included the raising of stage curtains accompanied by music and various other means to create a multisensory experience of art that was both pleasurable and instructive. Explanatory booklets were sold to visitors, which serve as key sources for understanding the contents of the exhibits and how they were staged. There are also some firsthand accounts by visitors to the Doolhoven, although these were exclusively written by well-educated European men and so do not give a full picture of the variety of audience reactions. The experiences of women, non-Europeans, and lower-class visitors are more difficult to track—we have to read between the lines of the existing primary sources to get a sense of how nonelites were situated and the types of hierarchies that were created by exhibits designed to change and improve. The instructive impetus of the displays thus signals what I think was the main purpose of the Doolhoven, which was to make inventive and exclusive forms of art and technology accessible to a broad mix of people, in the process attempting to transform them into civil, informed amateurs. The Doolhoven did not just seek to attract virtuosi and art lovers; rather, the itinerary through the sites was a rite of initiation into the secrets of the exhibits. This was the Silenus-like aim of these establishments: to convert those who were perceived as rough volk into refined liefhebbers.

Needless to say, this conclusion verges on the cliché. Seventeenth-century commonplaces about edification and amusement were elevated to a methodological approach by the 1976 art exhibition and catalogue *Tot Lering en Vermaak* (For learning and entertainment).[30] An important rethinking of Dutch genre painting, this study argued that the purpose of seemingly trivial or crude scenes of everyday life—domestic interiors, peasant festivities, landscapes, still-life paintings—was to instruct and delight. The emblems of Jacob Cats often are cited as key examples of this method: the visual image of something ordinary may seem incomprehensible, insignificant, or unsophisticated at first glance, but it contains valuable moral and religious truths. Cats himself explains this approach in the foreword to his first emblem book, *Silenus Alcibiadis, sive Proteus* of 1618.[31] As the title indicates, an emblem is like Alcibiades's Silenus box, a protean thing with hidden depths. Each emblematic image is given three levels of meaning relating to love, morality, and religion. In this manner, Cats draws on classicism to craft Calvinist culture.

In *Silenus Alcibiadis*, Cats likens the emblem to the apothecary's pot, which looks trifling on the exterior but when opened proffers curative balm. Drawing on the *Symposium*, François Rabelais similarly sums up the promise of the Silenus: "But had you opened that box, you would have found inside a heavenly and priceless drug: a superhuman understanding, miraculous virtue, invincible courage, unrivalled sobriety, unfailing contentment."[32] Pot and medicine are pictured on the Cats title print (fig. 4). Beneath the Silenus box in the niche on the left is the Latin phrase *meliora latent*: the secret it contains is betterment Also depicted are the people in need of this cure: a crowd that ranges from well-dressed burghers to ragged folk. The ideal expressed here is that the protean image delights and instructs in order to change protean people who, while rough on the outside, have inner potential. To muster the capacity of art for social change clearly was an expedient strategy for leading cultural producers like Cats, the supporters of the first Dutch Academy, and the urban patricians who founded the Doolhoven. Cats was notably appointed to one of the Dutch Republic's highest political offices: Grand Pensionary of Holland. His cultural efforts to reform and discipline the populace were more than theoretical notions; they were a form of practical politics, a means of creating an orderly republic.

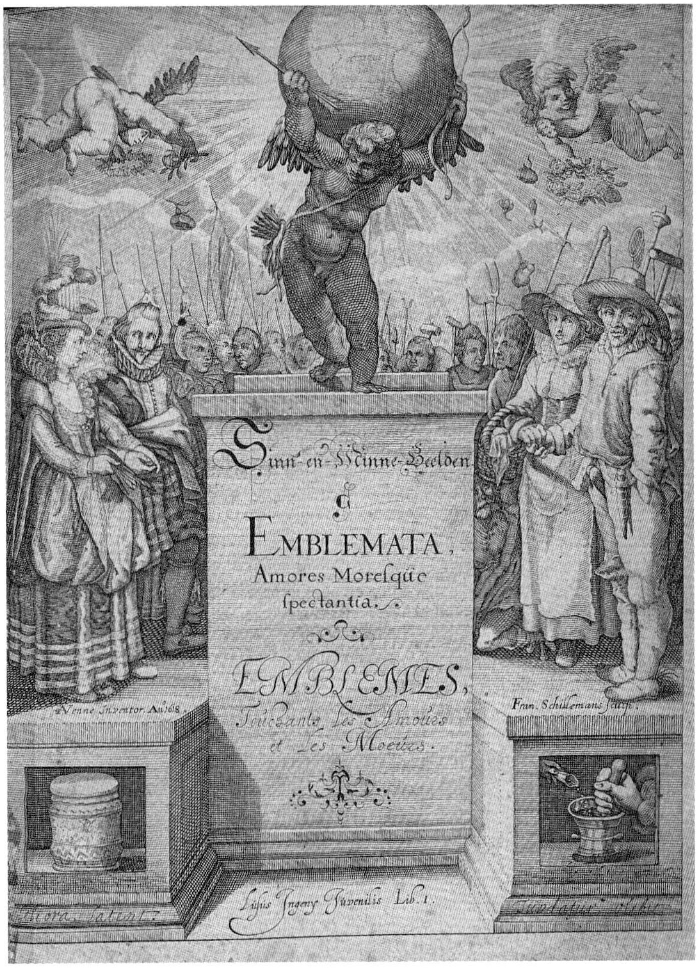

FIG. 4 François Schillemans, after Adriaen van de Venne, title page from Jacob Cats, *Silenus Alcibiadis, sive Proteus* (Amsterdam: Willem Jansz Blaeu, 1619). Photo: Beinecke Rare Book and Manuscript Library, Yale University, New Haven.

The idea of combining entertainment and instruction for the betterment of the general population was not invented by Dutch artists and writers; it was a well-established classical technique of persuasion. Classical authors, however, include a third term, which has not received as much art-historical attention. Cicero writes that the best orator teaches, delights, and moves his listeners: "To teach them is his duty, to delight them is creditable to him, to move them is indispensable."[33] With this, I return to the question of the impact of the Doolhof displays on their audiences. While the deciphering of symbols and iconographic themes was no doubt one route to moral edification and improvement, it was by no means experienced as a process of detached intellectual contemplation. Interpretation was a mobile, immersive, and transformative process at the Doolhoven, where protean artworks were devised to move visitors emotionally and physically as well as morally. The fountains, automata, and mechanical devices were surprising works, set up to startle viewers with unanticipated sounds and actions; the labyrinth was an ever-changing bodily experience that created conditions of sensory dislocation and visual uncertainty; and the waxworks, while deathly still, were perceived as "most liveley," in the words of one visitor.[34] These garden courtyards engaged audiences in multiple sensory experiences: confounding maze, spraying water, organ music, and Bacchic conviviality together enhanced the collective force of a gathering of strikingly lifelike moving artworks.

This kind of performative, embodied, sensuous, and emotive engagement with animated images has been largely repressed from art-historical methods like iconography, which tend to focus on the intellectual interpretation and contemplation of the work of art in relation to textual sources that convey similar moral messages. Moreover, the types of works that were showcased in the Doolhof exhibits—mechanical inventions, hydraulics, and wax portraits—have largely been excluded from accounts of a period mainly associated with the flourishing of Dutch oil painting. Frequently dismissed as a form of "primitivism," the lively artwork is most often encountered in popular culture studies or anthropologies of the image.[35] This academic compartmentalization explains, at least in part, why the Doolhoven have been relegated to the realm of folklore. Writing in the early twentieth century, art historian Julius von Schlosser aptly sums up the peripheral status of automata and waxworks as artifacts "that had lurked since earliest times on the threshold of art."[36] Bringing forward historical information about the production, display, and reception of experimental works, *Moving Statues* seeks to interrogate the boundaries of Dutch art history by turning to the marginalized objects that skulk there, which significantly alter our understanding of artistic priorities and cultural practices in the seventeenth century.

A reconsideration of the moving image has the potential to put art history in motion, animating it and dynamically opening it up to a consideration of different kinds of objects, modes of engagement, and methods of analysis. The legacy of *Tot Lering en Vermaak* was to inspire multiple studies of Dutch paintings that searched for hidden moral messages in everyday subjects, matching prosaic visual motifs to the improving texts of emblem literature especially. In her important critique of this approach, Svetlana Alpers emphatically asserts that Dutch art is descriptive, not prescriptive; instead of looking for disguised meanings, we should be more attentive to pictorial qualities.[37] Alpers rightly notes that text-based moral readings of artworks often overlook the striking visual and material characteristics of the works themselves, and her intervention productively shifted the field to considerations of vision and visuality and to the significance of images for scientific experimentation and new optical technologies. This turn to visual culture helped usher in the "New Art History" of the 1980s.

In so doing, however, Alpers devises a method that assumes trust in visual images. In her assessment, descriptive art is very close to actual optical experience: "Pictorial and craft traditions, broadly reinforced by the new experimental science and technology, confirmed pictures as the way to new and certain knowledge of the world."[38] Scholars have recently begun to question assumptions about the veracity of descriptive art and of vision itself. Stuart Clark, for instance, in his substantive overview *Vanities of the Eye*, demonstrates that there was widespread skepticism about the reliability of vision in the early modern period. Sensory information was perceived as uncertain, even deceitful. As Victor Stoichita and others show, much seventeenth-century art is characterized by an idiosyncratic "self-awareness"; artists employed a range of visual devices to draw attention to the illusory tricks—rather than the objective truth—of the mimetic image.[39] While Alpers posits a mode of looking at realism that verges on detached scientific observation, descriptive art has proven notoriously difficult to decipher, and the works themselves often repay attentive viewing by withholding information, creating puzzlement, uncertainty, and the suspension of any sure conclusions about meanings, morals, or empirical information. The art of describing thus cannot be apprehended as a way to certain knowledge; contemporaries saw it more as a confounding trap and characterized art making as a form of trickery, albeit an "honorable deception," in the words of artist and theoretician Samuel van Hoogstraten.[40]

The notion that scientific ways of looking are detached and objective can also be challenged. Alpers's *The Art of Describing*, with its emphasis on vision, tends to ally the descriptive pictorial mode with the development of scientific instruments, like the microscope and the telescope, and with viewing devices, like the camera obscura. Drawing on Alpers, Jonathan Crary, in his influential book *Techniques of the Observer*, claims that the camera obscura was a dominant metaphor for human vision across Europe in the seventeenth century. In Crary's account, this viewing technique signaled a new model of subjectivity in which the image is separated from the object and the act of seeing is sundered from the physical body of the viewer: "The observer's physical and sensory experience is supplanted by the relations between a mechanical apparatus and a pre-given world of objective truth."[41] Vision, in other words, was decorporealized.

However, alongside the telescope, microscope, and camera obscura, there were other sorts of artistic, scientific, and mechanical devices—automata, mechanical clocks, and hydraulic works, especially—that were equally prevalent in the seventeenth century and also were devised as tools to interrogate the workings of nature. As we will see, these stimulated viewing experiences were profoundly corporeal. If we reconsider the art of describing in relation to these types of artworks, we accordingly should rethink the mode of viewing prompted by highly illusionistic art, which does not simply convey visual knowledge of a pre-given world of objective truth. Many of the works displayed at the Doolhof gardens took mimesis to its extremes. Waxworks and automata, in particular, do more than just imitate nature—they look or even move as if they are alive. By engaging and deceiving the viewer, they pose questions about the contiguity of life and art, an issue explored further in chapters 4 and 6. The Doolhof exhibits also allow for a consideration of the interplay between various types of art objects, their environments, and their viewers. Instead of highlighting the discrete work of art as a locus of meaning or technical virtuosity, these pleasure parks immersed peripatetic visitors in a multimedia experience that unfolded as an engaged process of interaction. The moving image thus calls for a methodological approach that considers the impact of the work of art on viewers who cannot be characterized as detached observers.

Notably, this prompts us to reassess the now-ubiquitous term *visual culture* and shift to a consideration of multisensorial interactions with artworks. We cannot really bracket out a purely visual apprehension of things from all of the other sensory, cognitive, psychological, and emotive impressions that come into play. *Moving Statues* therefore contributes to growing art-historical interest in the moving image and experiential viewing. While Alpers's and Crary's narratives have long dominated visual culture studies, art historians have begun to reassess these claims, turning to primary source material as well as theories derived from disciplines such as philosophy and anthropology.[42] There has been a recent outpouring of studies that evaluate the affective and performative qualities of early modern imagery and reconsider aesthetic experience in terms of emotions, sensations, and embodiment. Important books by Caroline van Eck, Thijs Weststeijn, and Frederica Jacobs have definitively shown that classical and early modern art theory considered the moving and lively qualities of artworks as vital to their impact.[43]

Regarding approaches to Netherlandish art, a significant rethinking of the field is evident in special issues of flagship journals such as *Nederlands Kunsthistorisch Jaarboek* and *Journal of the Historians of Netherlandish Art* dedicated to exploring the importance of the passions and the sublime in encounters with early modern artworks.[44] There are also volumes devoted to particular reactions and embodied viewing practices like early modern horror, the erotics of looking, and theatricality.[45] Critical studies of the striking effects of automata and waxworks are appearing with more frequency, as are considerations of the labyrinth as an immersive art form.[46] What all of this current scholarship demonstrates is that—far from being supplanted—physical and sensory experience was integral to understanding early modern art. The camera obscura metaphor of detached observation therefore does not adequately characterize the period eye. Instead, I would argue that the works examined in this book—automata, waxworks, fountains, and labyrinths—offer more fitting models for understanding seventeenth-century visual culture as experimental, affective, embodied, and performative.

Seventeenth-century writers about art say very little about emblem-based moral meanings or the objective visuality of the new science. The art theorists also did not tend to oppose Dutch realism to Italianate classicism but developed sophisticated ideas about mimetic painting based on humanist knowledge of classical philology.[47] Indeed, pictorial realism itself was a form of classicism: the visual traditions of rhopography and the Silenus, as noted, were derived from ancient precedents, a fact that is usually overlooked in art-historical studies and exhibitions of Dutch pictures of everyday life.

Another key concept derived from classicism was *energeia*, used in rhetoric to denote the means of sparking emotional and physical audience engagement. The Dutch term for this process was *beweeglijkheid*. As Weststeijn and others explore, artists believed that every effective work of art should possess this moving quality, which connoted motion as well as emotion.[48] By persuading beholders that artificial figures could move, breathe, and speak, visual art could consume its beholders, immersing them in an intense illusory experience. Addressing his fellow artists, Willem Goeree summarizes the importance of infusing a work of art with beweeglijkheid:

> Since it is in the nature of figures depicted that they lack life and movement, this should stimulate us all the more to find the means to create the figures in the scenes in such a way that the spectator forgets all thoughts about canvas and panel, paint and oil, wood and stone and copper, be it cast or modelled, and to give him the idea that he sees before his eyes nothing painted or cast, but living human figures, so that he thinks they walk, move,

speak, shout, fight, hear, see, think, and do all other living actions of living people.[49]

Goeree describes a performative aesthetics in which spectators do not engage in detached observation of erudite iconography or optical effects but become intensely interconnected with living works that appear contiguous with their world.

Skillfully illusionistic realism was transformative because it was performative: it could solicit bodily and impassioned involvement with figures that appear to be "living people." In the words of one early modern commentator, the beholder thus confronted by the lively artwork "becomes another person."[50] The striking visual and sensory properties of art were never only for pleasure or edification: ultimately, the aim of beweeglijkheid was to rouse viewers to action. "To move them is indispensable," as Cicero insists; a work of art can only change people if it is able to move them, and intense emotive engagement with a truly compelling artwork could persuade onlookers to alter their ethical outlook.[51] This is why beweeglijkheid mattered. Although scholars have drawn attention to the significance of beweeglijkheid, there are few studies that demonstrate how the transformative force of the lively image was mobilized in the Dutch Republic to serve specific political, religious, commercial, and social agendas, a process that is the main focus of this book.

Controlled Confusion

Moving Statues interrogates many of the contentious debates that have polarized the field of Dutch art history by taking into account the complex functions of artworks that could be both prescriptive and descriptive, empirical and emotional, elite and popular, and even pagan and Protestant at the same time. Central to this approach is an emphasis on the early modern practice of thinking with paradox.[52] The art of paradox is the ability to hold incompatible opposites in tension. A key insight of the aesthetic theorists was that the fusion of opposing forces in an animated work of art could create a "terrifyingly confused inner struggle" in viewers.[53] At the heart of this struggle was the impossibility of reconciling the beneficial and harmful potential of art. The management and control of artworks and art lovers were of vital concern in the Dutch Republic because of the belief that art was dangerous. Protean works could proffer curative balm and deadly poison; they could operate as harmonious instruments as well as damaging weapons.[54] Authorities could never completely regulate the contradictory capabilities of animated art: it was a formidable tool and thus a locus of conflict.

With ultraorthodox Calvinism on the rise in the early decades of the seventeenth century—when the Doolhoven were first established—the lively artwork came under close scrutiny. The Doolhoven were semipublic sculpture parks, and sculpture was the art form most vilified in the Calvinist war against images. The iconoclasm of 1566 had cleansed the Dutch churches of much of their religious imagery, and statues were particular targets.[55] It was precisely the moving and lively qualities of the graven image that most offended. As one witness to iconoclastic destruction noted, "They hammer away mainly at the faces."[56] If only one part of an image was attacked, it was the eyes, which especially conveyed the lifelike or personlike qualities of the sculpted or painted figure. The iconoclasts struck at the moving image to stop it from stirring beholders to the false veneration of idols. Such concerns reescalated during the religious and political conflicts of the early seventeenth century. In 1613, to give a relevant example, Amsterdam's civic sculptor, De Keyser, was threatened with discipline by the Amsterdam church council, which claimed that his statues would move people to idolatry.[57] De Keyser was not the only cultural producer to come under censure. In 1622, Samuel Coster closed the Dutch Academy after it had been declared "anticlerical" at the repeated urgings of Calvinist church leaders.[58] Heinsius, to note another pertinent case, wrote an *Ode to Jesus Christus* in 1615 to offset the church's criticisms of his hymn to Bacchus, which was condemned by religious authorities because of its celebration of a pagan god.[59] In a similar move, Cats, too, thought it wise to distance himself from paganism: subsequent editions of *Silenus Alcibiadis, sive Proteus* were simply titled *Sinne- en Minnebeelden* (Images of morality and love).[60]

The Doolhof inns had an important advantage in this repressive context: as private businesses, they could foster cultural experimentation while evading the injunctions of the church. Even so, this was risky business. Eddy de Jongh states that after the iconoclasm, sculptors had to proceed cautiously, for they worked "in the shadow of Daedalus."[61] The legendary Daedalus was antiquity's most cunning artificer. He was credited with the invention of the automaton as well as the labyrinth, making him a fitting figure to preside over the Doolhof displays.[62] As explored further in chapter 4, early modern interpreters of the tale of Daedalus describe him as an artist who took human creativity to extremes. Renowned as the creator of clever devices, including the wooden cow used by Queen Pasiphaë to mate with the Cretan bull as well as the labyrinth built to imprison the monster born of this bestiality, Daedalus was famously punished for his transgressive ingenuity when immured by King Minos in the very labyrinth that he had so craftily designed. Francis Bacon's 1609 commentary is particularly eloquent about the paradoxical workings of Daedalus's creations. While he criticizes Daedalus as a "most execrable artist" and Daedalic artworks for their "mischief and destruction," Bacon also points out the social benefits of inventive ingenuity. He concludes that "unlawful arts, and indeed frequently arts themselves, are persecuted by Minos, that is, by laws, which prohibit and forbid their use among the people; but notwithstanding this, they are hid, concealed, retained, and everywhere find reception and skulking-places."[63] The Amsterdam Doolhoven, I am arguing, were just such skulking places. These labyrinth gardens contained, protected, and displayed the kinds of "unlawful arts" that had long been associated with Daedalus. Cast out of the churches, living statues were both sheltered and exposed as the "fine secrets" of the labyrinth.

Fostered by urban patricians, who were more invested in civic and commercial interests than the strictest doctrinal orthodoxy, the Doolhoven served as resourceful means to stimulate artistic and technological inventiveness in spite of religious prohibitions. The need for cultural innovation was particularly urgent in the late sixteenth and early seventeenth centuries, in the wake of the image wars. During this period, Amsterdam experienced its swift ascendancy to a major world trade center, largely due to the defeat of Antwerp in 1585 and the founding of the hugely profitable East India Company, the VOC, in 1602. The early decades of the seventeenth century were characterized by the large-scale immigration of merchants, intellectuals,

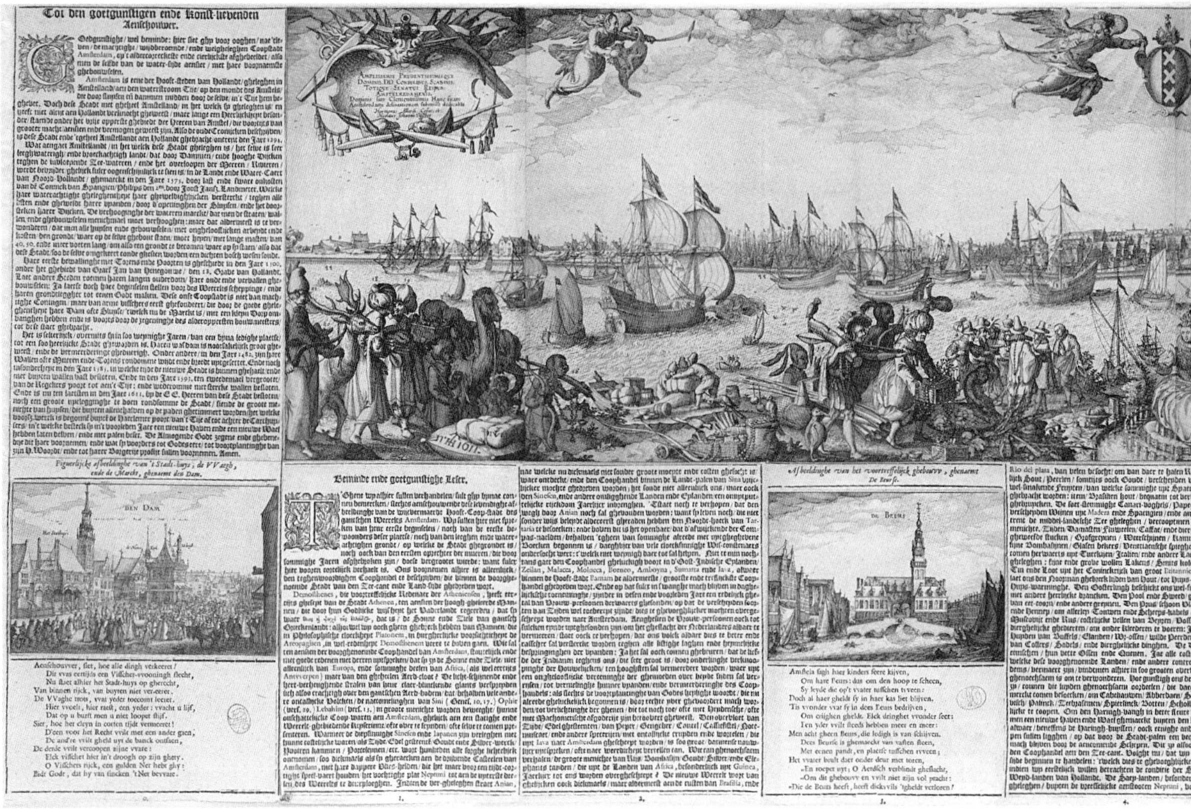

artists, and craftspeople, especially from the southern Netherlands and Germany. This influx of a foreign cultural elite, together with sudden economic dominance as Amsterdam eclipsed Antwerp in global trade, transformed the city in a remarkably short span of time. The population grew from thirty thousand in 1585 to one hundred and twenty thousand by 1632 and two hundred thousand by the late 1660s, an explosive pattern of growth exclusively due to immigration. Suddenly Amsterdam was the third largest city in Europe, after London and Paris.[64] The metamorphosis of Amsterdam from fishing town to overpopulated metropolis necessitated new cultural strategies to celebrate and make sense of—but also keep up with—the new situation.

The Doolhof exhibition sites were distinctive to the context of a republican city dominated by Calvinism whose wealth derived from mercantile capitalism and global trade. As an ambitious early attempt to picture Amsterdam's new place in the world, Claes Jansz Visscher's printed city profile of 1611 gives us a sense of the city's aspirations (fig. 5). The crowned Amsterdam Maid is seated in a small fenced garden at the center of the crowded composition, and the explanatory text proclaims that she is an empress who commands the whole

FIG. 5 Claes Jansz Visscher, *Profile of Amsterdam*, 1611. Photo: Rijksmuseum, Amsterdam.

FIG. 6 Claes Jansz Visscher, *Profile of Amsterdam*, 1611, detail. Photo: Rijksmuseum, Amsterdam.

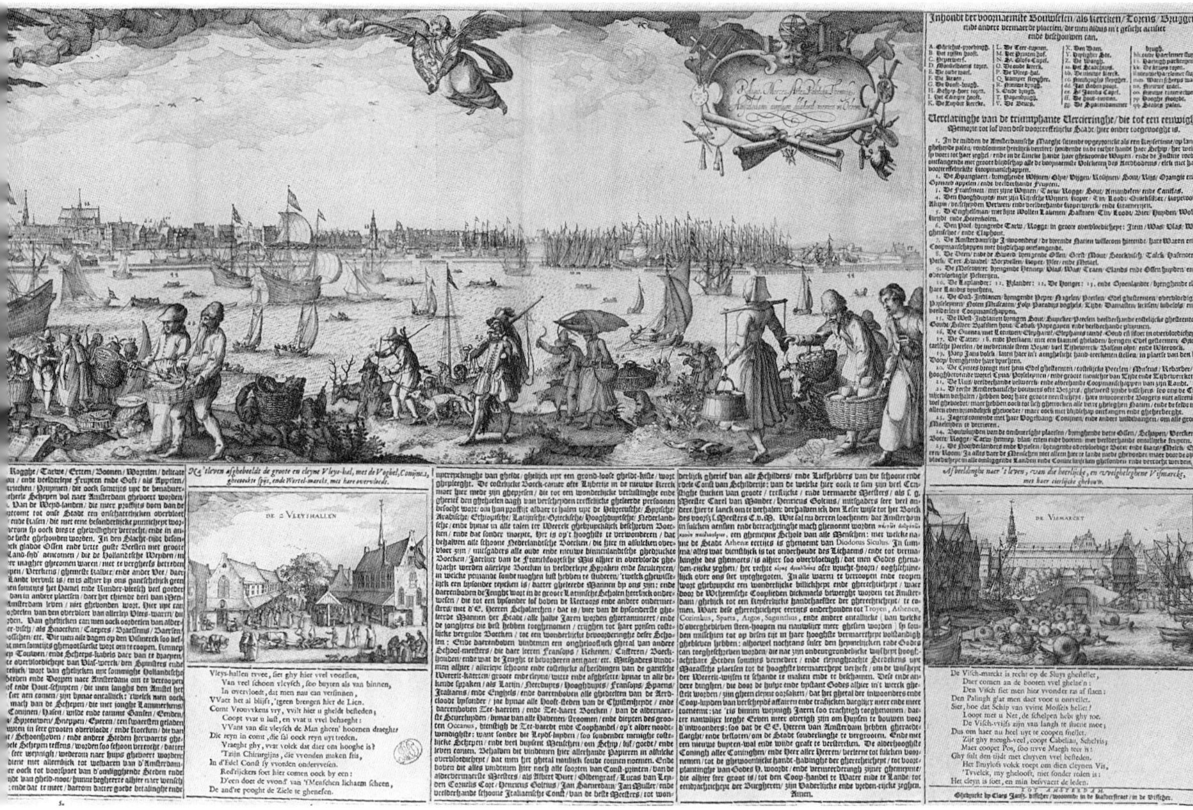
Inhoudt der voornaemste Bouwselen/ als Kercken/ Torens/ Bruggen/
ende andere vermaerde plaetsen/ die men aldus in't gesicht acrisier
ende beschouwen can.
Verclaringhe van de triumphante Verchieringhe/ die tot een eeuwighe
Na 't leven afghebeeldt de groote en cleyne Vleys-hal, met de Voghel, Conijnen,
ghevoegte spijs, ende Wortel-marckt, met hare overvloedt.
DE 2 VLEYSHALLEN
Vleys-hallen twee, siet ghy hier wel voorsien,
Van veel schoon vleysch, soo buyten als van binnen,
In overvloedt, dat men nau can versinnen,
VVaer het al blijft, 'tgeen brengen hier de Lien.
Comt Vrouwkens vry, wilt hier u gheldt besteden,
Coopt wat u lust, en wat u wel behaeght:
VVant van dit vleysch de Man gheen' hoornen draeght
Die reyn in comt, die sal oock reyn uyt treden.
Vraeght ghy, wat volck dat daer om hooghe is?
'Tzijn Chirurgijns, die wonden maken fris,
In d'Edel Const sy wonden onderweesen.
Red'rijckers soet hier comen oock by een:
D'een doet de wond' van 'tMenschen lichaem scheen,
De and're pooght de Ziele te ghenesen.
Afbeeldinghe naer 't leven, van die heerlijcke, en welgheleghene Vismarckt,
met haer cierlijcke ghebouw.
DE VISMARCKT
De Visch-marckt is recht op de Sluys ghestelt,
Daer comen an de booten wel ghelae'n:
Den Visch siet men hier wonder ras af slaen:
Den Palingh glat men daer voor u ontvellet.
Siet, hoe dat Schip van witte Mossels hellet!
Loopt met u Net, de schelpen hebt ghy toe.
De Visch-wijfs zijn van langh te staene moe;
Dus om haer nu heel uyt te coopen snellet.
Zijt ghy mengh-veel, coopt Cabelau, Schelvis;
Maer coopet Pos, soo uwe Maegh teer is:
Ghy sult den tijdt met cluyven wel besteden.
Het Bruylofs volck roept om dien cleynen Vis,
'Twelck, my ghelooft, niet sonder reden is:
Het cleyn is soet, en min beswaert de leden.
TOT AMSTERDAM,
Ghedruckt by Claes Jansz. Visscher/ woonende in de Kalverstraet/ in de Visscher.

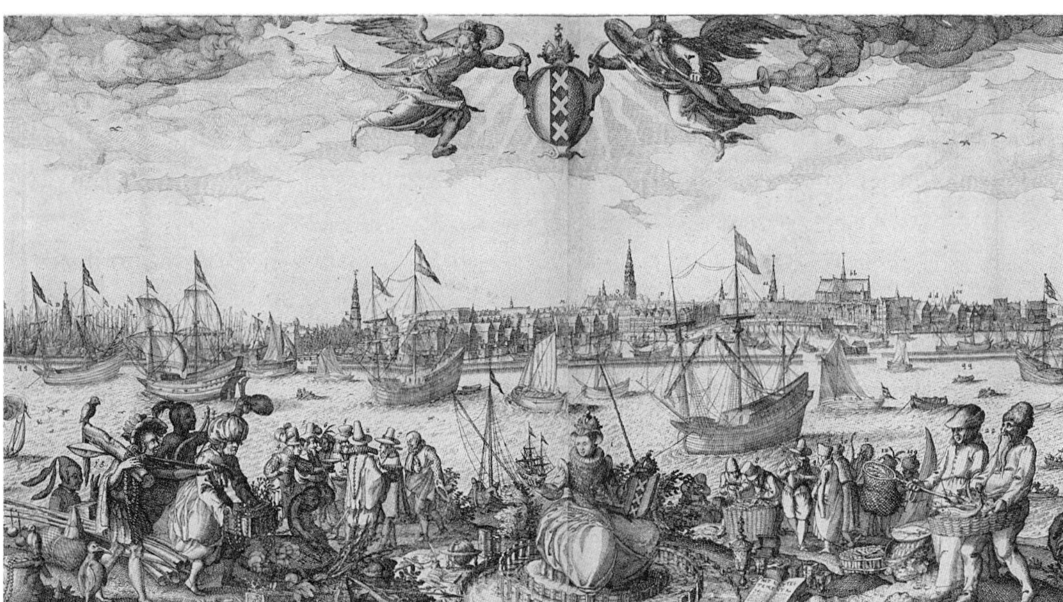

world. On her left are the traditional Dutch volk, bringing their contributions to civic prosperity: fishers, hunters, and farmers who carry the bounties of land and sea to the Maid. At her right are the peoples of the world, a long list that includes Spaniards, Frenchmen, Germans, Englishmen, Poles, the Amsterdam Jews, Danes, Swedes, Muscovites, Russians, Laplanders, Hungarians, Greenlanders, East Indians, West Indians, Guineans, Tartars, Persians, and the Chinese. They offer the abundance of their lands: spices, porcelain, Brazilwood, ivory, gemstones, grain, pearls, textiles, and many more commodities, which are pictured and inventoried by the print. The fiction of this image is that these astonishing riches are bestowed as gifts to the Maid. In return, she brings religion and law, represented by the rod of justice and the open Bible, and imparts learning and the arts, pictured as a globe, a quill, a painter's palette, two books, and some musical notation (fig. 6). The text explains that the Maid offers wisdom and knowledge to foreigners and residents; the city is "a common school for all people, as was formerly the city of Athens."[65]

It must be remarked that the Maid's offerings look rather paltry in contrast to the bounteous treasure she receives. In fact, this paean to Amsterdam comes up rather short when listing places where people could actually access the city's stores of learning, art, and wisdom. A library housed in the Nieuwe Kerk is mentioned, as are the city's map and print shops (promoting Visscher's own business). However, the Dutch Academy, Schouwburg Theater, and Athenaeum Illustre were not yet established at this time, nor was the New Town Hall, which later became an important expression of Amsterdam's cosmopolitan status. Artistic treasures that had formerly filled the churches had been removed or destroyed during and after the iconoclasm. In brief, there were few public places for the exhibition of art. Indeed, the accolades of Visscher's city profile may have seemed rather inflated in 1611. While Amsterdam was fast becoming a powerhouse global trade emporium, it had not achieved world dominance in the realm of culture. The Visscher print thus highlights, albeit unwittingly, what must have been a significant challenge for the urban elite: how to transform Amsterdam into an impressive global storehouse of learning and the arts in spite of considerable religious opposition to certain forms of cultural production, notably pagan classicism and potentially idolatrous graven images.

Visscher's elaborate printed profile stands near the beginning of a long tradition of civic panegyrics that flourished as Amsterdam continued to grow in size and status over the course of the century. Later city descriptions reiterated the claim that Amsterdam was a school for all peoples and directed readers (especially foreigners) to an expanding number of public places where they could access the arts and learning that the city offered. Notably, the Doolhoven are featured prominently in such publications. As the authority of the ultraorthodox Calvinists lessened, particularly around midcentury, Doolhof publicity began to proliferate more openly, advertising astonishing things for curiosity seekers. For instance, Caspar Commelin's civic history devotes a full chapter to four of the Doolhoven, describing various exhibits in detail and stressing that the sites offered "fine things for foreigners such as fountains, the

FIG. 7 Crispijn van de Passe, title page from *Dool-hof, Staende op de Roose-Gracht* (Amsterdam: voor Philips Lingelbach, 1660). Photo: Allard Pierson, University of Amsterdam Library Special Collections, OTM O 60–2758.

DOOL-HOF,

Staende op de Roose-Gracht, by de derde Brugh,

Alwaer vertoont worden verscheyden seer konstige Wercken, soo van Fonteynen als Astronomische Uurwercken. uytbeeldende eenige wonderlijcke Geschiedenissen: oock Herders Kluchten, met haer Beweging en Verroeringe, doende elck Beeldt sijn Actie of het leefde, tot Verwonderingh en Vermaeck der Aenschouwers, gepractiseert door *Mr. David Lingelbach* van *Franckfort*, als mede door sijn Soon *Philips van Lingelbach.*

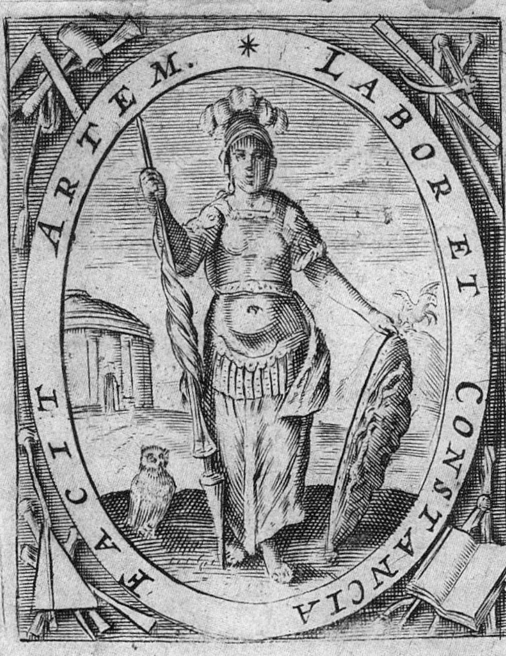

Voor *Philips Lingelbach*, Woonende op de Roose-gracht, by de derde Brugh.

c. 1660

performance of many moving histories, and many other similar things to see."[66] The sources proclaim that one of the main civic functions of these exhibition sites was to surprise and impress the city's many immigrants and foreign visitors with distinctive and inventive things that they could only see in Amsterdam, an audacious assertion that is interrogated further in chapter 5. In this sense, the Doolhof publicity substantiates the claim made by Visscher's print: Amsterdam was a new Athens, a center of learning for all the peoples of the world. The aim of educating civic populations was thus extended to the global populace. In this context, the republican city's art-exhibiting inns were important gathering places where travelers and locals could experience an esoteric collection of unusual artifacts. Situated as points of intersection in rapidly expanding networks of global transit, these distinctive venues mediated cultural diversity and knowledge exchange through pleasure and entertainment.

Further evidence that links the Amsterdam Doolhoven to the ideal classical culture of ancient Athens is found on the title page of a Nieuwe Doolhof pamphlet (fig. 7). The text introduces the "various artful works" displayed there, which include fountains, clockworks, and wonderful moving histories. These are described in terms of their beweeglijkheid: "Each image moves and acts as if it lives, for the wonder and delight of onlookers."[67] The frontispiece pictures Athena or Minerva, patron of Athens and goddess of wisdom, craft, and artistic practices. In Athena's left hand is her famous shield. Obliquely, we make out the Gorgon's snaking hair, open mouth, and wide staring eyes. The slanting view protects us, because this artful shield is a formidable weapon: the Medusa's gaze is so horrific that it petrifies those who look at her directly. Athena's shield is a particularly apt motif for the title print, since it is the moving image par excellence. The Gorgon's sinuous movements and petrifying stare can change humans into immobile stone statues. This Medusa effect is a reversal of usual viewing dynamics, as the animated work of art turns its beholders into inanimate artworks. A similar mode of viewing, involving the arresting effects of active artworks, was articulated in descriptions of visitors to the Doolhoven being "struck with wonder" by moving images so that they "could not even understand . . . but only gazed in wonder."[68] The blows of the iconoclasts seem to be reversed as reanimated graven images strike back.

This ability raises a key question about beweeglijkheid and the cultural strategies of the Doolhof displays. If the power of moving images was to transform viewers for the betterment of society, how exactly was this accomplished? It seems that the serpentine movements of the labyrinth led spectators into a place filled with lively artworks—a sort of lair where monstrous works could strike at them. If moving statues were mobilized to arouse strong passions like terror, desire, and wonder, would this not create chaos, confusion, and even idolatry rather than social order? The answer to this conundrum lies in the contradictory dynamics of passion and dispassion, motion and stillness, that characterized these exhibition sites and seventeenth-century aesthetic theory more broadly.

The Nieuwe Doolhof title print image shows Athena ringed with the Latin axiom: *Labor et Constancia facit Artem* (Labor and constancy make art) (fig. 7). The phrase resonates with civic ideals, specifically the practical implementation of Neostoic philosophy in urban life. A pivotal work was the 1583 treatise *De Constantia* by Leiden University professor Justus Lipsius. Described as a "practical guide to the art of living,"[69] this book revived interest in ancient Stoic philosophy across Europe and was extremely influential in the Dutch cities, where it was taken up as a working model for public life.[70] Neostoicism was widely adopted by the Dutch Republic's civic and political leaders as well as its foremost intellectuals, including playwrights and artists.[71] Lipsius counseled that public evils, especially the violence, upheavals, and uncertainties of war, were inescapable, and the only way to attain peace was by changing the self. The cultivation of immovable strength of mind calms the internal chaos that results when human emotions are stirred by unpredictable external events,[72] an insight succinctly expressed in the dictum of Michel de Montaigne: "Not being able to rule events, I rule myself."[73]

The Neostoic recommendation was to seek out the experience of strong passions with the aim of examining and moderating their potent effects. By learning not to be governed by fleeting emotions, the transformed person could meet every public evil with stoic constancy, dignity, and self-control. The difficult (if not unattainable) goal was to reach a dispassionate state of composure and contentment called *apatheia*. Heinsius, who was Lipsius's colleague at Leiden University, applied these ideas to aesthetic theory. In his important engagement with Aristotle's *Poetics*, Heinsius advises that the utility of emotionally stirring theater performances was to excite the passions in order to purge them, reducing them to the right measure and forcing them into order. Like Lipsius, Heinsius recognizes the futility of trying to use reason to completely overcome or suppress irrational emotions and advocates channeling passion's potent forces.[74] This approach resonates with the concept of beweeglijkheid: a performative work of art should have a forceful impact on viewers, stimulating physical motion, powerful passions, and ultimately transformation to a calm and balanced state of apatheia.

In keeping with the principles of Neostoicism, visitors to the Doolhof exhibits did not just attentively look at or study visual representations of classical myths. While the moral messages offered may have been similar to those discovered while probing emblems and paintings for deeper iconographic meanings, these labyrinth gardens solicited a fully embodied, eventful, and participatory understanding. Through their motions and emotions, participants physically took part in the narratives, at times even performing the actions, sensations, and decisions of characters like Theseus, Ariadne, Bacchus, and the Minotaur. By actively exploring ancient lore, audiences were challenged to examine and improve their own behavior and, ideally, to attain the equanimity of apatheia. This goal is what was promised by the inscription above the Doolhof door, which assured that lively performances, fountains, and music would reward visitors with "the purest joy and contentment" (see fig. 2).

Theories about scientific observation and learning also followed Neostoic precepts. In chapters 4 and 5, I will explore how philosophers like Aristotle and René Descartes advocate that emotive encounters with strange inventions can stir passions and arouse curiosity, moving viewers to philosophical contemplation and the acquisition of new knowledge. In a sense, then, the viewing of strikingly illusionistic and experimental works of art could potentially create the detached and objective observers that Alpers and Crary describe. If this was the goal, however, then it is important to recognize that in seventeenth-century aesthetic theory, the process of arriving at disengagement involved intensely engaged combat with the formidable emotive energy aroused by art.

Moreover, the position of the detached observer was one of rank and privilege. Neostoic cultural exercises have been described as emancipatory; instead of top-down political and religious control, entertainment and pleasure were employed to encourage individuals to assume active responsibility for the ethical cultivation of the self.[75] However, in what follows, we will see that Neostoic practices driven by Protestant zeal also produced a disciplinary apparatus intended to change others, especially the nonelite groups frequently disparaged in assessments of the Doolhof audience: women, youth, peasants, the urban poor, and non-European foreigners. Emotive, uninhibited, and embodied viewing practices were commonly associated with these groups, while calm, objective observation was most often posited as the cultural achievement of educated European gentlemen.

Moving Statues is structured in three parts. The first part introduces the proposed ritual route through the exhibitions, where drinking and art appreciation stimulated the arousal of passions, which were then curbed by the frustrating and difficult trial of the labyrinth, preparing peripatetic visitors for an intense encounter with the moving inventions inside the sculpture gallery. Chapter 2 begins with the first monumental artwork that visitors would have encountered at the Oude Doolhof: a multimedia sculptural fountain representing the *Triumph of Bacchus and Ariadne*. Situating this work in relation to innovations in hydraulic technology emphasizes its novelty; it was advertised as a triumph of Amsterdam. The intended functions of the work are assessed in relation to the classical accounts of Ovid and Catullus, which are put into dialogue with Heinsius's illustrated poem *Ode to Bacchus* as well as festive understandings of drinking and creativity. Primary sources indicate that engagement with the figural fountain in the tavern yard had the power to affect viewers so that they enacted some of the transformations that occur in the tale of Bacchus and Ariadne. Female visitors, for instance, were especially singled out and mocked for becoming like Ariadne in their sensuous submission to the wine god.

Spirited arousal was preparation for the next part of the itinerary, and chapter 3 follows the prescribed route into the labyrinth. Here we decipher how the challenging exercise of walking the maze was designed as a rite of passage that activated Calvinist understandings of original sin. The confrontation between courageous Theseus and the brutish Minotaur was experienced as an inner struggle to overcome the beastliness that lurks within every person, a battle that ideally stimulated self-awareness and an ethics of active participation in civic life. A consideration of the depiction of the Doolhof labyrinths on maps of Amsterdam suggests the place of these civic mazes within a wide-ranging program of urban reform, as the ambitious expansion and redesign of the city worked in tandem with disciplinary attempts to tame its populations. In this context, comparison to Amsterdam's prisons highlights what was distinctive about the Doolhof's maze gardens, which seductively used artistic invention and aesthetic pleasure as forms of correction.

The second part focuses on the Doolhof automata shows. Their particular modes of display are related to theories about bodies, passions, and machines articulated by philosophers like Aristotle, Bacon, and Descartes, all of whom specified how the exhibition of automata provoked strongly emotive reactions in order to stimulate thought and refine human behavior. With this understanding in mind, chapter 5 turns to Doolhof advertising that explicitly encouraged foreigners to visit these potentially transformative displays. Analysis of the publicity materials shows how foreign-looking automata were likened to slaves and how their foreign viewers were likened to automata. The forceful effects of the moving statue were thus deployed in support of the fiercely competitive and exploitative mercantile practices that brought about Amsterdam's precipitous rise. The chapter then introduces a case study that shows how early modern artistic invention was activated by the global mobility of people and moving art objects. The main focus is the inn of the scholar Theunisz, who honed his language skills and cultural knowledge in conversation with international visitors, including a Muslim diplomat from Morocco. I link the inventions displayed at this tavern with the innkeeper's access to rare Arabic manuscripts about making automata. These various interactions expose what the inns did not publicize: how their technological innovations were indebted to transcultural encounters and how foreign guests made significant intellectual contributions to the cultural life of the Dutch Republic.

The third part considers the Christian paganism of the exhibits. Chapter 6 turns to the display of waxwork portraits of Protestant leaders and infamous enemies. Like labyrinths, fountains, and automata, wax sculpture was an ancient and significant cultural form embedded in meaningful social, religious, and political rituals. The ability of wax to mimic life allowed it to channel the animated presence of historical personages into the present moment. Emotive responses to lifelike wax portraits were a means to provoke passionate political actions. Trading on the animated powers of wax portraiture and its previous ritual practices, the Amsterdam pleasure parks created a new kind of public ritual space for this highly efficacious form of political imagery, especially in order to generate active support for the Protestant cause in the religious wars that divided Europe.

The final chapter takes up the elaborate astronomical clocks and mechanical moving pictures that prophesized the coming domination of Protestantism over all the earth. Focusing on the clockwork created by Frankfurt immigrants David and Philips Lingelbach at the Nieuwe Doolhof, chapter 7 opens with the complex mechanized shows that performed the Old Testament prophecy of Daniel. Viewing these moving pictures, audiences could watch the rise and fall of various world religions and the eventual triumph of Protestant Christianity unfold in sequence. This visionary clockwork did not just represent or remind viewers of the end of time—their advanced technologies actively ushered it in. Presentations of ingenious inventions urged Reformed visitors to take decisive action in a critical age and to persuasively convert foreigners to Protestantism. Examining the connections between mechanical arts and evocations of a golden age, we come to understand how the Doolhof's clockwork was designed to play an apocalyptic role in imposing religious conformity throughout the world. In the manner of a labyrinth, the epilogue of *Moving Statues* returns to its beginning by assessing the global ambitions of these extraordinary exhibitions in terms of their failure and obsolescence.

The symbol of the labyrinth is a pictogram of controlled confusion, which best signifies the dynamic workings of a Doolhof. More than just taverns where art was displayed, the Doolhoven were designed as ritual spaces that revived pagan sculptural forms and cultural practices in an experimental program of Protestant reform. The rustic door of these Silenus-like pleasure parks opened into a realm of Bacchic excess designed to elicit squeals of terror, gasps of surprise, lascivious shouts of laughter, drunken stumbling, and the stunned silence of amazement as well as conversations and encounters among diverse peoples and inventions exhibited to foster specific Reformed ideals of social, political, and religious improvement. Surely among the most contradictory cultural experiments of early modernity, these were beer gardens for the cultivation of a Protestant self.

RITUAL ROUTES

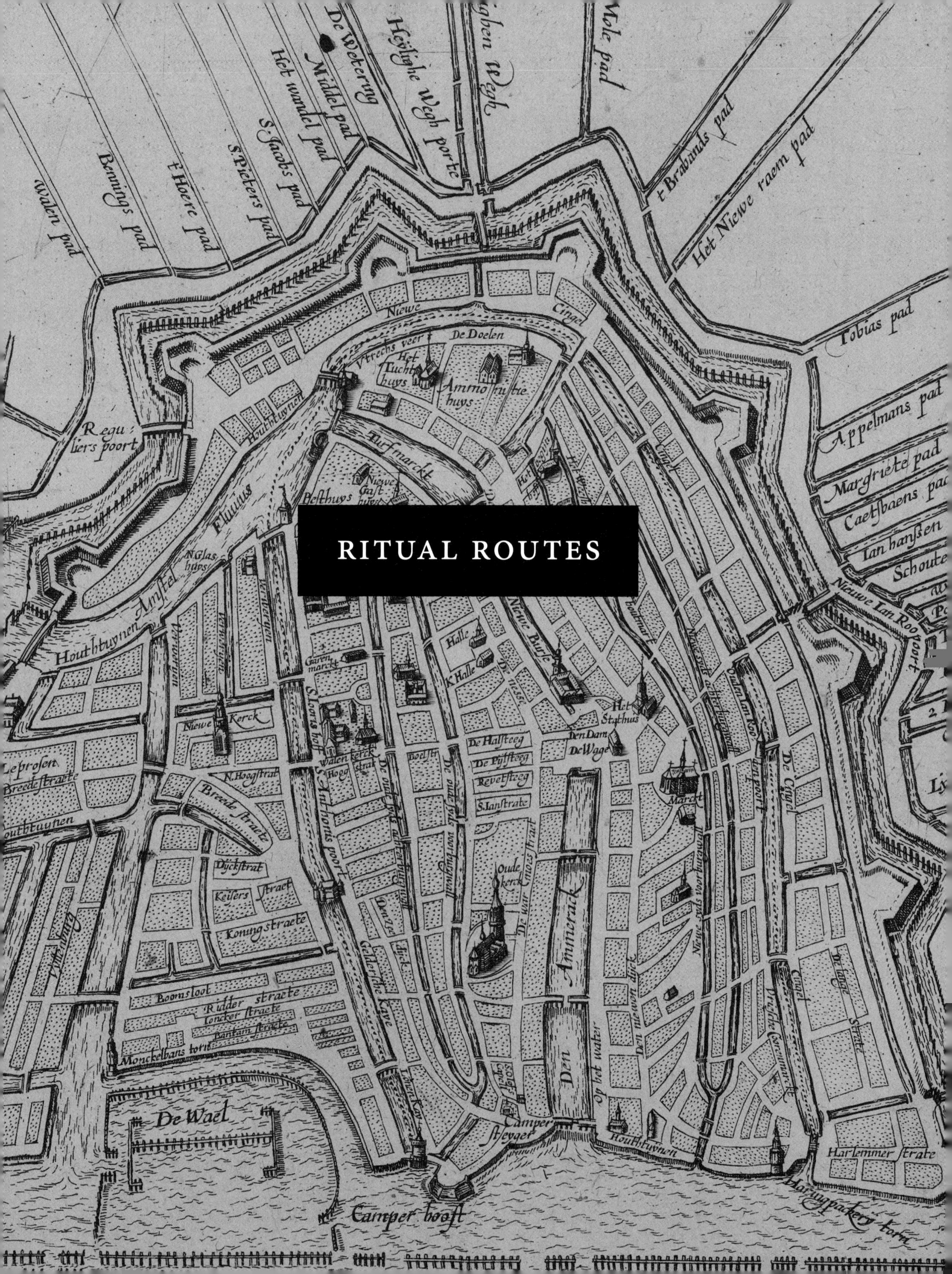

THE COURTYARD FOUNTAIN

Chapter 2

Bacchic Rites

Wine being drunk takes away Cares and Vexations from our Minds and adds a sort of Cheerfulness to them. . . . Wine both stirs up Wit and administers Eloquence.
—Erasmus, "The Profane Feast," 1518

For those traversing the bustling streets and waterways of seventeenth-century Amsterdam, the Doolhof gardens at the corner of Prinsengracht and Looiersgracht were heard before they were seen. The Oude and Nieuwe Doolhoven were located at the edge of the New Town, an urban expansion begun in 1610. Prinsengracht, the outermost of the newly dug ring of canals, the *grachtengordel*, linked the city to the countryside and connected the rivers Amstel and IJssel, the main transport routes in and out of the center. Traffic was directed along Prinsengracht in order to preserve the inner two canals—Herengracht and Keisersgracht—as quieter residential areas for affluent burghers. This made Prinsengracht one of the busiest arteries; the canal was filled with boats, and its quays were lined with markets, warehouses, and businesses, including numerous taverns and inns, many catering to foreign visitors.[1] Something extraordinary in this noisy soundscape was needed to interrupt quotidian circulations. A blast of trumpets, pipes, and drums at the corner of Looiersgracht assaulted the ears, directing attention to a portal there.[2] The inscription above the entryway to the Oude Doolhof mimicked the voice of a street crier whose job was to attract customers. Echoing his call, especially when read aloud, it commanded: "Don't just stand in front of this door, walk in."[3] Pausing at the threshold, the passerby heard, rather than saw, the promise of attractions within: the play of a fountain and organ music as well as tinny bleating and chirping punctuated by the chiming of bells. There was laughter and chatter but also startled screams, as the tinkling of delight mingled with shrieks that sounded like terror.

Those who interrupted their everyday expeditions in order to satisfy their curiosity would have passed through the portal into the spacious terrace of a tavern. Shaded by trees, the courtyard offered places to sit and enjoy glasses of wine or beer. A typical visit, then, might have begun with a drink or two. Souvenir pamphlets were on sale; the visitor could peruse one to get an overview of the site's attractions (fig. 8). The title page from 1648 features an image of a round hedge labyrinth and promises to explain various artful works whose movements were driven by clockwork, as well as a beautiful fountain—*The Triumph of Bacchus and Ariadne*—and some appealing life-sized statues of famed kings and princes of this century. All of this could be seen in the Oude Doolhof of Amsterdam on the corner of Looiersgracht, where "more and more improvements are made yearly."[4] Visitors thus discovered themselves in an art park and pub.

The exhibition of works of art in a drinking establishment was not a new practice. In fact, Dutch artist and theorist Karel van Mander praises this type of arrangement in his *Schilderboek* (Book of painters). First published in 1603 and reissued throughout the seventeenth century, it was no doubt familiar to the Doolhof proprietors and many of their art-loving visitors. Van Mander underlines the importance of art collections, notably in the wake of iconoclasm. He writes that during the troubles and the breaking of images, numerous artworks were stolen and secreted.[5] Private cabinets and galleries served as refuges for the arts during the war against images, hiding them away and protecting them. The work of art demands to be seen, however, and Van Mander was conscious of the need for public exhibition sites where liefhebbers, art lovers, particularly from abroad, could have access to the admirable arts of the Netherlands. Van Mander dedicated his *Schilderboek* to two art-loving Haarlem brewers who provided just such a space. He begins by lauding the manner in which these tavern keepers combined their passions for art and for beer. He then commends them for collecting "praiseworthy and important pieces by the most accomplished hands which you like to show in an amiable and generous manner to foreign gentlemen, strangers or other lovers of art—upon which occasions, moreover, you pour samples of your own art, namely the pseudo-Bacchus, or the god himself."[6] In the convivial enjoyment of art and drink, tavern visitors encountered the god himself. While this assertion may sound like hyperbole, in what follows, we will see how the animating powers of the moving statue activated Bacchic rituals in the Doolhof tavern yard.

Bacchic Triumphs

Exhibition sites are liminal spaces. They transport mobile spectators from the prosaic rhythms of everyday life into a ritual domain. As Carol Duncan shows, typically there is a prescribed route through an art exhibit. Following it, visitors animate a narrative that is designed to change them, usually by instilling particular social and political values along a transformative journey.[7]

FIG. 8 Title page from *Verklaringe van Verscheyden Kunst-Rijcke Wecken en haer Beweginghe: Door Oorlogie-Werck Ghedreven* (Amsterdam: Tymen Houthaeck, 1648). Photo: Allard Pierson, University of Amsterdam Library Special Collections, OTM O 62–2613.

VERKLARINGE

Van verscheyden kunst-rijcke wercken en hare beweginghe, door Oorlogie-werck ghedreven, benevens een seer schoone Fonteyn, zijnde een Triumph van Bachus en Ariadne.

Noch eenige aerdighe Beelden, levens-groote, van vermaerde Koninghen en Princen deser Eeuw.

Alles te sien in 't Oude Doolhof tot Amsterdam/ op de hoeck van de Lopers-gracht/ 't welcke jaerlijcks meer en meer verbetert wordt.

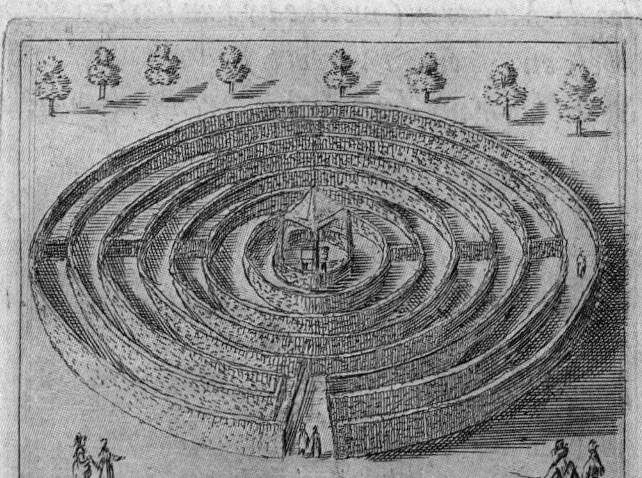

t'Amsterdam, Uit de Druckery van *Tymen Houthaeck.*

MDC XLVIII.

The labyrinth is specifically effective in this regard. In his *Natural History*, Pliny the Elder describes ancient labyrinths that contained "figures of gods, statues of kings, and effigies of hideous monsters."[8] Labyrinths have long been associated with the display of affective works, and to this day, many museums are experienced as compressed labyrinthine designs that take visitors from artwork to artwork along a twisting course that weaves back and forth through the building's windowless inner rooms and passageways. The ritual route highlights the most valued pieces as the curated experience unfolds.[9]

Extant publicity and commentary about the Oude Doolhof make it possible to recover the sequence of its exhibitions. From this, we know that the first work encountered on the proposed route through the site was the "extremely beautiful fountain of the Triumph of Bacchus and Ariadne" situated in the center of the main courtyard.[10] The fountain does not survive, and we mainly know about it from the guidebooks, which include an engraving by Pieter Holsteyn together with ekphrastic description (fig. 9); it also is discussed in some of the civic histories of Amsterdam and in various travel accounts. While the other Doolhof inns also had fountains, the Bacchus fountain at the Oude Doolhof attracted particular notice. Holsteyn's image shows an elaborate multifigural group arranged on top of a grotto surrounded by a moat. To either side of the main work are two basins that discharge vertical streams of water strong enough to suspend a ball and a crown. There is also a pronged candlestick that spurts water, and in the foreground, multiple jets of water spout up between the pavement stones. Standing at the center of the sculptural ensemble is Bacchus; naked except for a girdle and wreath of vine leaves, he holds a goblet and a bunch of dripping grapes. Ariadne kneels before the wine god; although she is fully clothed, high arcs of water appear to spray up from her breasts. Also featured is Bacchus's chariot pulled by two leopards; riding inside is a chubby putto, drinking from a flute, astride a wine barrel. Holding the reins of the wild cats is a crouching satyr who blows water from a long horn. Two more satyrs can be seen behind the main group: jets of water emerge from the mouth and from between the legs of the figure closest to Ariadne. While not visible in the Holsteyn print, some descriptions of this work state that the fountain group included Silenus on a donkey, which certainly is in keeping with the iconography of the triumph of Bacchus as well as the Doolhof's Silenus-box approach to colloquial classicism.[11]

Situated in the tavern yard before the entrance to the maze, this extravagant fountain was an effective transitional work between the two spaces, an animated presentation of Bacchus to stimulate the drinkers and of Ariadne to engage the maze treaders. Bacchus sets the tone. To combine the arts of brewers and vintners with art appreciation under the aegis of Bacchus must have appealed to the first proprietor of the Oude Doolhof, Vincent Jacobsz Coster. Coster was from one of Amsterdam's leading families: his uncle was Dr. Marten Jansz Coster, who had studied medicine in Bologna and later served as royal physician to King Frederick II of Denmark. Dr. Coster was appointed burgomaster of Amsterdam in 1578, the year of the *Alteratie*, or alteration, when the city established a new

FIG. 9 Pieter Holsteyn, *Triumph of Bacchus and Ariadne*, from *Verklaringe van Verscheyden Kunst-Rijcke Wecken en haer Beweginghe: Door Oorlogie-Werck Ghedreven* (Amsterdam: Tymen Houthaeck, 1648). Photo: Allard Pierson, University of Amsterdam Library Special Collections, OTM O 62–2613.

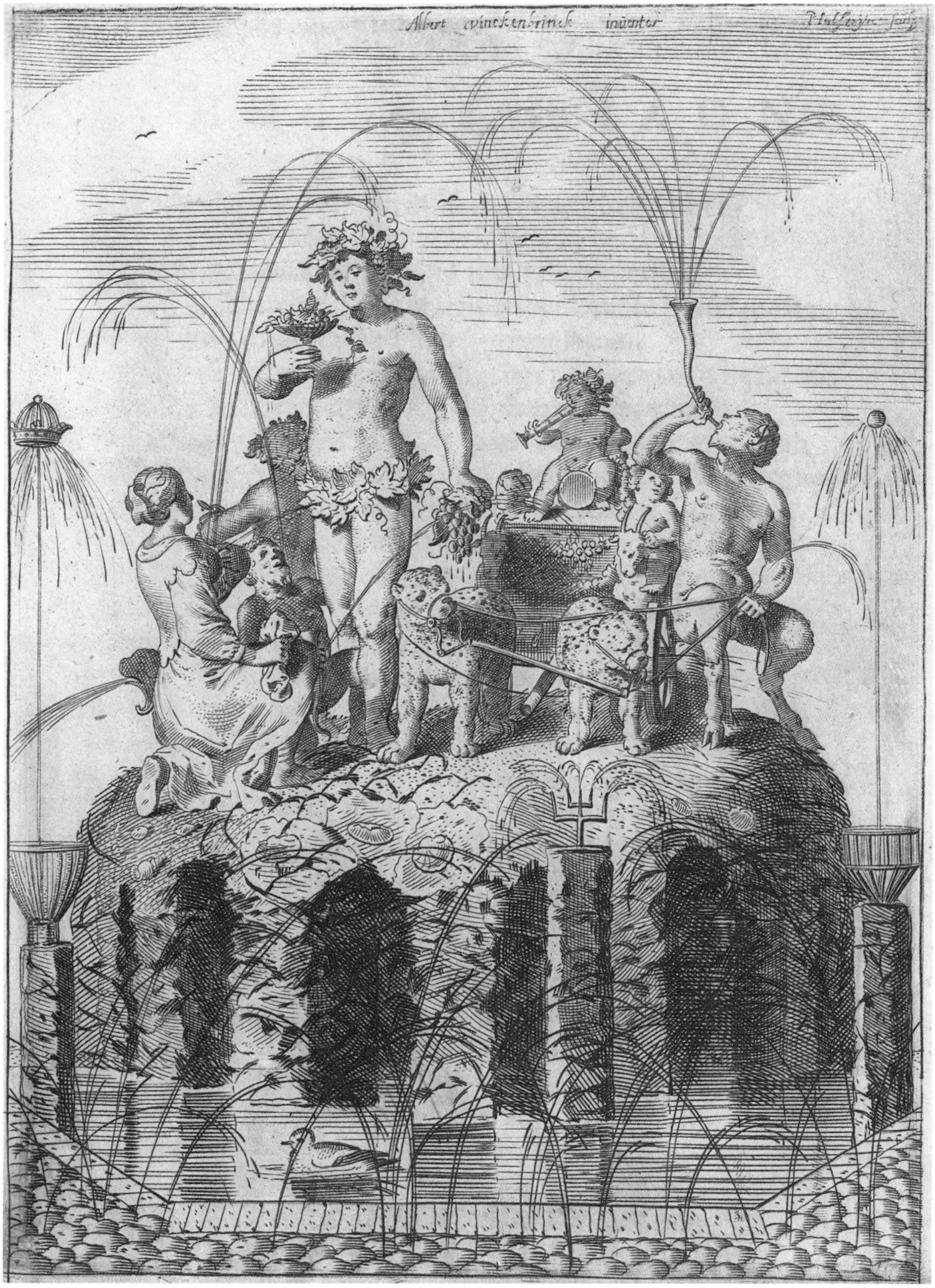
Albert cvinckenbrinck inuenter

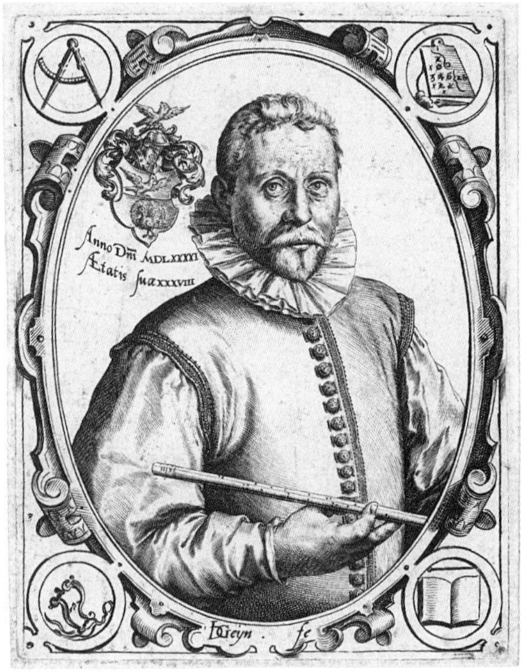

FIG. 10 Jacob de Gheyn II, *Portrait of Vincent Jacobsz. Coster*, 1591. Photo: Rijksmuseum, Amsterdam.

governing class composed of prominent citizens who adhered to the Reformed church.[12] Perhaps through his family connections in the civic government, Vincent Coster held the office of municipal wine assayer and was one of Amsterdam's main importers of wine: Bacchus was his god. A printed portrait of the publican by Jacob de Gheyn II highlights his Bacchic occupation (fig. 10). With the family coat of arms floating prominently above his right shoulder, Coster proudly holds the measuring rod used to gauge the contents of wine barrels. In the corners of the print are other tools of his civic office: a compass, a book, and a writing tablet marked with an equation, possibly in reference to the assayer's task of calculating excise taxes and duties on wine. There is also a bridle, serving as a reminder to curb the excesses of drink.[13]

Another portrait of Coster, a marble portrait bust, indicates his innovative artistic patronage (color plate 2). Made by De Keyser, this highly unusual work is the first known marble classical bust of a middle-class citizen to appear in the Dutch Republic, showcasing Coster's investment in reviving forms of classicism.[14] The portrait is enlivened by a draped cloak in the style of the ancient Romans, which is fastened in front with a small mask of a snarling lion. The base of the bust comes to a point above the open-mouthed face of a horned creature, possibly a satyr. Growling animal faces contrast with the calm, slightly upturned face of Coster, whose eyes look upward. The virtuoso formal balance of the marble bust on its small base conveys the equilibrium of the sitter's serene visage in juxtaposition with the agitated expressions of the beasts. A tension between precariousness and stability animate this skillfully executed work.

Like the publicans praised by Van Mander, Coster was a significant patron who owned a substantial art collection in addition to the Doolhof exhibits.[15] We might describe him as an innovative curator as well. His tavern was a peculiar kind of exhibition space similar to the one Van Mander described: a drinking establishment run by a leading citizen and avid art lover, where the enjoyment of various ingenious and admirable arts included local brews and fine imported wines.[16] Bacchus was obviously a fitting deity, and the god was doubly present in the Oude Doolhof tavern yard—in the drinks and in the fountain sculpture.

The figural fountain of Bacchus and Ariadne was a major artistic commission, and Holsteyn's print includes the name of the sculptor—Albert Vinckenbrinck. There is a 1648 engraved portrait of Vinckenbrinck, also by Pieter Holsteyn, done after a painting by his brother Cornelis Holsteyn (fig. 11).

Lying together with the sculptor's tools on the table before the sitter is a sketch of Vinckenbrinck's most renowned work, the elaborate carved wooden pulpit in Amsterdam's Nieuwe Kerk.[17] One of the sights of Amsterdam, it elicited much admiration, as in the description of English tourist William Mountague, who writes: "Nor did we ever see such a Pulpit in the World . . . built broad at the lower part, and so up like a Pyramid, all very curiously carved."[18] Vinckenbrinck's Bacchus fountain was equally curious and also attracted attention; numerous primary sources describe it as an extraordinarily intricate and artful work. The rockwork of the grotto solicited special notice: a collection of precious stones and seashells from far-flung shores, this costly display of exotic collector's items impressed viewers with the rarity of the design while transporting them to the distant isle of Ariadne's abandonment.[19] The blending of natural elements, like rocks, shells, and water, with human-made forms made fountains appear as wondrously paradoxical assemblages of art and nature. The sixteenth-century Italian humanist Claudio Tolomei eloquently sums up this merging by describing the pleasures of fountain craft "in which mixing art with nature, one can't judge if it is the work of the former or the latter; thus, one appears a natural artifact and another, man-made nature."[20] Artful fountains harnessed the forces of nature itself.

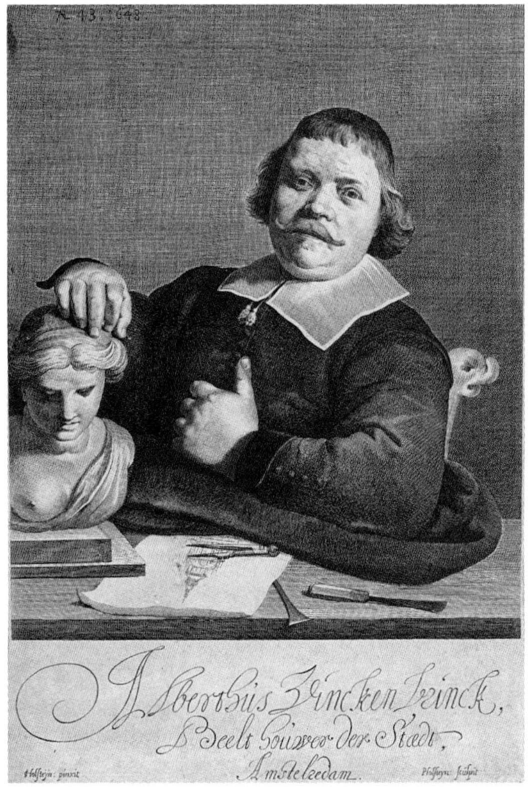

FIG. 11 Pieter Holsteyn, after Cornelis Holsteyn, *Portrait of Albert Jansz. Vinckenbrinck*, 1648. Photo: Rijksmuseum, Amsterdam.

The hydraulic technologies used to manipulate the flow of water were particularly noteworthy: "Nothing like this has ever been seen in the Netherlands before," boasts one of the advertising posters.[21] This inventiveness becomes apparent when we compare *The Triumph of Bacchus and Ariadne* to the earlier fountains made by Bargois featured on the 1622 Oude Doolhof poster (see fig. 3). Apparently, the guidebook's claim that "more and more improvements are made yearly" was no exaggeration. The sequence of Oude Doolhof advertisements reveals how fountains—which were extremely costly—were routinely replaced or upgraded. The fountains presented in the 1620s publicity do not appear in the later series of guidebooks, which began to be published at around midcentury. Archival records reveal that Bargois sold his first two fountains—which he called "the Jonas" and "the little ball"—to his beer dealer in 1626 for the princely sum of six hundred guilders. He then rented the garden adjacent to the Oude Doolhof and set up the Nieuwe Doolhof on

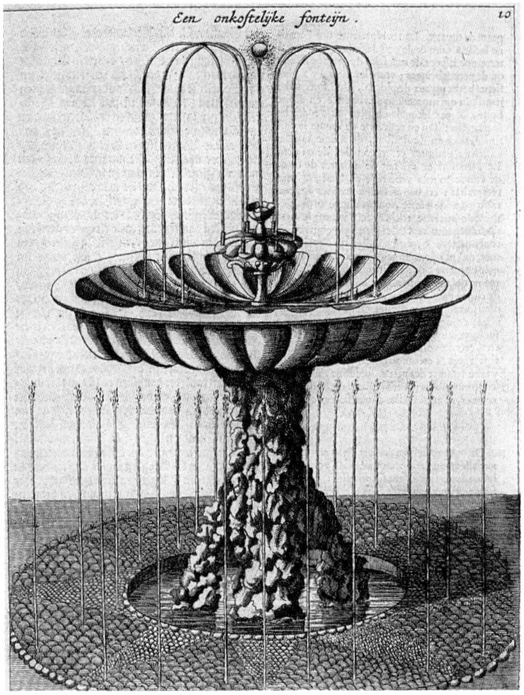

FIG. 12 Jan van der Groen, *Een Onkostelijke Fonteyn*, from *Den Nederlandtsen Hovenier* (Amsterdam: wed. Gijsbert de Groot, 1699). Photo: Bibliothèque nationale de France, Paris.

Looiersgracht, for which he created at least three new fountains.[22] Apparently no expense was spared as older installations were replaced or transformed into elaborate new assemblages.

The jets of water spurting up from the pebbled pavement in Bargois's fountains recur in Vinckenbrinck's work. Also similar are some aspects of the two water features, to the right and left of the Bacchus group, in which a crown and a ball are suspended by the force of water. A comparable device animates the fountain that Bargois named "het bolletje" (little ball). This work has a rocky base topped by a double basin configuration from which a high column of water elevates a ball. Water appears to pass through the orb, dividing into smaller sprays. A very similar, albeit rather scaled-down, version of the same type of fountain appears in one of the most popular Dutch books on gardening, *Den Nederlandtsen Hovenier* (The Netherlandish gardener), written in 1669 by Jan van der Groen, gardener to the Prince of Orange (fig. 12).[23] The illustration is labeled *Een Onkostelijke Fonteyn* (A priceless fountain). Like Bargois's design, it has a rock base, water jets that spout from a pebbled path, and a strongly powered central spray that pushes aloft a spurting sphere.

Such marvels were driven by new developments in hydraulic-pump technology. The journals of the Dutch philosopher and artisan Isaac Beeckman reveal that he was experimenting with pumps in the 1620s, using air pressure to make water circulate. Of interest for our consideration of pub fountains is the fact that Beeckman undertook this labor for beer breweries, businesses that relied on the supply of fresh, sweet water. Clarity of water was an important aspect of the aesthetics of fountains,[24] and this was a challenge in Dutch cities whose fetid canal water would not have been considered appropriate for garden fountains. Amsterdam's brewers brought in fresh water by boat from outside the town.[25] Possibly the tavern owners tapped into this supply for their fountains, and they may have used rooftop cisterns to collect rainwater.[26] The movement of fountain water, especially the trick of forcing it to rise and spray, was another difficulty due to the flatness of Amsterdam's terrain, making hydraulic pumps particularly efficacious. We know that Beeckman applied his practical experimentation with brewery pumps to artful fountain design: in his journal, he notes installing more than one "eeuwig spuitende fontein" (eternally spouting fountain) in the garden courtyards of wealthy regents.[27]

Knowledge of fountain making was disseminated in these decades by treatises like the one written by the French Huguenot and hydraulic engineer Salomon de Caus. De Caus traveled around Europe: the gardens he designed in Heidelberg for the Elector Palatinate Frederick V and his wife, Mary Stuart, daughter of James I of England, were celebrated as a wonder of the world. He also worked for Louis XIII of France and for the Prince of Wales in England.[28] De Caus's 1615 treatise, *Les raisons des forces mouvantes avec diverses machines tant utiles que plaisantes* (Principles of moving forces with diverse useful and pleasing machines), explains various hydraulic and pneumatic principles that could be employed in the working of fountains.[29] He acknowledges that he derived some of this know-how from the Hellenistic texts *Pneumatica* and *Automata* by Hero of Alexandria, to which I will return in chapter 5, along with the possibility that knowledge gleaned from Arabic manuscripts also influenced fountain making in Amsterdam.[30] De Caus's treatise is a very likely source for Bargois: both men were French Protestants and fountain makers.[31] There are several illustrations in de Caus's book that demonstrate how to use compressed air to force water from vessels and pipes into a vertical stream strong enough to raise a copper ball. In fact, this was de Caus's signature piece: he derived the method from the ancient treatise of Hero and described it as a "secret."[32] *Les raisons* includes illustrations and instructions about how to install water features that are very similar to those of Bargois and Van der Groen, in which a column of water emerges from a rocky base to suspend a ball so that it dances above the pavement.[33]

Notably, the *Triumph of Bacchus and Ariadne* fountain is much more elaborate in scale and grandeur than those illustrated in de Caus's treatise and Bargois's poster, or even the "priceless fountain" pictured in Van der Groen's book. With its multiple large sculpted figures, exotic shells and rockwork, and various types of inventive hydraulic effects, the fountain would have been a fitting ornament for a royal garden. Usurping cultural strategies that were common in princely collections and estates, this fountain—the opening artwork encountered on a visit to the Oude Doolhof—was designed to make a forceful first impression. The publicity asserts that it struck visitors with awe. *The Triumph of Bacchus and Ariadne* was the triumph of Amsterdam. With such a work, the city could vie with princely powers and compete with the splendor of major European cultural centers—particularly Rome, Florence, Bologna, and Paris—whose public squares boasted innovative classical fountains. The large-scale public figural fountains of these cities were inventions of the sixteenth century and most often included statues and motifs derived from classical antiquity. Such works enlivened urban space; the life-giving element of water created an atmosphere of prosperity, fertility, peace, and harmony. The ability to manipulate nature also expressed political power.[34] The Bacchus and Ariadne fountain sprang from this tradition of urban figural fountains. Indeed, water was arguably the most important element in Amsterdam's rise to power: its rivers, canals, and seaport were the sources of its successful economy, which was based initially on fishing and increasingly on lucrative overseas trade. The Oude Doolhof fountain's key difference from civic fountains in other European cities was that in the years when orthodox Calvinism held sway, the classical gods could not spout forth on the public square but had to take refuge in walled courtyards.

Becoming Bacchante

As garden historian Elisabeth MacDougall argues, fountains are the most important feature of a Renaissance garden; they are the focus of the overall design and usually provide an interpretive key for understanding the significance of the whole site.[35] Clearly, the ostentatious Bacchus fountain was a fitting introductory work, promoting as it did the cultural ingenuity and magnificence of Amsterdam as well as the artful inventiveness of the tavern's exhibits. However, the work poses a narrative challenge when considering the suggested route through the exhibition space, since visitors encountered this fountain before they entered the labyrinth. To start with Bacchus and Ariadne is to begin at the end of Ovid's tale of the Cretan labyrinth. After slaying Ariadne's half-brother, the Minotaur, and solving the maze with the help of her clew of thread, Theseus sailed with Ariadne to the island of Naxos, where the hero callously abandoned his new lover. Just as Ariadne awoke to realize that she had been left behind, she was rescued by the god Bacchus. This is the moment dramatized by Vinckenbrinck's fountain. The unwinding of the narrative in reverse offers, perhaps, a clue for deciphering this enigmatic exhibition site. Like a maze, the story of the Cretan labyrinth contains many twists and sudden changes of fortune. The theoretical term used by Dutch playwrights and artists to designate such reversals was *staetveranderinge*, a change of state.[36] A staetveranderinge is filled with the potential of transformation. As the most powerful moment in a performance, its emotive charge, or beweeglijkheid, could deeply move the audience members and thus affect an alteration in their inner states. As the seventeenth-century art theorist Franciscus Junius writes, a successful work of art is performative; it should affect us "not otherwise than if wee were present, and saw not the counterfeited image but the reall performance of the thing."[37] Vinckenbrinck accordingly chose an intensely moving scene for his affective sculpted fountain, and the impassioned impact of this work was the introductory experience of the Oude Doolhof exhibition.

Doolhof guidebooks explain that the fountain conjured the moment in which Ariadne wakes in confusion to discover that Theseus has set sail without her. In this fraught instant, the Cretan girl—keeper of the labyrinth's secrets and holder of its thread—knows that she has lost her way. Abandoned on route to Athens, she cannot return home to Crete. Ovid expresses the poignancy of her acrid despair as this realization dawns: "No, for my father and the land ruled by my righteous father—dear names!—were betrayed by my deed when, to keep you, after your victory, from death in the winding halls, I gave into your hand the thread to direct your steps in place of guide—when you said to me: 'By these very perils of mine, I swear that, so long as both of us shall live, thou shalt be mine!' We both live, Theseus, and I am not yours!"[38] There is a double twist here: Ariadne's betrayal of her family turns against her now that she has been betrayed. The ancient Roman poet Catullus encapsulates her remorse in a poignant lament: "Or can I hope for father's help? Whom I deserted / To follow a youth spattered with my brother's blood?"[39] Unable to go back or travel onward, Ariadne is stranded in a state of limbo.

The Doolhof guidebooks cite a poem by Dutch playwright Jan Vos that evokes Ariadne's reaction to this shocking staetveranderinge by describing her facial movements: "Her mouth is laughing still, but the eye begins to weep."[40] The face of a statue normally is frozen in expression, but the fountain's play of water animates immobile stone. A smiling face appears to weep as water drips from its eyes, conveying conflicting emotions. Spurting and flowing water constantly alters stone surfaces, invigorating the figures and conveying the motion and emotion (beweeglijkheid) of this charged transitional moment. Laughing and weeping, Ariadne moves from the giddiness of new love to the embittered grief of love's betrayal. The affective enlivening of immovable stone resonates with Catullus's evocative description of Ariadne's painfully indeterminate state, as she stands on the shore "bitter-eyed, / like some Bacchante's stone statue, watches, ah / Watches, and tosses inwardly on great waves of troubles."[41] Stone-still and billowing with sadness, Ariadne is so moved by her plight that she is immobilized. It is a dramatic moment that lends itself particularly well to the sculptor's art.

Just as the stricken Ariadne twirls into a stone statue on the shore, there is another staetveranderinge also captured by the fountain sculpture. The god Bacchus arrives in the midst of a storm: water sprays wildly over him and his rough retinue of satyrs, putti, and leopards. Vinckenbrinck may have derived his interpretation of this particular moment in part from the provocative Dutch poem *Hymnus oft Lof-sanck van Bacchus: Waerin 't Gebruyck ende Misbruyck vande Wijn Beschreven Wort* (Hymn or ode to Bacchus, in which the use and abuse of wine are described). Written in 1614 by Heinsius, one of the Dutch Republic's leading classical scholars, the poem first appeared in print in 1616.[42] An edition published by Willem Blaeu includes engravings by Jacob Matham. Like Vinckenbrinck's fountain, Matham's engraved image of Bacchus's triumphant procession condenses several moments of the story (fig. 13). Matham shows Ariadne at the shore as Theseus's ships sail over the horizon behind her. With clasped hands, she has just turned from horrified realization of her abandonment to apprehend the noisy arrival of Bacchus with his parade of inebriated followers. In the shaded foreground area in front of her is a somersaulting satyr. The shadowy upside-down figure signifies reversal, a sudden cartwheeling, as events turn head over heels. Acting as a *repoussoir* figure, the tumbling satyr directs attention to Ariadne's twisting form in the brighter background scene. The picture compresses multiple turnarounds that underscore Ariadne's pivot from passionate love to angry grief as she sees Theseus sail away. Crying out for his change of heart, she vainly shouts into the wind: "Turn about your ship, reverse your sail, glide swiftly back to me!"[43]

According to Catullus, as Ariadne laments on the shore, "spinning convoluted cares in her mind," there is another turn as she hears the arrival of her raucous rescue party. Undoubtedly a key source for Heinsius,[44] the passage in Catullus is worth citing at length:

> But in another part Bacchus in bloom flew by
> With rout of Satyrs and Nysigenous Sileni,
> Seeking you, Ariadne, and burning with love for you.
> For him the Thyads raved around with frenzied mind,
> Shrieking *Evoe*, *Evoe*, twisting their heads about . . .
> Others with open palms were banging tambourines
> Or clashing ringing clangour from thin rounded
> bronze.
> For many, horns blared out cacophonous *boom-booms*
> And barbarous reed-pipes screeched hair-raising
> music.[45]

In his retelling of this stirring moment, Heinsius explains that Bacchus—or wine—has always been a

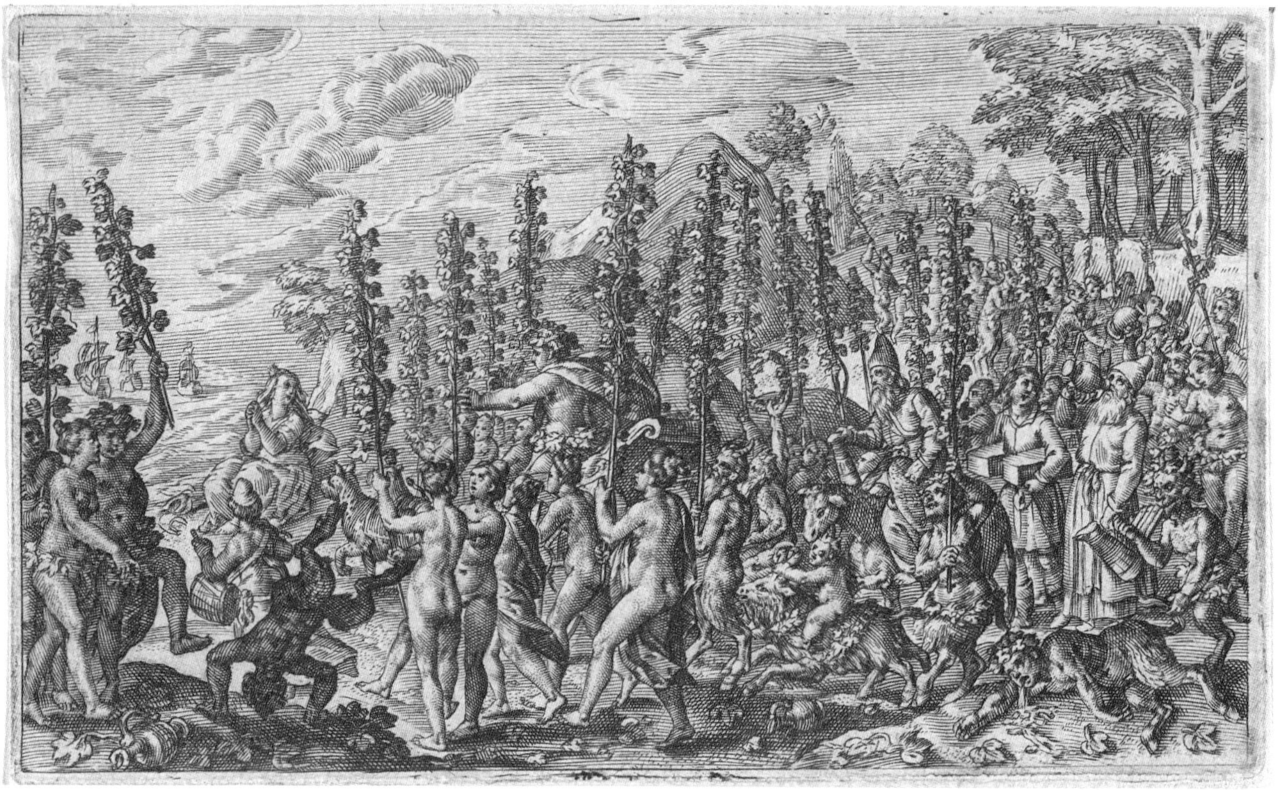

FIG. 13 Jacob Matham, after David Vinckboons I, *Bacchus Finds Ariadne on Naxos*, from Daniel Heinsius, *Nederduytsche Poemata*, vol. 3, *Hymnus oft Lof-sanck van Bacchus, Waerin 'tgebruyck ende Misbruyck vande Wijn Beschreven Wort*, ed. Petrus Scriverius (Amsterdam: Willem Jansz Blaeu, 1616). Photo: Rijksmuseum, Amsterdam.

helper in times of grief and misfortune, a refuge for men and women in need: "I am the great God, the father of wine, who drives away sorrow and pain. Minos's daughter swayed, and turning her sad face to him, rejoiced."[46] The fountain wittily expresses Ariadne's high-spirited assent to Bacchus's libidinous advance as spouts of water arc from her breasts to shower the wine deity, one spray hitting the vine leaves that girdle the god's genitals. If Holsteyn's engraving of the fountain is accurate, most of its water jets spurted from the bodily orifices of the figures and dripped from the grapes and cup held aloft by Bacchus (see fig. 9). This fountain thus appeared to transmute water into other fluids: saliva, breast milk, urine, semen, and, of course, wine. Its exuberant spraying together with the Oude Doolhof's noisy trumpets, pipes, and drums transformed the tavern courtyard from an everyday drinking hole into a mythic realm imbued with the forces of Bacchus.

By condensing the moment of Ariadne's abandonment with that of her rescue, the sculpture presented a number of turnings and multiple psychological states. It artfully compressed a great deal of motion and emotion so that the whole tale could be apprehended as if in the moving blink of an eye—what the art theorists called an *oogenblikkige beweeging*. This condensation is where a work of art differs from a theater performance or a written history. Whereas the performed or published story unfolds over time, the artwork captures a single loaded moment in which the full impact of the narrative is distilled.[47] The effect on the viewer is comparable to that of imbibing a concentrated draught of liquor.

Van Mander's observation that taverns could offer sanctuary to the beleaguered arts in the religious war against images thus aptly summarizes the purpose of the Oude Doolhof. Like a Silenus box, the semiprivate courtyard sheltered the Bacchus fountain, a transgressive sculptural work that certainly risked being condemned as pagan idolatry by the orthodox Calvinists, like Heinsius's *Ode to Bacchus*.[48] Defending himself against this religious censure in a letter to a fellow scholar, Heinsius emphasized that the intent of his controversial *Ode* was not to promote worship of pagan deities but to learn from classical antiquity. The preface of the *Ode* references Silenus with the claim that pagan myths may appear worthless on the outside but have great value inside, because they hide and contain profound wisdom and philosophy.[49] In his letter, Heinsius asserts that mythological deities metonymically stand for the powers of natural forces, like wine and love. He insistently repeats: "For they are no Gods, as I said before, but emotions, and they do not contribute to religion . . . but to terror."[50]

Surely a justification along the lines of Heinsius's protests could have been articulated for the intoxicating powers of the Bacchus and Ariadne fountain, a work like the Silenus box that secreted potent forces. As Heinsius insisted, the powers of Bacchus—or wine—are experienced as emotive energy, the arousal of strong passions. Bacchus is the god of *discordia concors*, a paradoxical state of balance between dynamic and dissonant elements.[51] While the many social ills associated with alcohol abuse were acknowledged by Heinsius and certainly were disparaged by many Calvinist moralists of the time,[52] the imbibing of wine also was lauded for its positive effects. The traveler Mountague, for instance, noted that the naturally phlegmatic Dutch stood in greater need of "strong and spirituous Liquors" because of their wet climate, which made them slow and heavy: "Their Country is a Bog, a Marsh, full of Vapours and Fogs . . . and therefore their frequent Tippling ought not to be charg'd on them as a Crime" for it "keeps out the Cold and the Fogs, quickens them, and makes them fitter for Action." In this account, frequent tippling brings about a beneficial balance: it makes a sluggish people livelier by offsetting the effects of a cold, wet climate on the humors that governed character. The lethargic wateriness of the Dutch needed the animation of strong spirits. Without the stimulation of alcohol, it would "have been impossible to have finish'd such great Works as they to their Honour, and the Safety of their Country have happily atchiev'd."[53] In Mountague's observations, Dutch ingenuity itself was indebted to the forces of Bacchus.

The benefits of Bacchus are conveyed in an etching by Cornelis Holsteyn (fig. 14), which depicts the same dramatic moment represented in his brother Pieter's print of the Doolhof fountain. The abandoned Ariadne turns sorrowful eyes up to the wine god, who arrives with his following of satyrs, bacchantes, putti, and leopards. The text along the bottom of this print provides a succinct summary that explains the prosaic significance of the story for every viewer, and indeed, for every

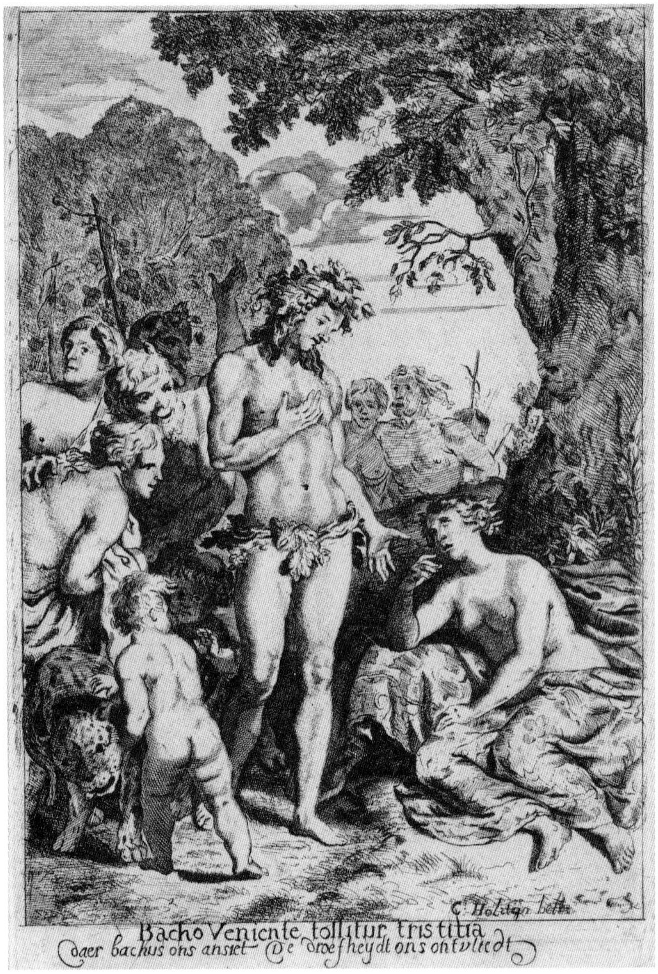

FIG. 14 Cornelis Holsteyn, *Bacchus Consoles Ariadne*, ca. 1628–58. Photo: Rijksmuseum, Amsterdam.

drinker: "daer bachus ons ansiet / De droefheydt ons ontvliedt" (when Bacchus looks upon us, sadness flees from us). Wine alleviates suffering; it effects a change of state. It especially triggers sensual pleasures, as with the bacchantes' strange and boisterous drunken rites and in Bacchus's seduction of the willing Ariadne.[54] Visitors at the Oude Doolhof, then, felt the joyous transformative force of Bacchus through their alcohol-enhanced sensory and emotional involvement with the artful fountain. In the process, they experienced something very like Ariadne's change of state—their troubles disappeared as they saw, heard, felt, smelled, and tasted Bacchus. Erasmus, whose insights run through Heinsius's poem, explains: "Wine being drunk takes away Cares and Vexations from our Minds and adds a sort of Cheerfulness to them. . . . Wine both stirs up Wit and administers Eloquence."[55] In this way, the drinking of alcohol prepared visitors for the Bacchic rites of this peculiar exhibition site. As cares dissipated into cheer, interpretive wits were aroused, priming drinkers for the next event in the arranged itinerary of interacting with the Doolhof's startling works of art.

As spectators stood near the Bacchus and Ariadne fountain, peering, perhaps, at its lubricious aspects, they unwittingly activated another set of artful waterworks. Trick fountains, installed under the pebbled walkway around the sculpture group, would suddenly spurt up jets of water, drenching those who stood there. The English baronet and member of Parliament Sir William Brereton visited both the Oude and Nieuwe Doolhoven in 1634. He later wrote about his experience of the courtyard fountains and noted with prurient delight that "those who stood within a circle, and near unto the rails, were well washed: a woman well washed under her clothes once or twice."[56] Visitors who sat drinking at the shaded benches and tables especially enjoyed the spectacle triggered by these cunning waterworks. Many sources assert how comical and pleasurable it was to see how the women and young girls who stood gawping at the fountain sculpture were surprised by cold water suddenly spraying up their skirts. They scream and they "hop about like new-born calves," which is very amusing for the bystanders, chortles Tobias van Domselaer in his 1665 history of Amsterdam, which includes a lengthy description of the Doolhof's attractions.[57] The commentators described the soaking of female victims; they did not recount their own unguarded reactions and undignified movements or those of other men, even if their lusty observations do reveal how their passions were stirred.

How was it that women in particular were the targets of hydraulic devices that were hidden underfoot, unnoticed until they spouted up to wet the unwary? Water jokes of this sort were featured in elite and princely private parks across Europe.[58] There is evidence dating back to the medieval chateau gardens at Hesdin in Flanders to confirm that this sort of water hoax had long been aimed at women. The 1432 account books of Philip the Good, Duke of Burgundy, catalogues the garden's mechanical artifices, which included: "a machine for wetting ladies when they step on it," "eight pipes for wetting ladies from below," and "everywhere under the pavement other pipes and 'engiens' to wet the ladies from below."[59] In the sixteenth century, the mechanisms of comparable devices were described in the travel journals of Montaigne. Visiting the estate gardens of the wealthy Fugger family outside of Augsburg in 1580, he observes: "While the ladies are busy watching the fish play, you have only to release some spring: immediately all these jets spurt out thin, hard streams of water to the height of a man's head, and fill the petticoats and thighs of the ladies with this coolness."[60] Montaigne encountered a similar trick at the famed gardens of Cardinal Ippolito d'Este at the Villa d'Este of Tivoli, outside Rome. Again, Montaigne notes that the hydraulic attack was "launched by a single spring that can be worked from far off."[61] Montaigne's remarks reveal how a person positioned at a distance deliberately chose female victims and then activated the seemingly capricious and indiscriminate waterworks. This practice indicates the likelihood that in the Doolhof courtyard, as well, an operator watched and waited for female marks to stand just above the concealed pipes before releasing the spring that triggered the sudden emission.

As Montaigne describes it, this wetting from below was an erotic experience—the discharge of forceful squirts of water went up the ladies' petticoats to fill their thighs. The commentaries of both Brereton and Domselaer about the Doolhof waterworks imply that in this setting, hydraulic ejections were directed at female spectators who had already been aroused by gaping at the lascivious sculptural group of Bacchus's pursuit of the overly ardent Ariadne. In his theoretical treatise on Aristotle's concept of tragedy, Heinsius expressed a concern that may have troubled the Doolhof proprietors as well: while certain things can "excite and move similar passions in everyone, they do so even more in emotional people" who tend to "lose control" of themselves.[62] Heinsius followed Aristotle in claiming that women were among those who were most prone to lose command of their bodies and emotions: they consequently did "not have a correct understanding of public affairs, or ordinary life, or wisdom, or learning."[63] Ariadne is an exemplar of female inconstancy. Her betrayal of father, brother, and homeland was driven by an overwhelming passion for Theseus, and in the end only Bacchus, the god of excess, could fully slake her immoderate desires.

In this way, the ritual itinerary of the Doolhof exhibition park made distinctions between different kinds of visitors. The broad audience who enjoyed these pleasures clearly included women, but they were singled out as particularly prone to inappropriate and overly passionate responses. The trick fountains operated as a sort of shaming rite, using sensory stimulation to both trigger and mock uncontrolled female behavior. Indeed, part of the joke was how it played with theories of the humors that governed human bodies and temperaments, specifically the idea that a typical female body was cold and wet. Heinsius's colleague Hugo Grotius summarizes this view in his 1631 *Introduction to the Dutch Jurisprudence*: "The female sex, being colder and moister, customarily has lesser ability for affairs that require understanding."[64] Women's supposed coolness and wetness apparently made them impressionable; they were characterized as acting largely in reaction to external forces and thus were governed by inconstancy rather than reason. Trick fountains that aimed to drench women from below accordingly marked the cold and wet female body by setting off unchecked behavior. The association of women with fountains lent itself to many misogynistic jests. As William Shakespeare put it in *The Taming of the Shrew*, "A woman moved is like a fountain troubled."[65] Such jokes had serious consequences, since they dismissed women as unfit for full participation in public life.

This classical fountain, therefore, did not merely represent or allegorize a moralized ancient story; it actuated both the physical movements and the various emotional states that occurred in the narrative of Bacchus and Ariadne. Overcome by drunkenness, licentiousness, and sensuality, the cavorting, screeching, and guffawing spectators were initiated—like Ariadne—into Bacchus's rowdy retinue. The sudden sensational shock of water on flesh prompted unthinking physical reactions and heightened emotions—starts and shrieks of surprise; mocking, joyful, or embarrassed laughter; and drunken lasciviousness. In this way, the fountain activated multiple Dionysian aspects of the Doolhof by triggering unrestrained bodily impulses in both female targets and male onlookers, intensifying emotions and arousing passions. Visitors became bacchantes through the uninhibited behavior stimulated by the fountain. As Heinsius emphasizes, Bacchus was a loosener of limbs: "He unbinds our senses and understanding and devours our cares."[66] The pranking waterworks activated a fluid release of body and mind, freeing visitors from conventions of civility and the constraints of proper public manners. This artful work affected the visitor like heady wine. In his 1609 commentary about Bacchus in *The Wisdom of the Ancients*, Bacon concludes: "Of all things known to mortals, wine is the most powerful and effectual for exciting and inflaming passions of all kinds."[67] By releasing the passions, the fountain produced beholders who, like bacchantes, "raved around with frenzied mind."[68]

In the manner of discordia concors, or discordant accords, the fervors brought on by these hydraulic works could also have the opposite effect. A good wetting with cold water was considered a means of bringing unruly female passions under control. As another male commentator put it, "The *Element of Water* . . . doth allay and cool them [women], and therefore, they are . . . not so hot and *desirous of Copulation as Men*."[69] In following the logic of the ritual itinerary of the Oude Doolhof, it seems that immoderate desires were aroused by the salacious Bacchus and Ariadne statue and then quenched—literally, with a cold shower squirted directly up women's skirts. The Bacchic spraying of the fountain thus induced frenzy while also bringing the humors into balance. The strong passions of those considered overly emotional were unleashed and then doused as a means of teaching excitable people—women especially—to gain self-control.

Spirits

Taverns had always been dedicated to Bacchic rites. In their praise of the wine god, Heinsius and Erasmus, as well as François Rabelais and Pierre de Ronsard, draw on widespread ancient cultural traditions that were practiced across Europe. The carnival, or *kermis* in Dutch, was in many ways a Bacchic festival with roots in the pre-Christian past.[70] As the classic study by Mikhail Bakhtin shows, carnival celebrated the lower parts of the body and porous bodily orifices through the regenerative cycles of drunkenness, purging, feasting, and fertility. Carnival rites engaged all residents of a city or town and were enacted in public spaces like marketplaces, town squares, and, of course, inns and taverns, where Bacchus presided over local festivities.[71] Carnival signified temporary release from the rules that governed everyday behavior but could also cruelly mock the socially marginalized and reinforce the status quo. While scholars have disagreed about whether carnival was primarily a form of social protest or of social control, its operations are probably best understood in terms of paradox: it functioned in the manner of discordia concors. Turmoil and equilibrium work in tandem in the realm of Bacchus.

The celebration of kermis—perceived as both pagan and Roman Catholic—was banned in Amsterdam after its Alteratie to Calvinism in the late sixteenth century. In spite of its repression, carnival themes were adapted and reinvented in comic culture and imbue the pictorial traditions of depicting rowdy peasant festivities and boisterous tavern scenes, forms of rhopography designed to reveal the hidden value of the seemingly mundane.[72] Clearly, carnivalesque practices also found refuge in the Bacchic courtyard of the Doolhof inns. Pubs had long been at the center of kermis, and when age-old festive practices were prohibited, taverns provided spaces of resistance to restrictive policies and regulations.[73] Van Mander speaks of taverns giving shelter to the arts in the aftermath of the Reformation while also harboring the god himself. If fountains provide the interpretive key to the larger program of a garden, then the

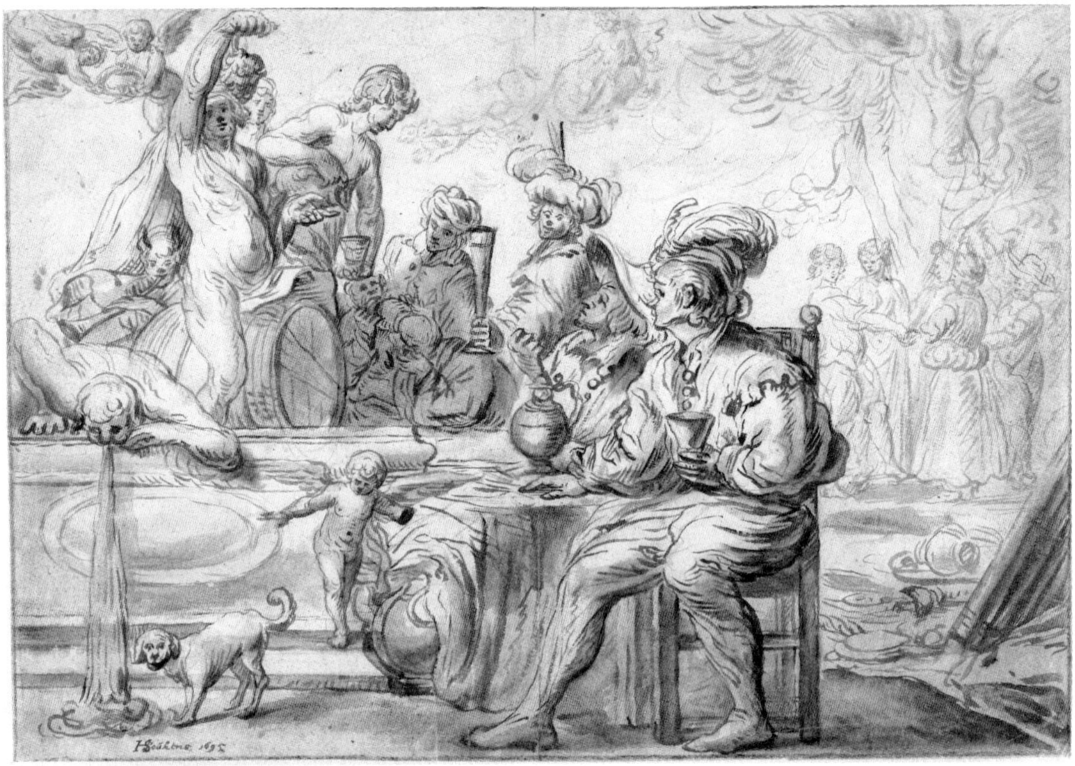

FIG. 15 Hendrik Soukens, *Bacchanaal*, 1695. Photo: Rijksmuseum, Amsterdam.

Oude Doolhof's *Triumph of Bacchus and Ariadne* animates a space of Bacchic license where artistic creation and festive behavior were reasserted in a triumphant reinvention of pagan rites in the here and now.

A drawing by Hendrik Soukens shows how a Bacchic fountain could turn ancient myth into lived experience (fig. 15). Drinkers in a tavern yard are seated at a table next to the base of a sculpture group. The statues appear to come to life as the corpulent Bacchus, riding on a barrel, raises a cluster of grapes while his satyrs serve wine to tavern patrons. Hunched over the base of the fountain is a vomiting drunkard. This spewing of bodily fluid signifies how overindulgence in the pleasures of Bacchus acts as a purgative. The experience of the work of art cannot be disentangled from the embodied experience of drunkenness in this scene. The statues intoxicate the participants, altering their physical and emotive states. In his *Hymne de Bacchus* (Hymn to Bacchus) of 1554, French poet Ronsard elucidated the higher purpose of seemingly degraded drunken behavior:

> Thanks to you, Father, sated with your sweet
> ambrosia
> We raise up to heaven the human fantasy;
> Borne in your chariot and no longer sinful
> Now that we are purged by your liquor, we dare to
> climb to the heavens.[74]

In Soukens's sketch, Bacchus's chariot, pulled by two leopards, swoops through the sky to the right of the wine-pouring satyr and above the heads of the inebriated crowd. Heinsius was a close reader of Ronsard and considered the purging of passions an important part of aesthetic experience, a means to arrive at deeper understanding and higher truths.[75] A similar understanding runs through the influential 1586 treatise *Ethics or the Art of Living Well* by Amsterdam artist and moralist Dirck Volkertzoon Coornhert. The book opens with an anecdote about a drunken boatman who staggered through the city streets until he vomited; thus sobered, Coornhert asserts that he was made "ready for the salutary draught of pure truth."[76] Soukens's drawing intimates something similar by associating crude and earthy bodily experiences with spiritual interactions.

With the double vision of a drunkard, Soukens pictures the transformative powers of the ancient classical god operating in the here and now of the local tavern yard. As a pagan satyr fills the drinker's cup, olden times flow into the current moment. The anachronic confluence of the divine and the mundane in the Bacchic fountain operated with immediacy; its impact was instantaneous and direct, as if Bacchus really was present in the work, and his ancient godly powers flowed from the fountain into its contemporary audience of tavern goers, thus transforming them.

Indeed, the fountain's mysterious potency was its creative force. As Van Mander writes in the *Schilderboek*, the lively spirit or *gheest* of a gifted artist was "like joy-inducing wine."[77] Gheest was a heavenly substance: the soul of the gifted artist received an outpouring of gheest at birth. Overflowing with spirit, artists were endowed with inventive powers. In a transfer of spirits, artists' gheesten inspired and imbued their work, enlivening and animating it, infusing the artwork with the moving qualities of beweeglijkheid. According to this theory, the spirit of the lively artwork could, in turn, enter its viewers, penetrating through bodily orifices to infect them with the energy of the work.[78] The creative spirit of the artist and the transformative capacities of the artwork were thus like the inspirational qualities of "joy-inducing wine"—taken into the body to stimulate spirited behavior. Ronsard's verses—"Sated with your sweet ambrosia / we raise up to heaven the human fantasy"—encapsulate this set of beliefs, linking human ingenuity with divine inspiration. This spirit doctrine also informs Mountague's more prosaic observation that without frequent tippling, it would have been impossible for the sluggish Dutch to achieve their great works.

The process of spirit transfer was facilitated by the liquidity of fountain sculpture, particularly a Bacchic one dedicated to wine. Alcoholic drinks (or "spirits") were in fact thought to contain a measure of liquid gheest. An excess of spirits, like an excess of alcohol, inspires wit and laughter, important to the inventive processes and creative pleasures involved in the making and interpretation of works of art.[79] Gheest was also thought to be present in bodily fluids, especially semen; human procreativity was a spiritual force.[80] The salacious sprays of the Oude Doolhof's Bacchic waterworks thus signified and stimulated creative release through the ecstasy of sensual gratification. Fertility and creativity commingled as the spurting fountain penetrated its viewers. No doubt this is why statues of Bacchus, Ariadne, Silenus, bacchantes, and satyrs commonly adorned classical fountains. Hero of Alexandria's *Pneumatics* includes a Bacchus fountain with figures of maenads and wild cats; the god's thyrsus sprayed wine.[81] When these forms were revived in the Renaissance, they resuscitated the spirit of ancient fertility rites.[82] In the Dutch tavern courtyard, therefore, the imbibing of alcohol and the wild spraying of a fountain that encapsulated a stirring classical story incited the spirited responses of visitors. Form and content merged as the ancient forces of Bacchic inspiration flowed between artist, artwork, and viewer.

High-spiritedness can quickly run out of control, however, degenerating into the undisciplined drunkenness, frenzy, madness, violence, and debauchery of a bacchanalia. The spirit is a natural force, but, as Van Mander urges, its potency needs to be trained, mastered, and disciplined—kept in balance.[83] A fountain's harnessing of nature through human artifice strives to achieve equilibrium. This balancing act is also achieved by Bacchus: "He keeps the peace in us between discordant accords," writes Olivier de Magny in his Ronsard-inspired *Hymne de Bacchus* (Hymn to Bacchus) of 1559.[84] While steadiness and harmony may have been the aim of the Doolhof displays, this noble goal did not mean that they escaped censure. In 1629, there was a municipal bylaw that specifically targeted inns with fountains for disturbing the peace with the racket of rowdy customers and the blasting of trumpets, pipes, and other noisy instruments.[85] Visitors, too, did not always appreciate the higher purpose of these sites. A satirical French booklet about Amsterdam remarks that the celebrated Doolhoven were frequented by "des sots païsans Hollandois" (drunken Dutch peasants).[86] Clearly a mocking reference to boorish behavior, the word "païsan" also denotes "pagan," an apt descriptor of the Doolhof's Bacchic excesses.[87]

If we think of the Oude Doolhof as an early modern theme park, then its main theme is metamorphosis. When Domselaer joked that bawling visitors sprayed by water tricks hopped about "like new-born calves," he was describing the transmuting powers of Bacchic waterworks powerful enough to create fledgling Minotaurs, whose wild cavorting was part-human, part-bovine. The operations of this fountain were both playful and degrading; there was terror in its joyful pleasures. This stimulating artwork conveyed the god himself, alleviating cares and making visitors caper like noisy bacchantes; in the process, they might have seen themselves and others as maenads, satyrs, or even Minotaurs. The spirited arousal of human beastliness was preparation for the next part of the itinerary, which was to move into the garden's hedge labyrinth. In winding their arduous way to the center, participants took on the role of another character: Theseus, the hero who confronts the monster in the maze. In the ritual performance prescribed by the discordant accords of this strange exhibition site, the loosening of limbs and unleashing of passions prepared initiates for the next part of the route, their entry into the controlled chaos of the labyrinth, where they would encounter the beast within.

INTO THE LABYRINTH

Chapter 3

Containing the Human Monster

He is not me but he is more than me: his stomach is the labyrinth in which he has lost himself, loses me with him, and in which I discover myself as him, in other words as a monster.
—Georges Bataille, "The Sacred Conspiracy," 1936

The ritual itinerary of the Oude Doolhof unfolded with ever-increasing sensory disorientation. Spatially, this bewilderment was experienced as constriction. Visitors were lured from the broad streets and canals of the New Town into a boisterous tavern yard where their senses were assaulted by the liquid refreshments of the cunning Bacchus. Then (for a small admission fee), they entered a second enclosure to be entrapped by the tight constraints of a tricky maze. Translated literally, a doolhof is a wandering courtyard—a bounded space that encompasses the possibility of straying endlessly along meandering paths. Walkers are at liberty to roam where they will while imprisoned within its green walls. While enjoyable, there is also an element of madness or *dolheid*: the winding ways of a doolhof are designed to create confusion, which is heightened for wanderers already tipsy with drink.

Part architectural and part sculptural, the maze is a dynamic art form. It is not experienced as a static container or as a discrete crafted object but as sensory immersion in ever-changing mobile and emotive spatial practices. There is something machinelike about a maze, and people are its moving parts; the participants are integral components of its complex structure.[1] The body is affected by the space, compelled to feel its way through the interstices and passageways offered, and the body affects the space by activating its various possibilities. The deeper one goes into a labyrinth, the more difficult it can be to get out again. Sensations intensify with ensnarement. Pleasure mingles with panic, delight with dismay; each step is an exercise in paradox. Perhaps the Dutch poet, politician, and gardener Cats had this sort of maze in mind when he wrote in 1625 of a bittersweet doolhof that was both pleasant and terrifying, "a garden full of fearful joy and a thousand strange blows."[2] As with the Bacchus fountain, the discordant accords of the hedge labyrinth were designed to transform visitors who followed the Doolhof's ritual route.

The Labyrinth of the World

The Amsterdam Doolhoven are the first known civic labyrinths that were multicursal.[3] Previous public labyrinths—chiefly designed for churches and town halls—were unicursal: there was a single way, and by conforming to it the seeker was led to the center in a contemplative quest for truth, self, and God. The maze at Chartres Cathedral is the best-known example of the unicursal variant. The main difference between unicursal and multicursal mazes is that the single path is authoritative—it offers no choices—whereas the multipath type is activated by the will of the wanderer. The unicursal maze does not need walls, barriers, or enclosures—the pattern can simply be indicated on the ground, and pilgrims largely rely on their sense of sight to follow the route to the center. The path is coiled tightly within a small space so that maze walkers take the longest way possible to reach the middle, but if they stay the course with patience, perseverance, and faith, they will eventually reach the goal. The interaction with this type of maze is experienced as meaningful action.[4] In the context of the cathedral, it prompts meditation on a life centered on trust in God and faith in a comprehensive design.[5] Maze treaders proceed in the secure knowledge that their passage unfolds within the artful plan of a higher power.

While unicursal mazes appeared mainly in religious contexts, multicursal hedge labyrinths, such as the ones in Amsterdam, were primarily features of private pleasure gardens and were associated with the myth of Daedalus and the Cretan labyrinth.[6] One of the earliest mentions of a *maison Dédalus* (house of Daedalus) occurs in fourteenth-century accounts of the famed park of the Dukes of Burgundy at Hesdin, whose trick waterworks were equally renowned.[7] By the mid-sixteenth century, garden mazes—often termed *dédales* or *maisons de Dédalus*—had become fashionable in the landscape parks of Italian, French, and Flemish palaces and villas. The sophisticated patterns of the Amsterdam labyrinths were probably derived from these sorts of private garden designs. The architectural treatises of Sebastiano Serlio and Hans Vredeman de Vries, which include various maze plans, may also have been sources of inspiration.[8] Like the Bacchus fountain, then, Amsterdam's bifurcating mazes were highly inventive art forms that were probably unfamiliar to many visitors.[9] The multicursal sculpted garden maze had hitherto been the preserve of the elite.

In contrast to the unicursal type, the multicursal maze issues a perplexing challenge. While you can put your trust in the plan of a unicursal maze, the bifurcating type is designed to outmaneuver you. Its artful twists and turns might lead you nowhere in a contorted enigma that can arouse intense frustration. To solve the puzzle, maze walkers must pit their wits against the ingenuity of the maze maker. The desire to win this contest by conquering the maze and finding the center conflicts with fear of what may lurk at the core. The forking maze requires courage as well as perseverance: a maze walker is a warrior who fights a battle against the maze's "thousand strange blows."[10] This involvement is very different from the contemplative experience of passively following the unicursal path. The multicursal maze can only be resolved by actively trying and failing in an arduous process of wandering, erring, hitting an impasse, turning back, choosing a different track, and so on in the repetition of a process devised to make walkers lose faith in their own misguided efforts.

In his civic history of 1664, Philipp von Zesen writes that in the Amsterdam Doolhoven "the senses were lost and the eyes unable to comprehend."[11] To walk this doolhof was an exercise in cognitive and sensory dissonance. The maze, as Bernard Tschumi points out, affects sense before reason.[12] In terrestrial encounters with dark corners and blind alleys, vision becomes an unreliable guide. And while one may hear sounds, voices, or a "chorus of idle footsteps,"[13] these sonic cues only augment the deceptive techniques of the closely encompassing hedge walls. The multicursal maze thus prompts a trial and error method of experiential investigation—it involves questioning faulty sensory information, battling strong emotions, and acknowledging personal failings. While the unicursal maze assumes faith in a visible scheme, the design of a multicursal maze is not evident and demands an active process of interrogation.

The inability to trust the senses may cause frustration and failure, but this misguided experience can lead to deeper understanding. As Erasmus explains in his essay *The Sileni of Alcibiades*: "The real truth always lies deeply hidden, so that it cannot be easily attained, not by most people. Most people are stupid and have a distorted vision of the world. They judge everything according to the first impression made on their senses. Over and over again they make mistakes, they go astray, they are misled by false images."[14] In the manner of the rough exterior of a Silenus box, a multicursal maze inculcates skepticism about surface appearances. To be led astray by unreliable images and perceptions was to realize that the human sensorium only provided what Erasmus recognizes as "a distorted vision of the world." Intense awareness of being lost in a misleading world ideally would prompt the search for a different path and recognition that "the real truth always lies deeply hidden."

In the battle to decipher a complicated maze, it is very useful to have the help of a guide, like Ariadne's clew of thread. Recall Ovid's account of Ariadne's grieving reproach to Theseus: "To keep you, after your victory, from death in the winding halls, I gave into your hand the thread to direct your steps in place of guide."[15] This aspect of the classical labyrinth—the need for guidance—was given moral meanings by scholars such as the Reformed pedagogue Jan Amos Comenius. Comenius's popular work *The Labyrinth of the World and the Paradise of the Heart* was published in the 1620s. In it he exhorts that without a guide, the defeated maze warrior is "doomed to wander and grope about it without ever finding his way out."[16] The long subtitle of Comenius's didactic book explicates how the world is like a labyrinth: "In this world and in its works there is nothing but confusion and staggering, floundering and drudgery, deceit and fraud, misery and sorrow, and finally weariness and despair with all." Maundering in the labyrinth of the world is a performance of human sin and error.

In Comenius's influential interpretation, as foolish maze treaders come to recognize the vanity of all worldly human endeavors, they learn to

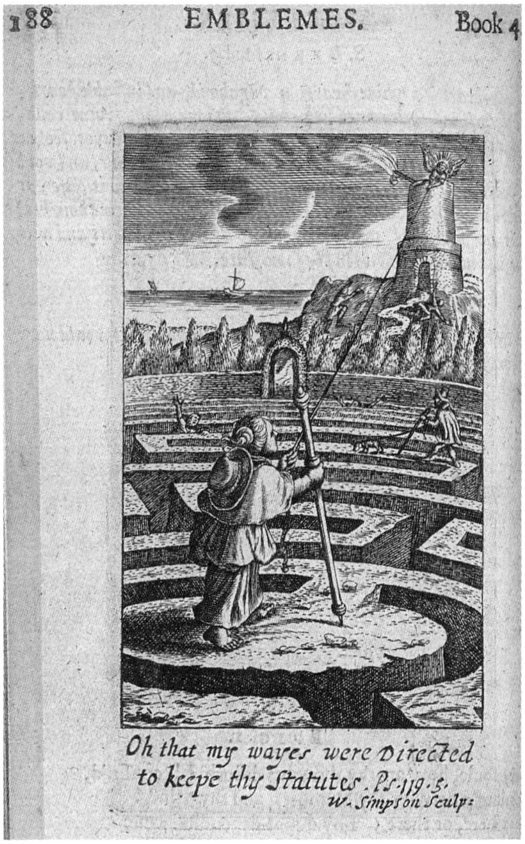

FIG. 16 Frontispiece to emblem 2, book 4, from Francis Quarles, *Emblemes* (London: G. M. [George Miller?], 1635). Photo: Beinecke Rare Book and Manuscript Library, Yale University, New Haven.

follow the will of God and to be directed by the guidance of the Bible—the essential clew of thread. This understanding of the labyrinth was predominant in the early modern period. For instance, Claude Paradin's 1551 emblem combines an image of a multicursal labyrinth with the motto "Fata viam invenient" (Fortune shall show the way). The explanatory text emphasizes that it is God who provides the thread: "The symbol of the labyrinth may perchance signifie that we are led by the grace of God to find the way that leadeth to eternall life, the same giving the thread as it were of his holy precepts into our hands, which when we have once taken hold of, and do follow, we turne away from the dangerous wanderings and fearful by wayes of this world."[17]

The need to trust God's guidance rather than worldly efforts while struggling through the labyrinth of life was an evocative theme for Reformed moralists. Francis Quarles's 1635 English edition of the popular emblem book *Pia Desideria* by Antwerp Jesuit Herman Hugo adapts the interpretation of the labyrinth to emphasize Reformed understandings. The emblem's image shows a pilgrim walking atop the (iconographically unusual) raised path of a maze (fig. 16). The motto exhorts that when human will and ingenuity fail, we must turn to God to find our way. In the image, the Christian wayfarer holds a thread proffered by an angel who offers assistance from the vantage point of a high tower outside and above the labyrinth. Others are also walking the maze but without this divine guide; some have tumbled off the true path into the trenches of sulfur that surround it. Small background figures attempt to climb to the heavenly tower unaided and fall, like Icarus, with arms outstretched. The accompanying poem explains, "Where if his footsteps trust their owne Invention / He fals without redresse, and sinks beyond Demension."[18] Human invention cannot be trusted. Those who rely on their own wits are bound to fall. They will never make their own way out of the maze of worldly error.

While Hugo's earlier Jesuit publication advocated perseverance, faithfulness, and effort, Quarles's Protestant version focuses on human error and original sin.[19] Quarles's interpretation stresses that although it may appear that multicursal maze walkers have choices, in fact, human labor, free will, and invention do not lead to salvation. The monster at the center of this earthly maze is human depravity. The Reformed conclusion is that we cannot be saved by our own works but must rely solely on the grace of God. God alone can comprehend the full design of the maze as seen from on high; he has laid out the pattern of each life, and only he can provide the clew.

The Amsterdam mazes no doubt were open to the sorts of Reformed interpretations offered by Comenius and Quarles. Indeed, this religious view would have been an effective way to answer potential critiques from church leaders about the revival of paganism at these sites. Those well versed in Calvinist theology would know that the reformer himself often compared the seductive pleasures and errors of the world to a labyrinth and advocated reliance on God. The difficulty inherent in Calvinist doctrine is that believers must trust in what they cannot see—God is outside and above the worldly maze. An invisible agent who works in mysterious ways, God cannot be apprehended by the human sensorium. In John Calvin's view, God is as incomprehensible as a labyrinth: "For it must be concluded, that the light of the Divine countenance . . . is like an inexplicable labyrinth to us, unless we are directed by the line of the Word; so that it were better to halt in this way, than to run with the greatest rapidity out of it."[20] Believers must accept on faith that an inscrutable God is leading them in the right way, even if all earthly struggles to discern the truth are suspect because of human shortcomings. While walking the unicursal church maze is experienced as meaningful action that leads to God, the multicursal maze generates futile action—a realization that worldly efforts alone do not lead to salvation. The faithful can only limp along blindly without the secure knowledge of their ultimate salvation, which is only known by an unfathomable, labyrinthine God. For Reformed believers, the labyrinth thus encapsulates the anxiety-provoking doctrine of predestination.

The Quarles emblem concludes, "The Road to Resolution lies by doubt."[21] In the Calvinist creed, believers must doubt their distorted vision of this world and trust in the unseen and extra-sensual: the will of God is a labyrinth of such beautiful intricacy that it cannot possibly be understood. This belief generates a profound sense of world-alienation, which is expressed by the pilgrim soul in the Quarles emblem: "Thus I, the object of the worlds disdaine . . . onely relish what the world counts vaine. . . . Her Truth, my Error; Her freedom is my Jayle; and her delight, my Terror."[22] The earthly labyrinth is a fearful, error-filled prison. Its many delights should inspire terror, as they are deceitful guides that entrap the unwary in worldly ways. By confounding earthly experience, the maze prompts awareness of the failings of sensate knowledge while concurrently forcing reliance upon this uncertain way. We may yearn to shuffle off this mortal coil, but as long as we live we can only stumble along in the worldly maze of human life. As Calvin asserts, the world is filled with error, and every individual mind is a labyrinth also error-filled.[23] Humanity was conceived and born in sin; to find truth, we must turn against the world and against ourselves. The believer's true home is outside and above the world. And yet, in our lifetimes, we cannot leave the world or the self. The wayfarer thus rejects and depends upon the labyrinth of the world. This paradoxical state of being is expressed in the poignant lament of Quarles's pilgrim, who says to the labyrinth: "With thee I cannot live, I cannot live without thee."[24]

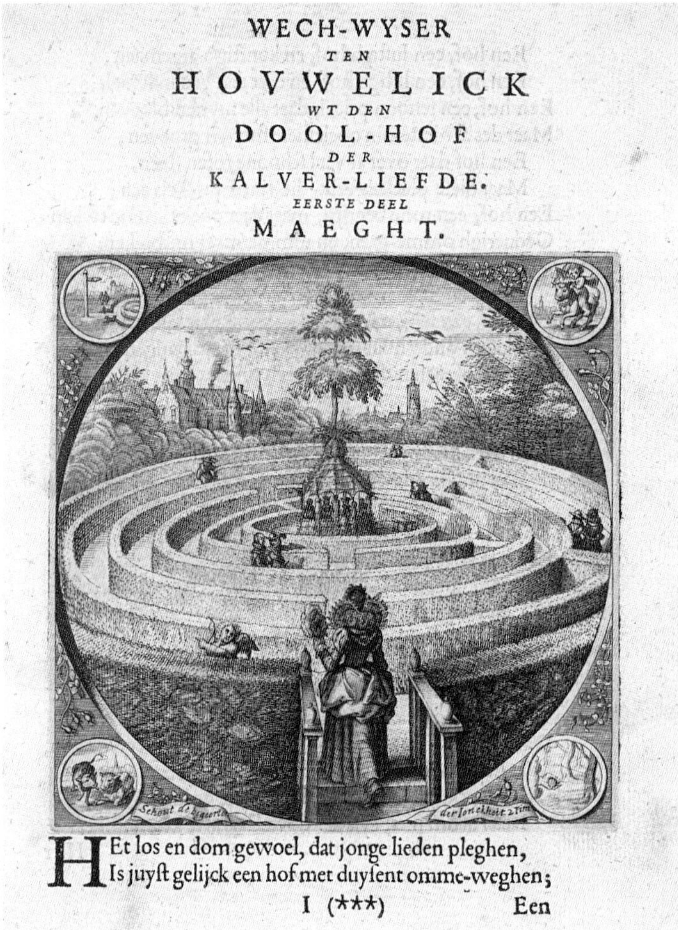

FIG. 17 Adriaen Pietersz van de Venne, *Guide to Marriage Through the Maze of Calf-Love*, from Jacob Cats, *Houwelijck dat Is de Gansche Gelegth-eydt des Echten Staets* (Middelburg: Jan Pietersz van de Venne, 1625). Photo: Rijksmuseum, Amsterdam.

Indeed, worldly delight can never be entirely repudiated, or there would be no earthly life. In one of the best-selling Dutch books of the seventeenth century, the Calvinist moralist Cats confronted the problem of sex. *Houwelijck* (Marriage), first published in 1625, opens with the image of a labyrinth (fig. 17). The engraving by Adriaen van de Venne shows a young maiden following Cupid into a round hedge maze filled with courting couples who wend their way to a pleasant pavilion in the middle. A two-tiered tree, like the Maypole of spring fertility rituals, tops this central bower.[25] The emblem draws on an oft-repeated dictum: the labyrinth is like sensual pleasure or voluptuous worldliness; it is seductively easy to enter into, but the deeper in you go, the more difficult it is to get out.[26] Green hedge mazes like those pictured in Cats's book—and like those planted at the Amsterdam Doolhoven—had long been associated with the complexities, entanglements, joys, and ambivalences of erotic love. Labyrinth historian Hermann Kern traces the origins of northern European "labyrinths of love" to pre-Christian fertility rites, most likely Mayday celebrations at the start of spring. In these pagan rituals, the form of the labyrinth is like a uterus, penetrated by a narrow path. Frustration and difficulty arouse desire, and the conquest of the maze generates rebirth.[27]

In *Houwelijck*, Cats describes entry into the maze as a rite of sexual initiation. The *doolhof der kalverliefde* designates the early phase of love, when young people are stirred by the strong and conflicting emotions of *kalverliefde* (calf love). Cats's poem compares what he calls the loose and irrational impulses of young love to the confusion of a labyrinth with a thousand byways.[28] The challenge, as articulated by Cats, is to negotiate a path through this overly passionate stage of life into the safe haven of marriage at the center. Cats offers his poem as a guide to lead young women—whom he finds especially vulnerable—through the tricky tangle of burgeoning sexual urges.[29] The monster in this maze is sexuality; its Minotaur is the animality of calf love. The moralist concludes that the sweet pains of erotic love should only be let loose and confined within the bounds of Christian marriage. In this way, the journey through the maze brings about a *staetveranderinge*, a change of state. As an initiation rite, it is a test, a battle against vigorous desires. To emerge victorious signifies the beginning of a new phase of life, in which the pleasures of passion can be enjoyed within well-defined limits. In this interpretation, the labyrinth both provokes and constrains worldly sensual pleasures. As the English poet Thomas Wyatt writes,

Alas, I tread an endless maze
That seek to accord two contraries;
And hope still, and nothing haze,
Imprisoned in liberties. . . .[30]

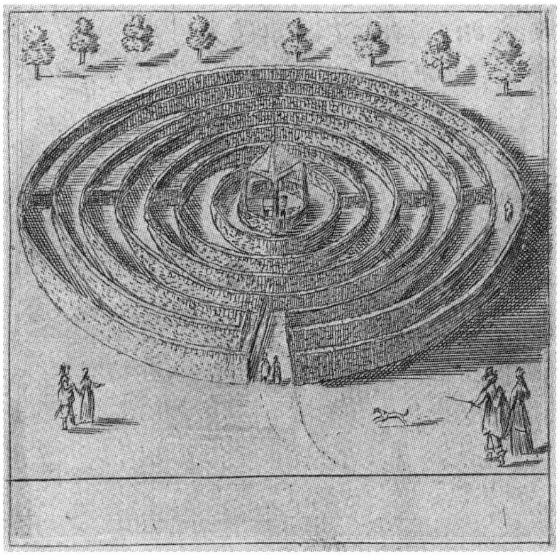

FIG. 18 Title page (detail) from *Verklaringe van Verscheyden Kunst-Rijcke Wecken en haer Beweginghe: Door Oorlogie-Werck Ghedreven* (Amsterdam: Tymen Houthaeck, 1648). Photo: Allard Pierson, University of Amsterdam Library Special Collections, OTM O 62–2613.

To tread the endless maze is to be "imprisoned in liberties." The maze's pattern of paradox is a tension between license and control; its multiple turnings were designed to bring contrary extremes into balance.

Descriptions of how the Oude Doolhof Bacchus fountain made women "hop about like newborn calves" no doubt were in reference to Cats's well-known writings about the doolhof of calf love.[31] The title print image of an Oude Doolhof guidebook depicts a round maze with a pavilion at the center, similar to the one in Cats's book (fig. 18). A number of strolling couples walk toward or into the maze, and behind it is a row of distinctive two-tiered trees that evoke the Maypole. After the unleashing of passions in the tavern yard by Bacchus, god of discordia concors, visitors entered the pleasurable constrictions of the green labyrinth. Cats's evocation of the maze of calf love thus suggests a conceivable interpretation of the Doolhof mazes, which reinvigorated ancient fertility rites within tight moral restraints.[32]

Theseus Fights the Man-Bull

As fertility rites generate life, they initiate the mortal struggle that ends in death. Arriving finally at the center of the Oude Doolhof maze—marking its innermost transitional locus—the wayfarer encountered a sculptural group of Theseus battling the Minotaur, the man-bull that fed on the blood of those who penetrated the labyrinth's hidden heart. Plutarch described this mythic monster: "A mingled form and hybrid birth of monstrous shape / Two different natures, man and bull, were joined in him."[33] The Theseus and Minotaur statues do not survive, and the only extant image is a small detail on the title page of one of the Oude Doolhof guidebooks (fig. 19). The booklet describes the figures as larger than life-sized, and Domselaer's civic history notes that some dead bodies lay strewn about them.[34]

The image indicates that the Theseus and Minotaur group was displayed on the roof of a small garden pavilion or grotto. At the culmination of their difficult wanderings, visitors entered a dark, cramped, isolated space. A Nieuwe Doolhof guidebook quips that as in the maze, as in life, everyone wanders as long as he can, until he drops into the grave.[35] Enfolded in the middle of the worldly labyrinth is death; every crooked route eventually brings you there. Following from the pleasures of the Bacchic courtyard, the constrictions of the maze might have initially felt like a sensual embrace. But with each false turn and every failed attempt, frustrated maze walkers might have begun to feel that they were caught in the deathly coils of original sin. Voluptuous worldliness is easy to enter into but difficult to exit. Another guidebook explains that the doolhof represents "the pattern of human life, which after many difficulties comes to its resting place and thus travels back to where it came from."[36] In a labyrinthine move, the enclosed endpoint signifies a reversal as the turning point back to the origin: the entrance of the maze is also its exit. The enigmatic labyrinth thus signifies another well-established paradox: *in my beginning is my end.*[37] To be born is to die. Indeed, if the labyrinth is the world, then there is only one way out. Playful and fearful, womblike and tomblike, the Amsterdam mazes pulled people out of quotidian experience into a ritual space that aroused embodied awareness of their own mortality.[38]

Turning back to beginnings, the terminal point of the maze also led to mythic origins. The Theseus and Minotaur sculptures identified the Amsterdam labyrinth as a creative reinvention of the fabled Cretan maze. Arriving at the center, visitors understood that—like Theseus—they had followed Ariadne's thread. The complicated route that they had walked delineated and revived an ancient pattern of bodily movements that stretch back to antiquity.[39] The multicursal maze was recursive: the choreographed actions of its mobile spectators revitalized a dynamic mythic structure. This potency was harnessed to larger civic narratives as maze walkers performed the promise of the maze.

The relevance of the myth of Theseus for Amsterdam had been brought forward by one

FIG. 19 Title page from Jan Vos, *D'Oprechte Aenwijser* (Amsterdam: Alexander Iansz Lintman, 1674). Photo: Allard Pierson, University of Amsterdam Library Special Collections, OTM O 62–2616.

d'Oprechte Aenwijſer,

tot verſcheyde Kunſt-rijcke Wercken,

En haer wonderlijcke beweegingen, gedreven door Oorlogie-werck, nevens eenige aerdige Beelden levens groote, van eenige voorname Koningen en Princen onſer Eeuw.

Oock ſeer uytſteeckende Fonteyn-werck, nevens een luſtigh Doolhof.

Welcke alles kan geſien worden op de hoeck van de Lopers-gracht/ in het Oude Doolhof/ tot Amſterdam.

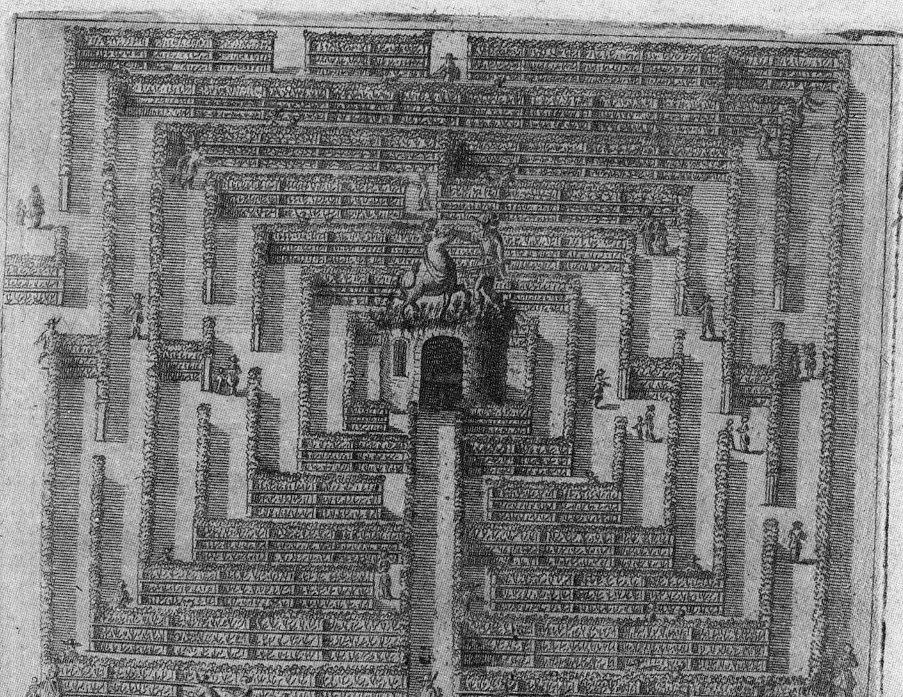

t'Amſterdam, gedruckt by *Alexander Ianſz. Lintman*, in de Beuling-ſtraet, 1674.

of the city's foremost playwrights and leading citizens, P. C. Hooft, who had personal and professional connections to the Doolhoven. The son of a wealthy Amsterdam burgomaster, Hooft was sent on a tour of several French and Italian cities as a young man and returned inspired to develop vernacular classicism in his home city. The first theater play that he composed after coming back from his travels was *Theseus and Ariadne*. Written in 1601, the work appeared in print in 1614 with a second edition in 1628.[40] During these decades, it was performed multiple times by the Amsterdam Chamber of Rhetoric called De Egelantier.[41] This civic society was the city's main center of literary production before the founding of the Dutch Academy by Hooft and a small group of others who broke from De Egelantier in 1617. De Egelantier was conceived as "a school of the mother tongue, open to anybody, without exceptions."[42] Its activities centered on the writing, recitation, and performance of plays and poetry and the staging of *tableaux vivants* at public festivities and civic competitions.[43] It was open to all citizens and drew its membership mainly from the broad middle classes, including many artists and skilled artisans. Notably, it also included members of the civic government and thus had the semiofficial public role of deliberately crafting cultural politics for the city. As Spies argues, the main aim of De Egelantier was to create urban harmony.[44] The Neostoic philosophy of Lipsius provided a moral foundation that de-emphasized religious controversies. The society included members of different Christian confessional groups and deliberately avoided dogma or zealotry, which were seen as disruptive to urban life. The aim was to heal the divisive religious, political—and artistic—conflicts of the Dutch Revolt and to generate civic accord through performance and creativity.[45] The Amsterdam Doolhoven, as I am arguing, were designed along similar lines.

The foreword of *Theseus and Ariadne* declares that the play was written for "the honor of the fatherland and our mother tongue."[46] Hooft's particular aim for De Egelantier, and later for the Dutch Academy, was to advance a new kind of Dutch classicism, reviving ancient forms in the vernacular and making them accessible for the betterment of the urban population. The emphasis of Hooft's *Theseus and Ariadne* is the bravery of Theseus, son of the king of Athens. Proving himself an ideal leader, the hero virtuously offers his own life to bring an end to the sorrow and troubles of the Athenians, who were forced to send a tribute of young men and women to Knossos, where they were imprisoned in the labyrinth to wander until they died of starvation or were devoured by the Minotaur. Joining this sacrificial group, Theseus determined to fight the monster for the deliverance of his people. The play includes a chorus, which sings the praises of the virtuous leader, a true servant of the community.[47]

Hooft drew from "The Life of Theseus" in Plutarch's *Parallel Lives*, which details the prince's founding of the lovely and famed metropolis of Athens. Theseus is credited with gathering the scattered people of Attica into one town and establishing "government without a king and a democracy, in which he should only be commander in war and guardian of the laws."[48] He also enlarged Athens, issuing to strangers the famed invitation: "Come hither all ye people." While the "promiscuous multitude" of foreigners threatened the new democracy with disorder and confusion, under Theseus's guidance, order was maintained in the expanding city.[49] For audiences, the comparisons to the Dutch stadtholder, Prince Maurits, son of William of Orange, would have been evident. The model of government proposed by the legendary Theseus was similar to the idiosyncratic political structure of the United Provinces, a republic with a strong central ruler—the stadtholder—who was not a monarch but was charged with command of the military and the maintenance of order. Moreover, like Theseus, Prince Maurits encouraged urban expansion; Amsterdam's civic magistrates consulted with him before proceeding with plans for the New Town, built to accommodate legions of immigrants.[50]

In his praise of the prince, Hooft followed the precepts of Lipsius: the just ruler is constant and governs with reason; he is an exemplar to the people. Extolling steadfast Theseus, the chorus in Hooft's play concludes: "Prudence usually goes along leveled ways / God is against passion's roundabout ways."[51] This passage echoes the advice of Lipsius, who holds up Theseus as a model of stoic behavior: "For whoever listens to reason has tamed the desires and the riotous impulses of the mind, and he is safe from losing his way in the labyrinths of this life, if he follows it as the thread of Theseus."[52] Reason, explains Lipsius, always turns toward God, its origin. To be caught in the maze of the world is to be entangled in troubles and buffeted by fortune and changing circumstances. Prudence, faith, and fortitude are needed to negotiate its many trials. Like the exemplary Theseus, good leaders—and good citizens—must master the confusions, wild emotions, and sensory distractions of life's maze. They should not be led by their passions but submit to God.[53]

Van Mander's commentary on Ovid's *Metamorphoses* similarly articulates the significance of Theseus's battle. Rewriting the ancient myth along the lines of Protestant morality, Van Mander urges readers to conquer their fleshly, sinful, beastly, irrational natures—to escape the crooked entrapments of Daedalus's doolhof. They should slough off the doolhof's voluptuous coils like a snake sheds its skin in order to enter a new life of honest piety. He concludes by advising readers to become like Theseus, who unraveled the labyrinth, slayed the monster, and followed "the straight path of reason."[54] This is the right path for the exemplary citizen: to emerge from the labors of the labyrinth a transformed subject, in control of the self, the beast, and the city.

In Hooft's play, the steady, masculine Theseus stands in contrast to the flighty, feminine Ariadne, who fails to take control of her passions. In the second act, Ariadne's nurse Corcyne reproaches the young woman's immoderate love for Theseus, admonishing: "You can calm this frenzy if you are willing." Ariadne responds: "If you are willing, that's true, but I can't and I won't will it."[55] Like her mother, Pasiphaë, Ariadne allows herself to be governed by unrestrained and excessive desires; she refuses to exert her will over her frenzied emotions. Pasiphaë's shameful lust for the Cretan bull engendered the hybrid Minotaur, symbol of humanity's beastly nature. Such a brutish human had to be incarcerated in the labyrinth, a prison for monstrous cupidity.[56] For this reason, Bacchus, rather than Theseus, is the more fitting mate for Ariadne, as only the god could satisfy her voluminous passions.[57] The Doolhof guidebooks contain a similar warning. The description of the Bacchus fountain concludes that Ariadne's unbridled lust ultimately caused her death, as she was slain by the chaste goddess Diana for failing to guard her maidenhood.[58]

In a republican city (like Athens or Amsterdam), male citizens were exhorted to become stoic leaders like Theseus: by achieving mastery of themselves, they secured civic harmony. Citizens' wives and daughters, as urged by Cats and others, were also expected to attain self-mastery, especially by containing their passions within the bounds of marriage. The vernacular classicism of early seventeenth-century theater and of the Doolhof sculpture parks thus sought to apply ancient wisdom to everyday life. The purpose was to extend the virtues of stoicism beyond the ruling elite with the aim of educating and reforming as many as possible. Amsterdam was to be a new Athens, "a common school for all people."[59] The people included Dutch men and women, both old and young, as well as the "promiscuous multitude" of foreigners in the city, who were strongly encouraged to visit the Doolhoven. To walk the labyrinth was to perform the utopian promise of civic concord.

Yet the maze always discloses human shortcomings as well as human capabilities. As maze warriors emulated Theseus's heroic efforts, the battle they fought was against their own inherently flawed selves. Within the labyrinth, we discover that—like the Minotaur—we all have two different natures, human and beastly, conjoined in us. Doubling back to a point already glimpsed in passing, the true monster in the maze is the self. Those caught within the labyrinth's twisting intestinal routes must go inward to grapple with their own alterity, their secreted Minotaurs, their contained carnality. As Van Mander advocates, one should turn upon the maze of the self in solitary stoic combat against the brute within.[60] This admonition resonates with Calvin's doctrine of original sin and human depravity: "Hence that immense flood of error with which the whole world is overflowed. Every individual mind being a kind of labyrinth."[61]

The self-analysis prompted by the labyrinth requires that the subject move to regard itself as an object in need of reform. *Nosce te ipsum*—Know thyself—was a much-repeated classical maxim.[62] The 1599 poem by Sir John Davies expresses the inadequacies of self-knowledge:

> My self am center of my circling thought
> Only myself I study, learn, and know . . .
> I know my life's a pain and but a span;
> I know my sense is mock'd in ev'ry thing:
> And, to conclude, I know myself a man,
> Which is a proud, and yet a wretched thing.[63]

To know the self is to know that the self cannot be known by the self.[64] Within the labyrinth, "sense is mock'd in ev'ry thing," and constant circling reveals the self's contrary ways. This notion is precisely what the seventeenth-century commentators assert: as long as one walks the maze, an ongoing, moving battle within the hybrid self continues, for this is the pattern of human life. The labyrinth thus described is a site of perpetual conflict, a field of opposing forces. As the poet Wyatt writes: "Alas, I tread an endless maze / That seek to accord two contraries."[65]

As a paradigm for human action, the labyrinth celebrates human effort and potentiality, but it also generates painful awareness of human impotence—of what we cannot be, or do, or make, or know.[66] To repeat Wyatt's elegant précis: to tread the endless maze of the world is to be "imprisoned in liberties." The wondrous pleasures and possibilities of the human can never transcend mortal limitations. In this way, the multicursal labyrinth maps the contradictory space of early modern subjectivity. To be human is a proud and yet a wretched thing. It is thus a multicursal subject that traverses the forking ways of the labyrinth—tangling with inner monsters in an impossible quest to establish calm, undisturbed reason within the self and the city.

The Urban Maze

Having walked inside the labyrinth and penetrated its secret heart—which is to experience the inspiring capabilities and distressing inadequacies of the mortal self—we now take a view from above, which situates the labyrinth gardens within projects of urban reform. The earliest Amsterdam doolhof is pictured in miniature on Balthasar Florisz van Berckenrode's elaborate nine-sheet engraved wall map of 1625 (fig. 20). This map is the first detailed cartographic depiction of the newly completed urban expansion, which doubled the size of the city. Taking us out of our earthbound bodies so that we hover over the urban fabric like birds or celestial beings, Berckenrode's map especially proffers the voyeuristic pleasure of peering into all of the hidden spaces normally concealed behind street facades. Based on information garnered from recent surveys, each street, canal, and bridge is rendered in painstaking detail, down to the trees planted along the quays and the variety of boats that traverse the busy waterways. Not only do all of the city's public buildings, courtyards, and market squares appear, but every single house and back garden is likewise laid out for the eye to see.

The courtyard gardens along the recently constructed canals—Herengracht, Keizersgracht, and Prinsengracht—are particularly intriguing. The cartographer has rendered with care the various patterns of geometrically designed plantings and parterres. These backyards belong to Amsterdam's most prosperous citizens—mainly merchants who had profited from the city's expansion into global trade. The New Town, completed in the first decade of the seventeenth century, offered spacious lots that allowed for the construction of larger homes and gardens than those in the old and cramped city center, where precious little open space was attached to private dwellings. Berckenrode's detailed depiction of these sizeable city gardens gives a glimpse of the large amount of inaccessible green space enclosed by the newly constructed blocks of exclusive canal homes. In fact, the map offers a forbidden view: buildings on the transversal side streets (the *dwarsstraten*) that overlooked these backyards were prohibited from having windows that faced onto the secluded gardens of the wealthy.[67] Berckenrode thus caters to the inquisitive desire to look into private enclosures that could not be accessed—physically or visually—from the street.

As we peruse these walled gardens, an anomaly in the urban fabric appears in the large courtyard at the corner of Prinsengracht and Looiersgracht (fig. 21). There, a denser patterning of lines alters the texture of urban space, attracting and confounding the eye. Endowed with the voyeuristic abilities of airborne Icarus, we regard the design of a labyrinth from the sky. This is the garden of the Oude Doolhof inn. So detailed is the rendering that it includes a small splashing fountain as well as tiny sightseers who stroll through the main courtyard. The maze pictured on the map is square, and at its center is a pavilion with pointed roof. This depiction is very similar to the labyrinth and pavilion on the 1622 Oude Doolhof poster discussed in chapter 1 (see fig. 3). The cartographer must have been aware of this advertisement, as it was designed by his brother Cornelis Florissen van Berckenrode.[68] As argued above, the street poster enticed tourists and local amateurs with the contradictory strategy of publicizing secrets: presenting just enough information about the labyrinth garden, with its fine fountains and first-rate wines, to lure with the promise of further "fine secrets for art lovers." The 1625 map employs a similar strategy—it makes secrets public by revealing just enough to tantalize. Offering a privileged view, it exposes the city's hidden places while keeping viewers at a distance.

Designed to celebrate Amsterdam as the primary trade emporium of all Europe, Berckenrode's work created a new standard for Dutch city maps.[69] Intriguingly, the small labyrinth pattern at the corner of Prinsengracht and Looiersgracht became a common feature on subsequent maps of Amsterdam. For example, the Oude Doolhof maze appears in Joan Blaeu's important 1649 atlas, *Toonneel der Steden van de Vereenighde Nederlanden* (Theatre of Netherlandish cities) (fig. 22). While indebted to Berckenrode, Blaeu updated the map of Amsterdam by integrating changes, especially recent urban expansions. Curiously enough, this new map also included an extension of the Oude Doolhof garden (fig. 23). Next to the Oude Doolhof courtyard and maze that were depicted on the Berckenrode map, Blaeu added the Nieuwe Doolhof's circular maze and courtyard with fountain. These new features had been built sometime after 1626, when the proprietor expanded the business by purchasing an adjacent lot on Looiersgracht. By incorporating the labyrinth gardens in official maps of the city, the mapmakers highlighted the civic importance of these distinctive cultural sites. The delineation of tiny labyrinths also enhances the truth claims of the city maps. Attending to the smallest details that textured the urban fabric, these maps present themselves as scrupulously accurate depictions.

The inclusion of miniature mazes on maps of Amsterdam relates to urban design and cartography. The maze was, in fact, an ancient symbol for a walled city. Comparable to the labyrinth, the city is a brilliantly complex structure of human making whose intricate network of multiple intersecting and bifurcating streets and byways cannot be fully comprehended when walking within it. An ancient apotropaic device, the labyrinth's twisting internal paths, similar to a dense network of urban streets, could create confusion and confound invaders, while its external boundary, like the city wall, offers protection to those within.[70] The complicated and perplexing configuration of a labyrinth or a city is only revealed by a view from the sky—it is the illusory God's-eye view of the mapmaker that exposes its artful design. For the cartographers of urban space, the symbol of a labyrinth—a small map—may have appealed as an emblem of their own practice.

Joan Blaeu's 1640 map of Crete, or Candia, indicates the mapmaker's engagement with the ancient history and symbolism of the labyrinth, for it includes a small circular maze labeled *laborinto*.[71] The term evokes the etymology of the word: a labyrinth—*labor-intus*—contains human labor, both in its making and in its deciphering.[72] Experienced from within, the labyrinth exposes the failings of human efforts, but when seen from above, a complex labyrinth design reveals the inventiveness of what Pliny the Elder calls "the most stupendous works, perhaps, on which mankind has expended its labors."[73] In relation to cartographic practices, the linear intricacies of depicting a labyrinthine configuration showcase the sophisticated engraving techniques used in mapping. The emblem of the labyrinth encapsulates the mapmaker's ingenious labor in compressed form.

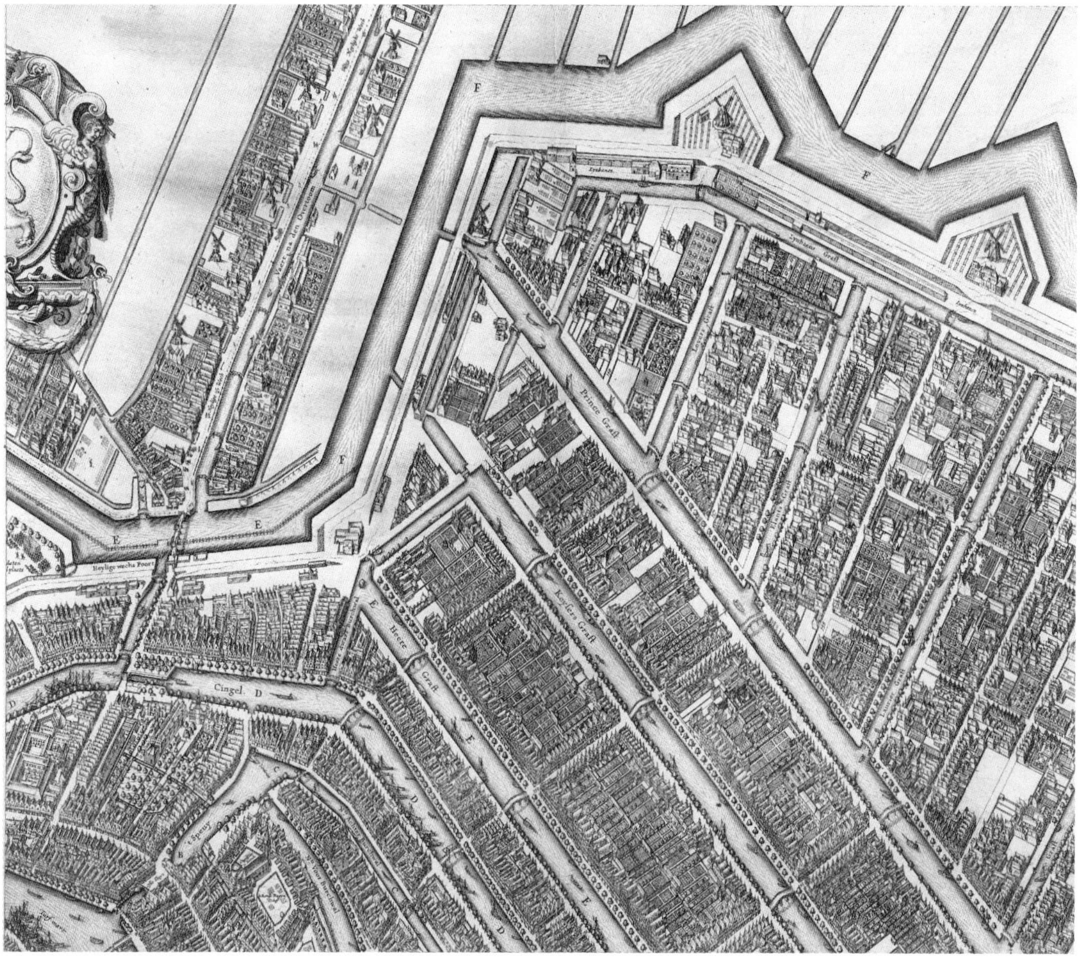

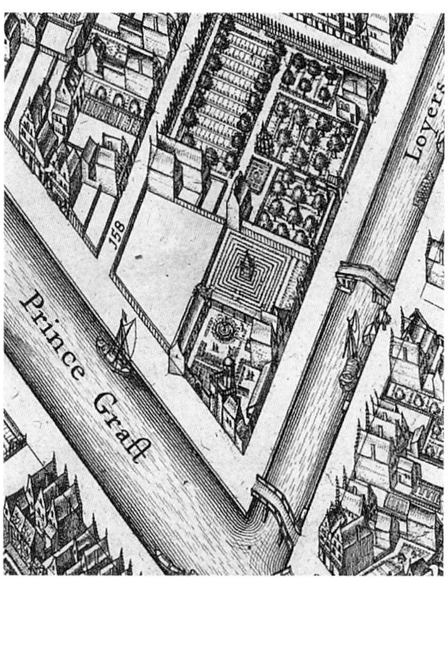

FIG. 20 Balthasar Florisz van Berckenrode, *Map of Amsterdam*, 1625, top middle sheet. Photo: Rijksmuseum, Amsterdam.

FIG. 21 Balthasar Florisz van Berckenrode, *Map of Amsterdam*, 1625, detail showing Oude Doolhof. Photo: Rijksmuseum, Amsterdam.

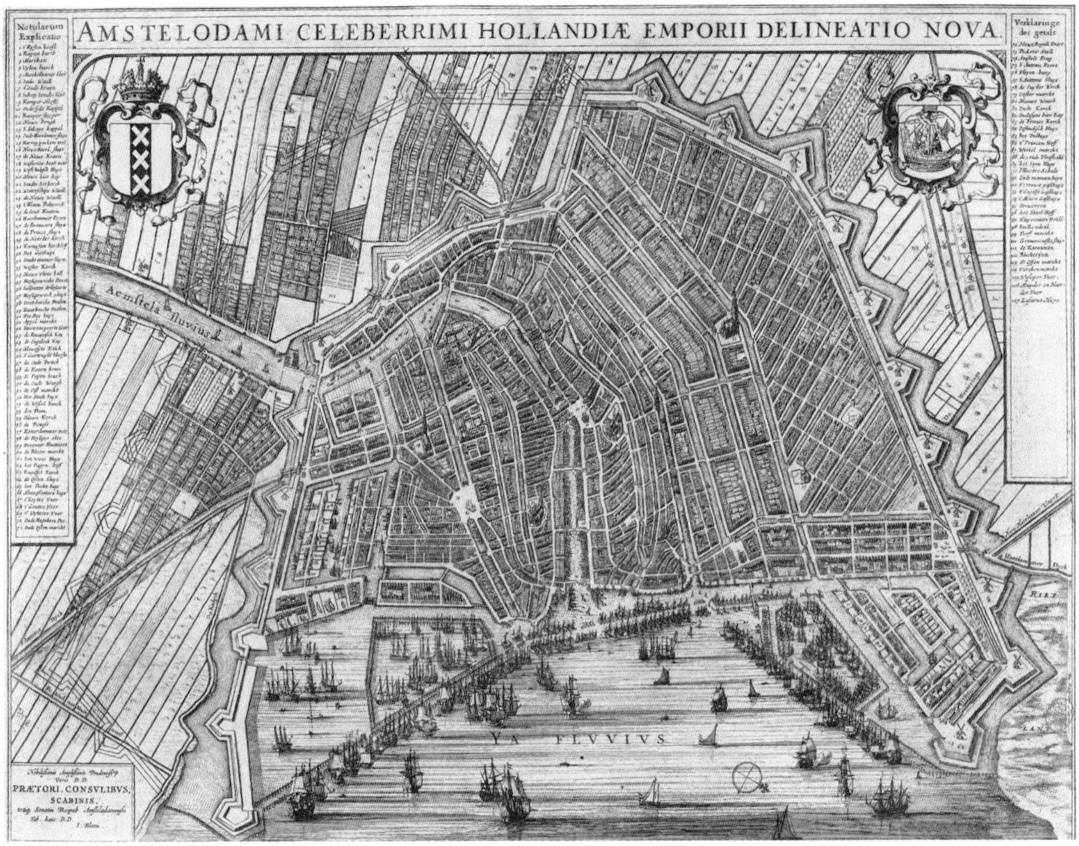

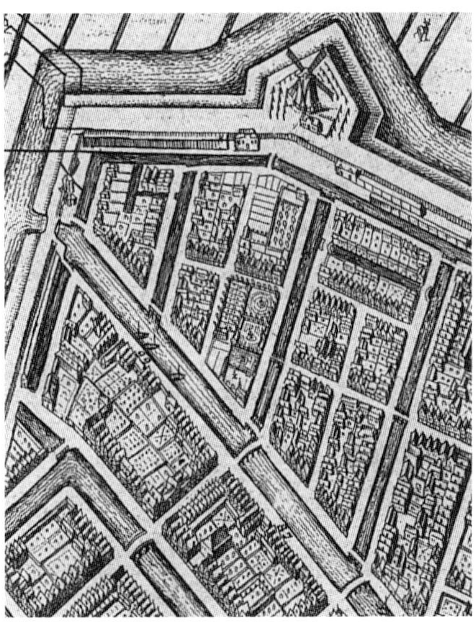

FIG. 22 *Amsterdam*, from Joan Blaeu, *Toonneel der Steden van de Vereenighde Nederlanden, met hare Beschrijvingen* (Amsterdam: Blaeu, 1652). Photo: Rijksmuseum, Amsterdam.

FIG. 23 *Amsterdam*, detail showing Oude and Nieuwe Doolhoven. From Joan Blaeu, *Toonneel der Steden van de Vereenighde Nederlanden, met hare Beschrijvingen* (Amsterdam: Blaeu, 1652). Photo: Rijksmuseum, Amsterdam. http://hdl.handle.net/10934/RM0001.COLLECT.226661

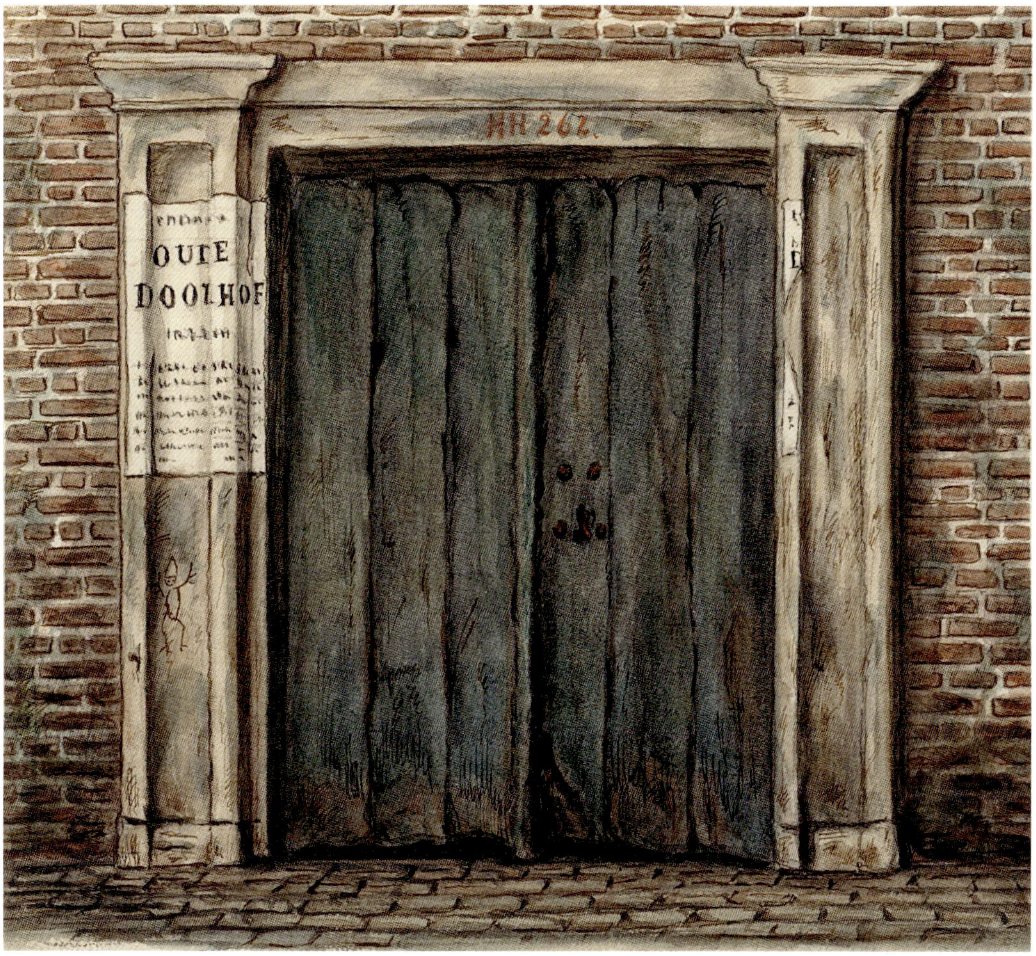

COLOR PLATE 1 Johannes ter Gouw, *The Door of the Oude Doolhof*, 1864. Photo: Amsterdam City Archives (010055000321).

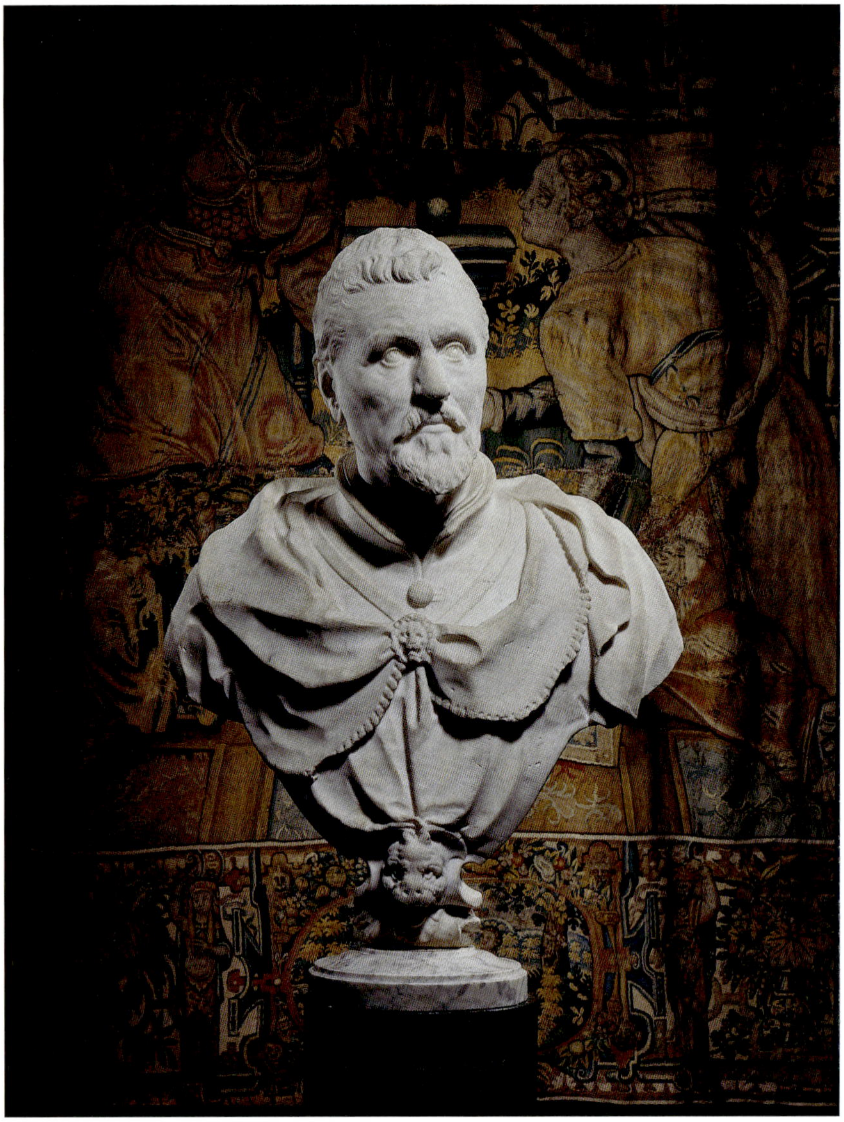

COLOR PLATE 2 Hendrik de Keyser, *Bust of a Man, Probably Vincent Coster*, 1608. Photo: Rijksmuseum, Amsterdam.

COLOR PLATE 3 (OPPOSITE) Albert Jansz Vinckenbrinck, *David and Goliath*, ca. 1648–50. Photo courtesy of Amsterdam Museum (Richard de Bruijn).

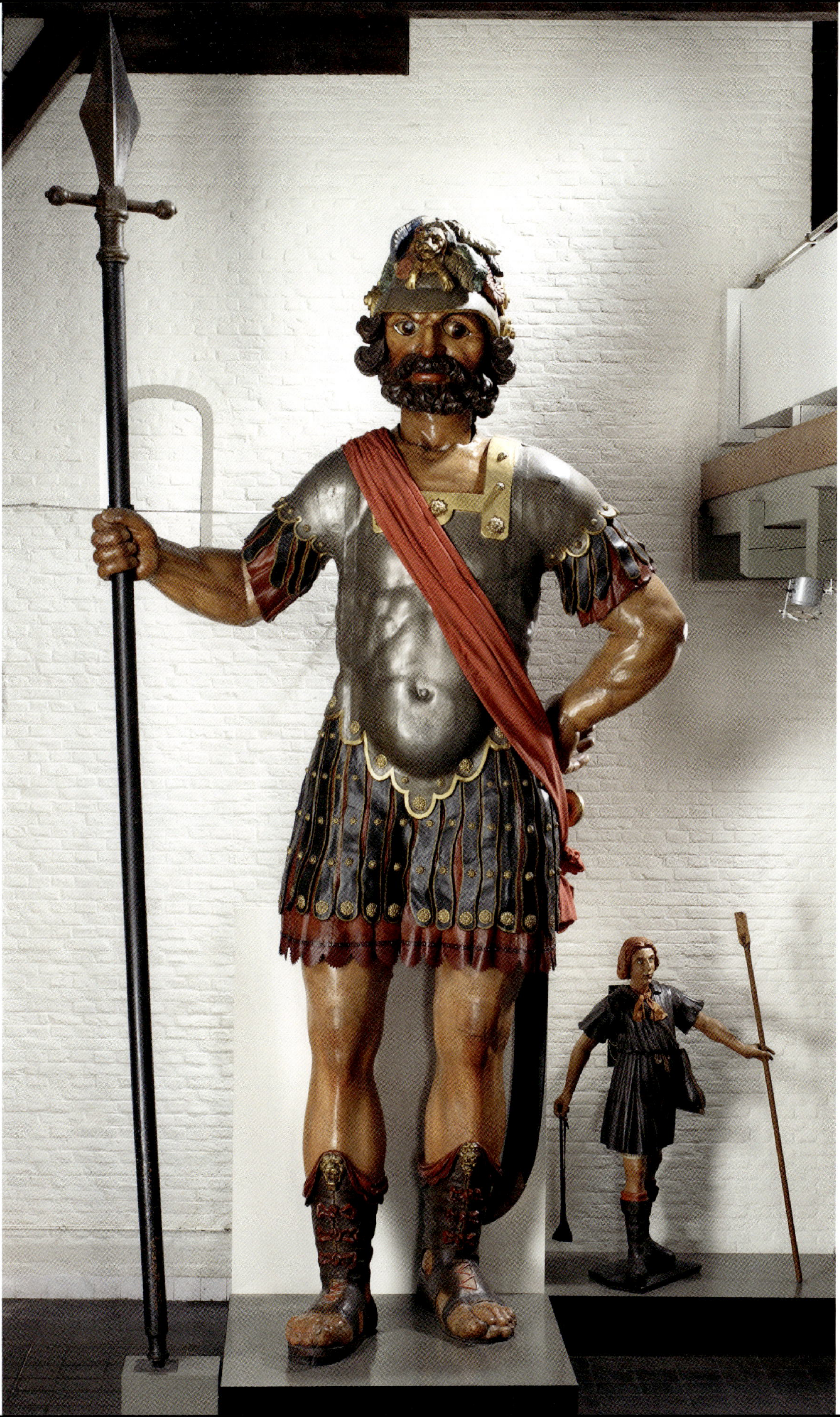

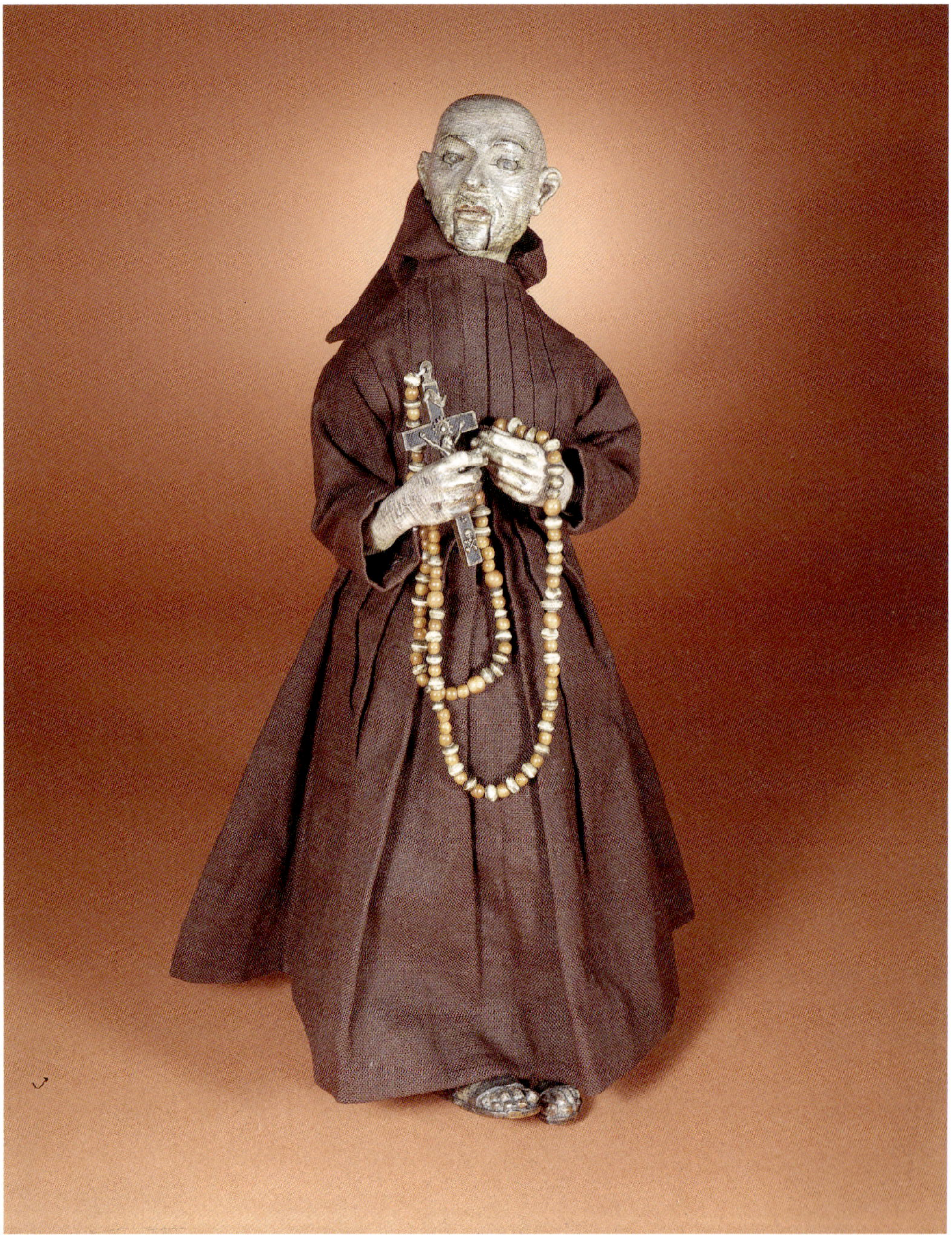

COLOR PLATE 4 *Automaton of a Friar*, mid-sixteenth century. Photo: Division of Work and Industry, National Museum of American History, Smithsonian Institution, Washington, DC.

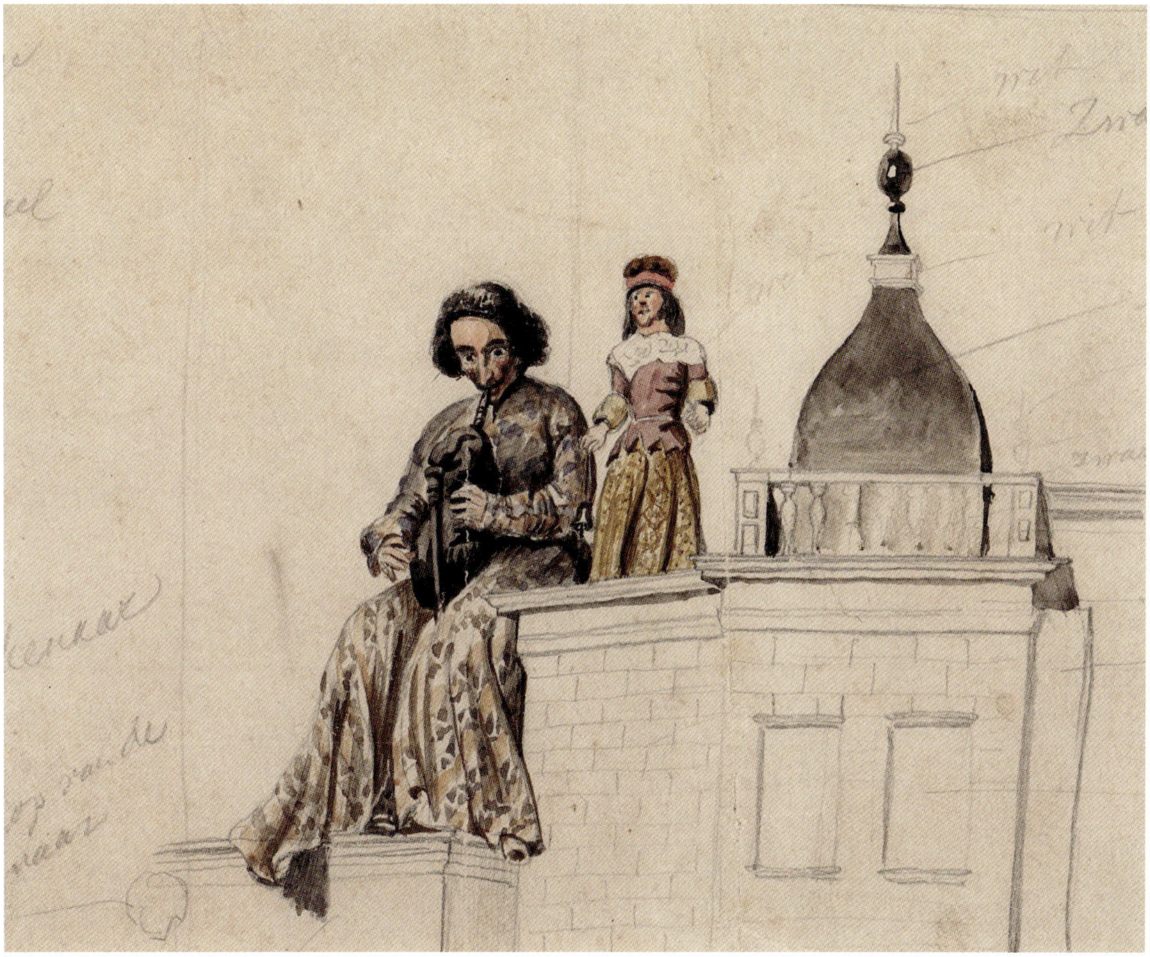

COLOR PLATE 5 J. G. L. Rieke, *Oude Doolhof, Diverse Sketches*, ca. 1861. Photo: Amsterdam City Archives (010097014918).

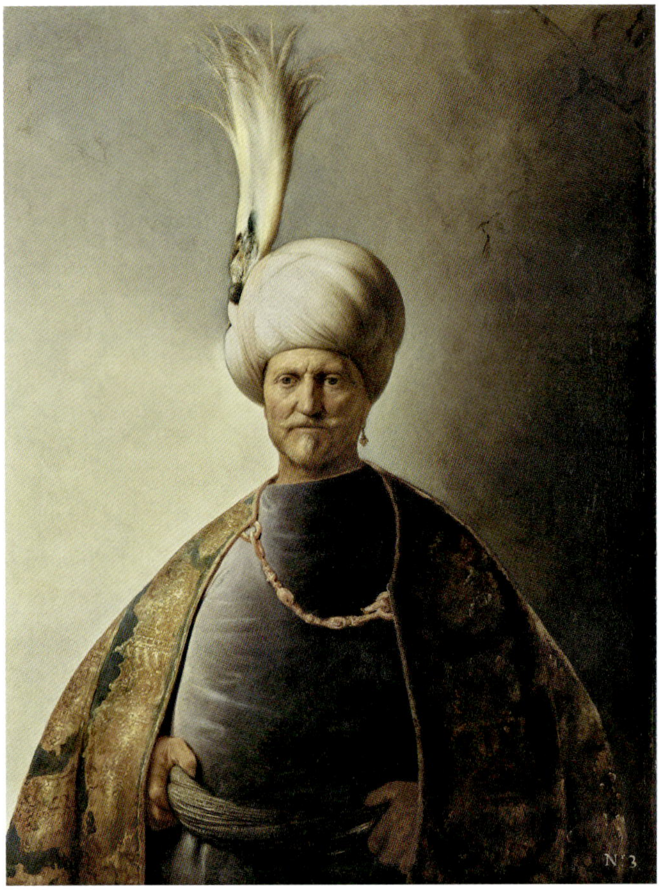

COLOR PLATE 6 Jan Lievens, *Old Man in Oriental Garb*, 1628–30. Photo: Stiftung Preußische Schlösser und Gärten Berlin-Brandenburg / Wolfgang Pfauder.

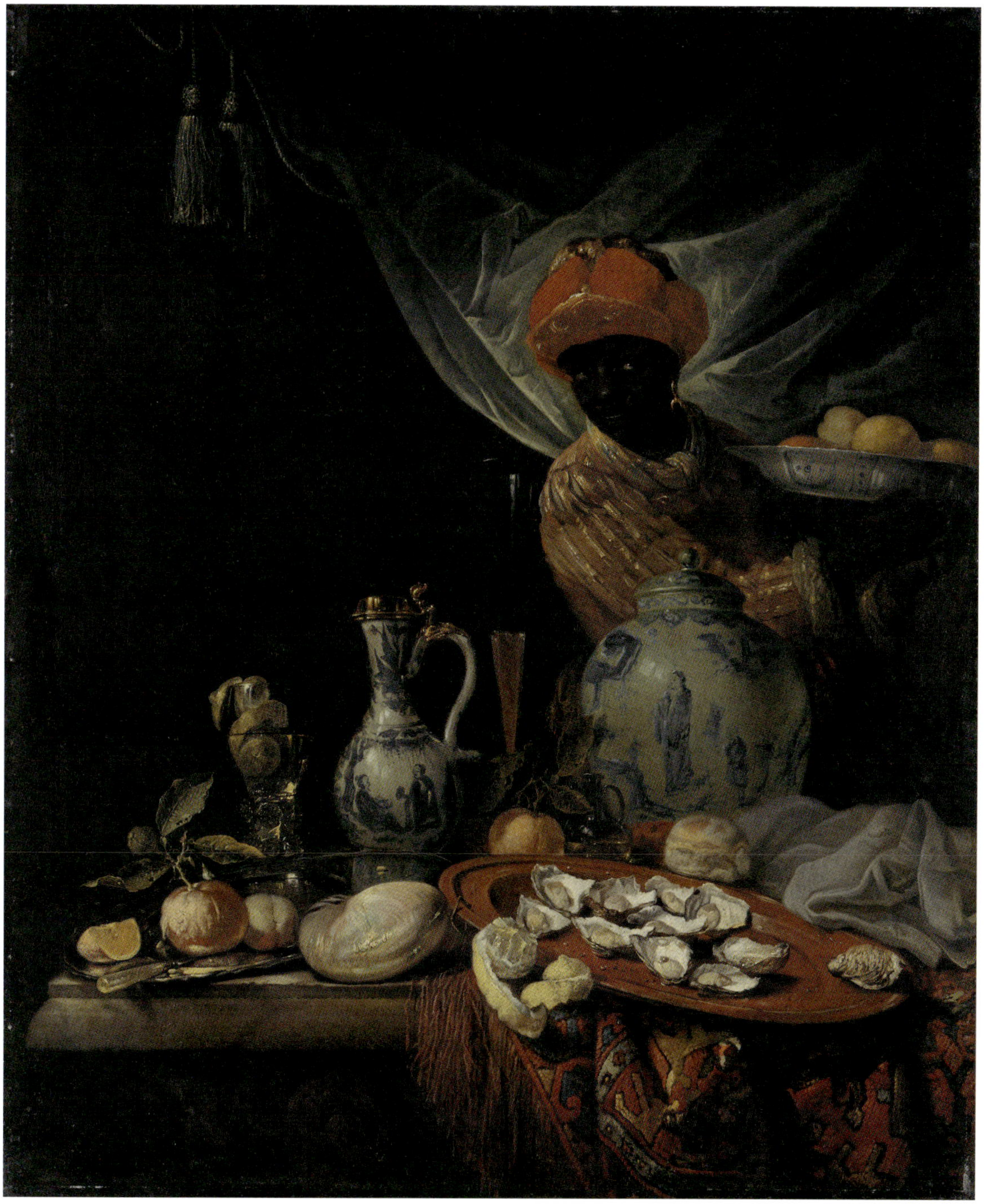

COLOR PLATE 7 Juriaen van Streek, *Still Life with Moor and Porcelain Vessels*, ca. 1670. Photo: bpk Bildagentur / Alte Pinakothek, Bayerische Staatsgemäldesammlungen, Munich / Art Resource, New York.

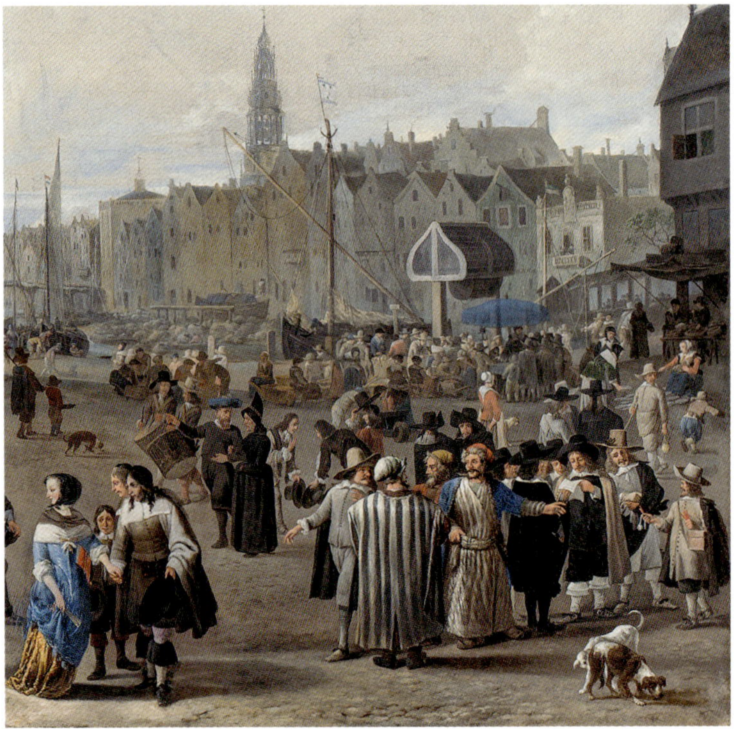

COLOR PLATE 8 Johannes Lingelbach, *The Dam*, 1656, detail. Photo courtesy of Amsterdam Museum.

COLOR PLATE 9 (OPPOSITE) *Peacock Basin*, folio from a copy of Ismāʿīl ibn al-Razzāz al-Jazarī, *Book of Knowledge of Ingenious Mechanical Devices*, Mamluk, probably Cairo, Egypt, 1354. Hervey Edward Wetzel Fund, Museum of Fine Arts, Boston, 22.1. Photograph © 2022 Museum of Fine Arts, Boston.

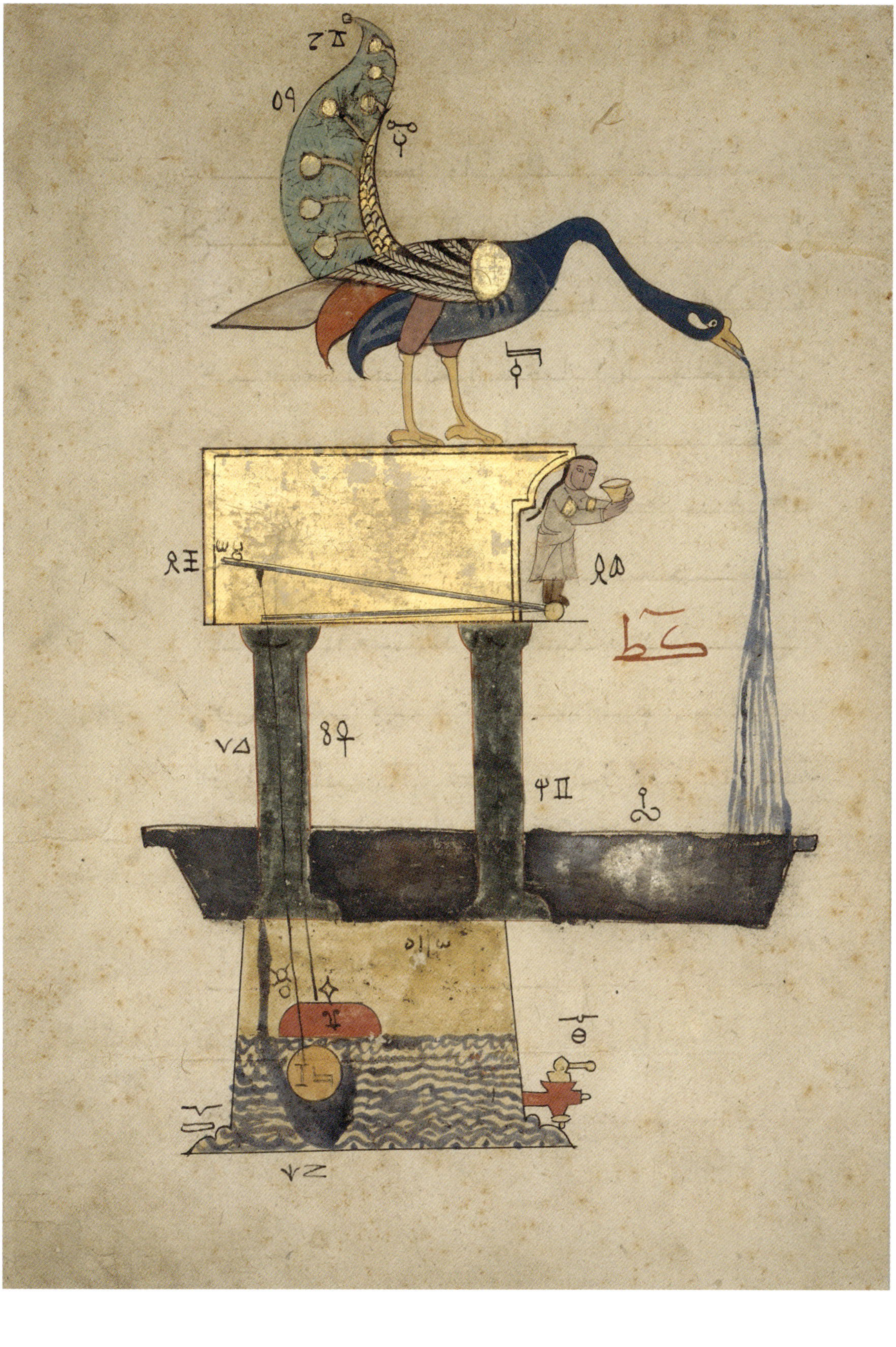

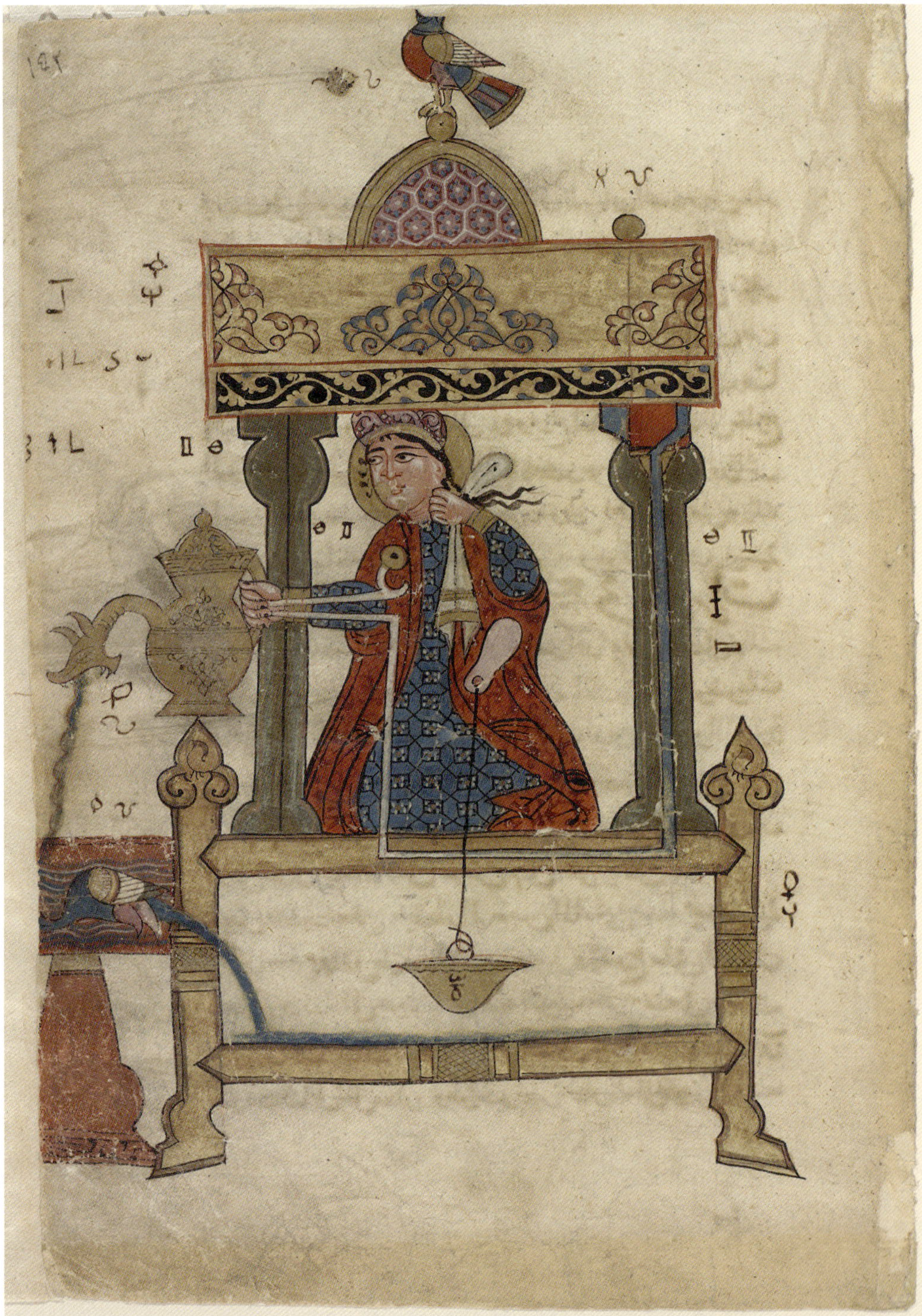

COLOR PLATE 10 *The Basin*, folio from a copy of Ismāʿīl ibn al-Razzāz al-Jazarī, *Book of Knowledge of Ingenious Mechanical Devices*, Mamluk, probably Syria, 1315. Calligrapher: Farruq ibn Abd al-Latif. Photo: Freer Gallery of Art, Smithsonian Institution, Washington, DC: Purchase—Charles Lang Freer Endowment, F1930.75.

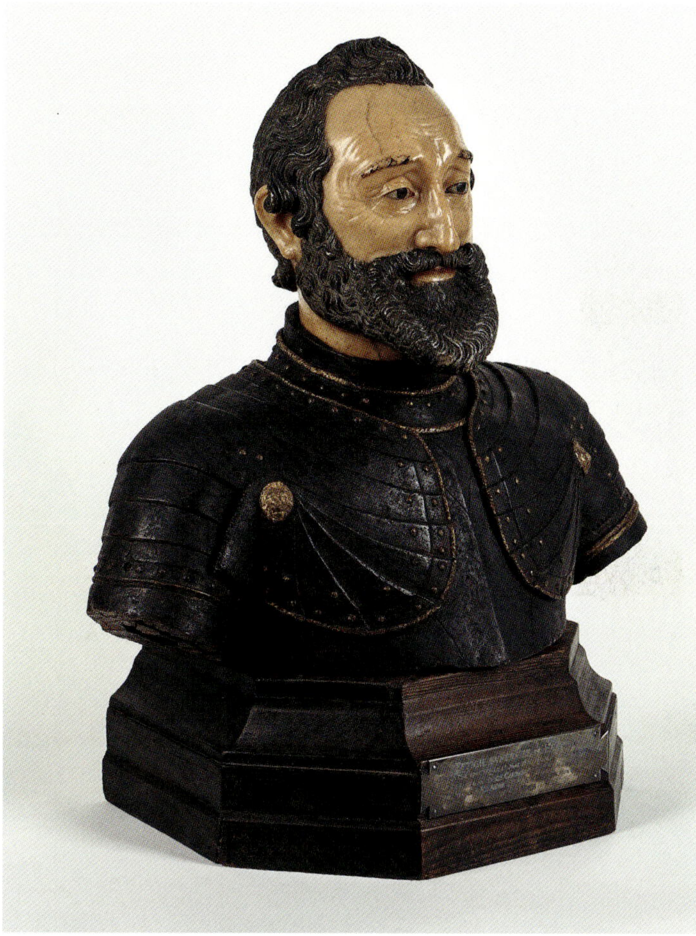

COLOR PLATE 11 Michel Bourdin, *Wax Bust of Henry IV*, early seventeenth century. Photo: CCo Paris Musées / Musée Carnavalet.

COLOR PLATE 12 Hendrik de Keyser, *Terra-cotta Bust of William of Orange*, ca. 1615–20. Photo: Rijksmuseum, Amsterdam.

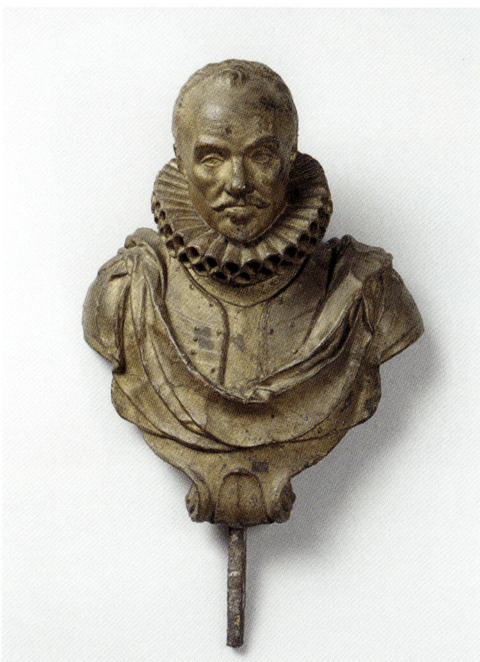

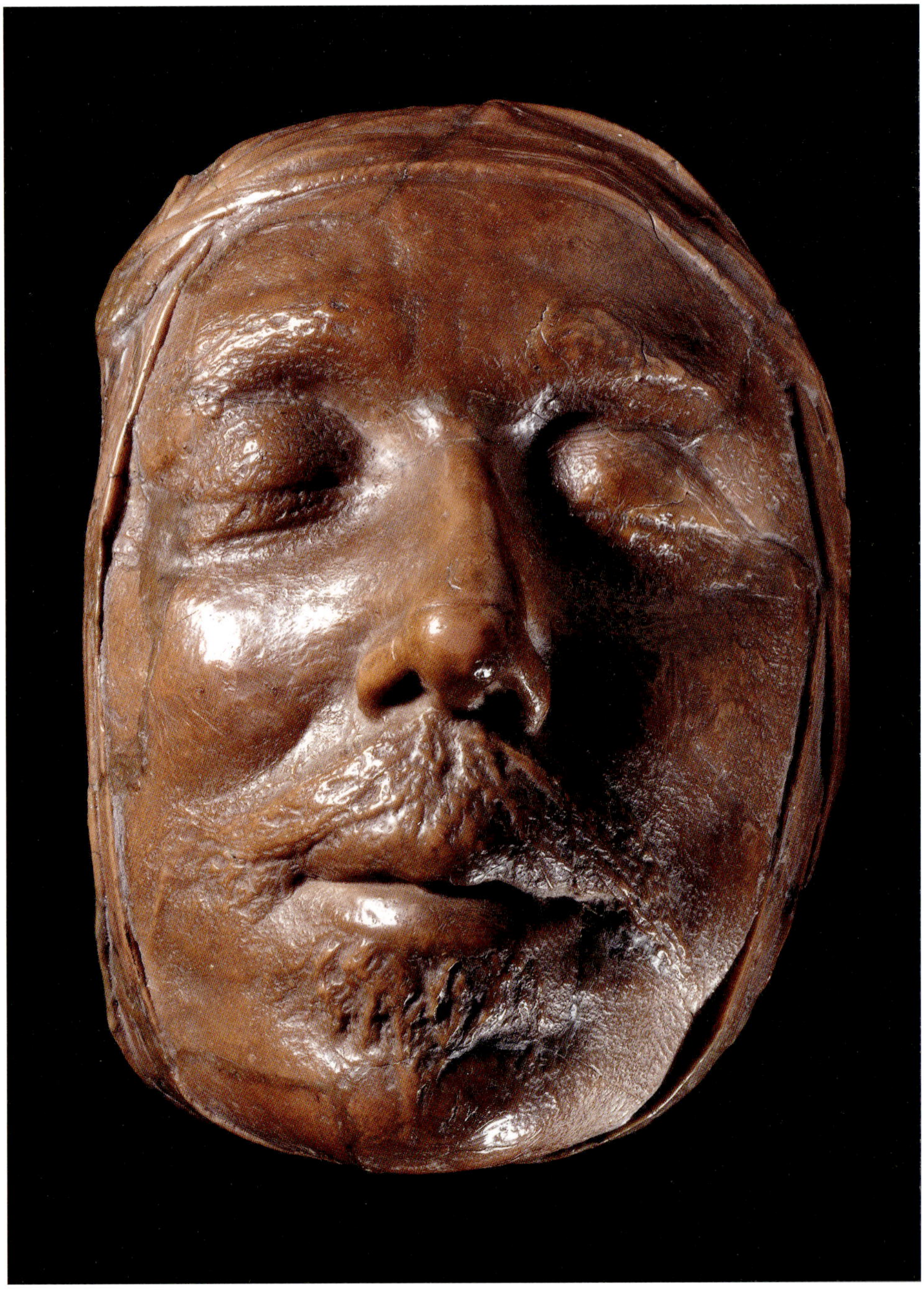

COLOR PLATE 13 Wax copy of death mask of Oliver Cromwell, seventeenth century. Photo © Trustees of the British Museum, London.

By including the labyrinth on his map of Crete, Blaeu was, in fact, engaging with a well-established cartographic convention derived from Ptolemy's *Geographia*.[74] Significantly, the depiction of the Cretan labyrinth was understood to be paradoxical: its presence on maps of Crete both supported and undermined cartographic truth claims. This is because the labyrinth of Crete—while famed—was not an actual geographical site but a mythic place. Recognizing the contradictory character of this Ptolemaic convention, one commentator writes: "Many do mention the Labyrinth in Crete which was built by Daedalus, of which . . . Pliny (lib. 36. cap. 13) witnesseth that there is now nothing remaining. Wonderfull therefore is the stupidnesse and ignorance of the Inhabitants, who doe still shew at the foote of the Mountaine Ida, a new Labyrinth instead of that old one: but it is no wonder that the Cretensians should Cretissare, that is, should be Lyars."[75] A telling slippage occurs here, since it is not the Cretans themselves but European cartographers who were showing the labyrinth at the foot of Mount Ida. The reference to Cretan liars connects this distinctive cartographic convention with the famed paradox of The Liar, which is the model of self-contradiction in logic. The paradox is this: the Cretan Epimenides claims that all Cretans tell lies. If he is telling the truth, then as a Cretan, he must be lying. If he is lying, though, then he must be telling the truth.[76] A similar self-contradiction occurs when a map that claims to be accurate details a site known to be fictional. The Cretan labyrinth is thus a sort of Cretan liar, a perfect equivocation. Through its inclusion, the map of Crete claims that the map of Crete is a liar.

Returning to the small labyrinths on maps of Amsterdam, one wonders if the same contradictory logic is at play. If the labyrinth was long recognized as a symbol of a walled city, do its paradoxical dynamics also characterize urban space? The proximity of the Amsterdam mazes to the new city wall may be noteworthy in this regard; the Oude and Nieuwe Doolhoven were constructed at the outer edge of the New Town, at its expanding frontier. The significance of the new city wall can be perceived in Georg Braun and Frans Hogenberg's 1617 map of the Amsterdam extension in progress, included in their influential atlas of cities, *Civitates Orbis Terrarum* (fig. 24). This map takes a multitemporal approach to indicate the changing contours of the city. The old city walls and moat are shown still intact. Outside of them, especially to the west, the cartographers have included a series of footpaths and ditches that radiate from the outer moat. This area was called *de paden* (the paths). By the late sixteenth century, Amsterdam was attracting thousands of immigrants, and many of the poorest lived in *de paden*. In 1608, there were more than three thousand houses there, all of them illegal. It was forbidden to construct buildings this close to the city walls; it created problems for civic defense. Moreover, thousands of inhabitants were evading taxation. There also were concerns about crime, disease, indigence, and other difficulties associated with unregulated urban overcrowding.[77]

When the truce with Spain in 1609 brought a hiatus to war, the municipal government seized the opportunity and embarked on a plan to alleviate the overpopulation problem. As noted, the magistrates consulted their Theseus, Prince Maurits, about this urban renewal plan. The strategy was to build a new city wall that would encompass the unregulated area called *de paden*. Construction began in 1610: on Braun and Hogenberg's 1617 map, the parts of the new city wall that had already been constructed are indicated by solidly rendered,

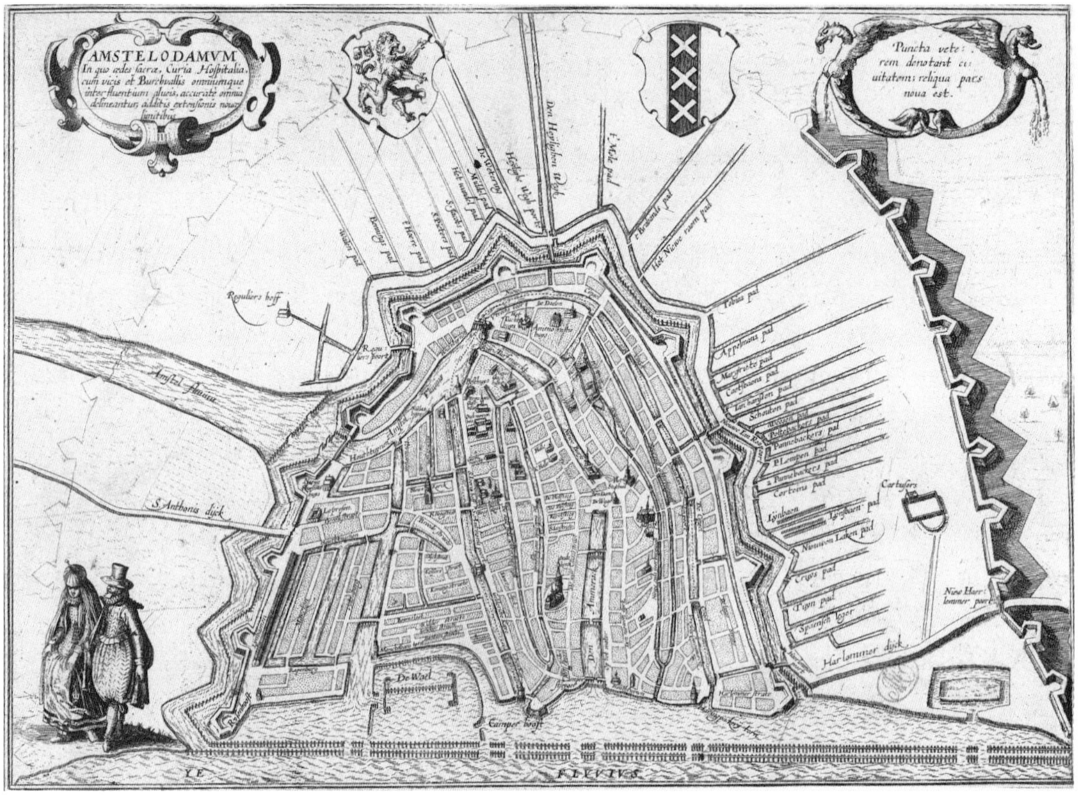

FIG. 24 Anonymous, *Map of Amsterdam, 1612*, from Georg Braun and Frans Hogenberg, *Civitates Orbis Terrarum* (Keulen, 1617–18). Photo: Rijksmuseum, Amsterdam.

three-dimensional fortifications that throw crenulated shadows onto the waters of the broad moat. The projection of the further extension of the wall, which would transform the city into its half-moon shape, is sketched with a single line. The cartographers hence indicate three moments in the lengthy and laborious process of urban transformation by picturing the old city, the new city, and a future expansion. Their map clearly shows how the new wall was designed to incorporate *de paden*.[78] The narrow jumble of paths in the disorderly district outside the old city walls was literally being contained: brought inside the walls, it was also brought under civic jurisdiction. The neighborhood of the Doolhof gardens had formerly been in this chaotic zone, recently restricted by the new city wall. This location suggests why cartographers mapping this newly constructed part of the city may have chosen to emphasize the Doolhof labyrinths. The plan of a maze elegantly pictures a bounded space that contains complexity and chaos. A pictogram of controlled confusion, the labyrinth represents the walled city in miniature.

The highly detailed topographic urban maps developed by Berckenrode and Blaeu record and celebrate civic expansion. They present the myth of a perfectly ordered city, in which unruly elements have been brought under control with the extension of the city wall. Looking again at Blaeu's map of the expansion, we note that, like Braun and Hogenberg, the cartographer included palimpsest lines to project how future city walls would come to encompass suburban sprawl as well as displace cows, haystacks, fields, and windmills, regularizing the areas outside the existing walls (see fig. 22). The geometric orderliness of the metropolis expands in space and time, as the planned city extends its uniformity into the countryside and into the future with subsequent enlargements throughout the seventeenth century. Civic histories of Amsterdam repeatedly praised the organized appearance of the New Town, lauding Amsterdam as the most meticulously planned European metropolis. The rational layout of the semicircular canal ring was contrasted to the oldest part of the city, which had developed haphazardly over the course of centuries, resulting in dark, cramped, labyrinthine streets and alleys overhung with crooked buildings.[79] Urban chronicler Von Zesen describes the new streets and canals as broad and straight—designed, he says, for the foremost citizens.[80] He and other authors of Amsterdam's civic histories drew on Ovid's *Metamorphoses* and spoke of the city enjoying a new golden age of civic harmony. In fact, the myth of the Dutch Golden Age is derived from these texts, which connect ideals of social and political concord with the remarkable cultural achievements of Amsterdam, an issue to which we return in chapter 7.[81]

The New Town thus represents a new way to manage urban space and populations.[82] Its geometric pattern appears abstract, methodical, and organized when mapped. The harmony of this visual order was designed to create social order. The wealthiest citizens lived on the broad new canals, while *de paden* area was named Jordaan and remained a neighborhood for the urban poor and working classes.[83] In addition, many disorderly aspects of urban life were relegated into separate enclosures within the city, also indicated on the maps: the madhouse, the lepers' house, prisons, workhouses, orphanages, and hospitals. These locales served to confine elements that were potentially disruptive to civic harmony: madness, crime, homelessness, indigence, violence, and disease. As Anthony Vidler argues, rationally designed spaces need to contain within them areas of unreason to serve as asylums for the forbidden, the disorderly, the frenzied, and the wanton. These elements, which disturb the ordered principles of planned space, must be given a place within the plan to serve as the tangible antagonists it fights against. The preservation of contradictory forces is thus a necessary part of defining rational urban space.[84] For reason to prevail, monsters must be kept.

The imprisonment and protection of dangerous forces had long been the paradoxical purpose of the labyrinth, a bounded space that encompasses bestiality, violence, horror, and death. Recall that the mythical purpose of the ancient maze made by Daedalus was to incarcerate as well as protect the monstrous Minotaur, the offspring of Queen Pasiphaë and the Cretan bull. The Minotaur was human, even princely, but also beastly. Such duality similarly characterizes the labyrinth: it preserved the queen's son from harm, and it imprisoned a murderous brute. A perfect paradox, the Cretan labyrinth was doubly apotropaic: it kept danger out, and it kept danger in. An analogous paradox typifies Amsterdam's new city wall: designed to protect the city from foreign incursions, it also encompassed the labyrinthine paths and ditches of the densely overpopulated *paden* area, bringing its poor and largely foreign outlying population into the city. It kept migrants out, and it kept migrants in.

Indeed, it is possible that the first public multicursal mazes initially appeared in the New Town because the restructuring of urban space and the reform of urban populations were closely intertwined during this period of intensive city planning and growth. Amsterdam's prisons were particularly innovative, and it is instructive to compare their principles to those of the Doolhoven.[85] The Spinhuis—literally, a house of spinning—was the women's prison, while the men's prison was called Rasphuis, as its inhabitants were set to work rasping red Brazilwood into powder for dyes. Both workhouses were designed to correct deviant behavior among the urban populace.[86] Their aims were in part derived from the treatise *Boeven-tucht* (The discipline of misbehavior), written in 1587 by Coornhert, a literary figure and artist who was influential for De Egelantier. Coornhert addressed pressing social problems of his day. His treatise explains that the displacements of war and religious discord were bringing legions of migrants into the ever-expanding Dutch cities. Masses of them, according to Coornhert, were idle layabouts, indigents who were turning to crime. Houses of correction were proposed as a productive solution to this urban disorder. Amsterdam's city fathers put Coornhert's recommendations into practice and established the two workhouses for men and women in the late 1590s. P. C. Hooft's father, the burgomaster C. P. Hooft, was one of the founders of these new civic institutions. Interestingly, the tax revenues paid by Amsterdam innkeepers for liquor licenses were an important source of funding for the Spinhuis, so there was a direct connection between the tavern and the house of correction.[87] If the sellers of drink contributed to social ills, they also paid for their amelioration.

The idea of the workhouses was to detain prostitutes, beggars, and other criminals and to teach them productive work, thus training them to make industrious contributions to public life.[88] The women confined at the Spinhuis, accordingly, were set to the tasks of spinning, weaving, sewing, and lace making. Above the portal of the Spinhuis was a verse by P. C. Hooft, which succinctly addressed this method of reform: "Don't be afraid, for I exact no vengeance for wrong but force you to be good. My hand is stern, but my heart is kind."[89] Another inscription at the Spinhuis stated that the aim of this benevolent correction was to teach indigent girls and women not to take the crooked way—the *doolwegh*—of begging and idleness.[90] Thus the forced labor of spinning produced a sort of Ariadne's thread: a "straight thread of reasonableness, to come from the confused ways of this world into the path of righteousness," in Van Mander's words.[91] The spinning and weaving of thread into intricate patterns guided wayward women to the right path.

A key idea was that labor and incarceration could tame the inner beast. Coornhert likened the "urban scum" to savage bears and wolves—their depraved nature made them like dangerous wild animals that had to be disciplined and made useful for the greater advantage and convenience of society.[92] The portal of the men's workhouse includes a sculptural relief by De Keyser, also a member of De Egelantier (fig. 25).[93] The work resonates closely with Coornhert's description of wrongdoers. A group of beasts—including lions, wild swine, and a wolf—are harnessed to a wagon driven by a foreman with a whip. The cart is laden with logs—the red Brazilwood that was rasped by the

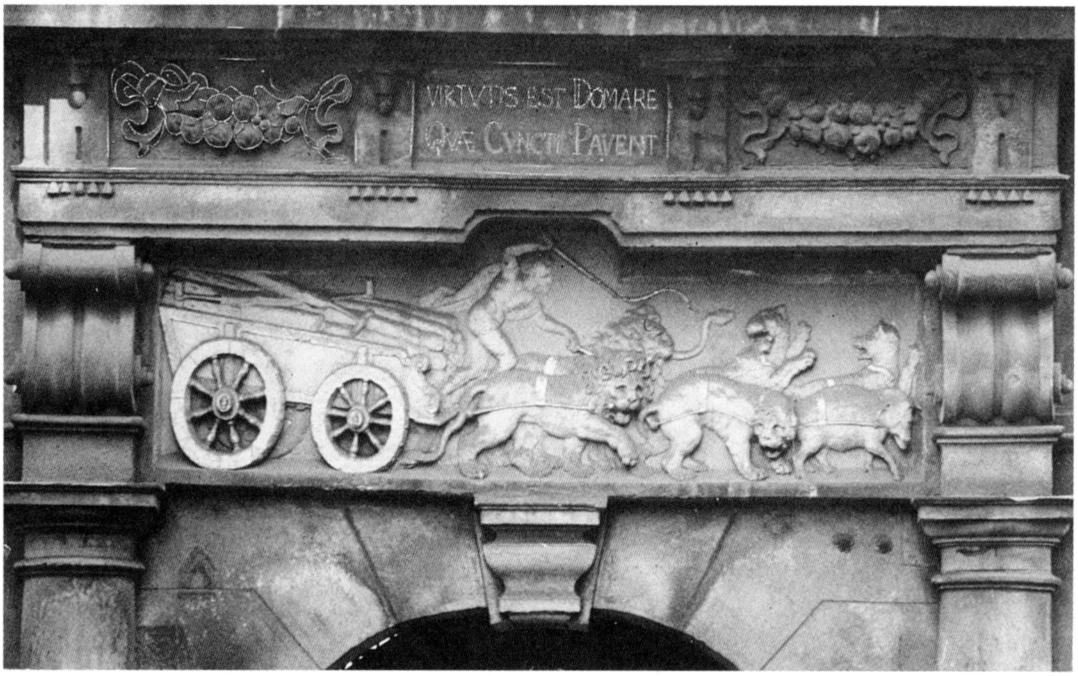

FIG. 25 Hendrik de Keyser, *Portal of Amsterdam Rasphuis*, 1604, detail. Photo: Amsterdam City Archives (BMAB00013000019_004).

inmates of the Rasphuis. The Latin inscription is from Seneca: "It is virtuous to tame those whom all fear."[94] The message is evident: if urban miscreants refused to modify their beastly behavior by their own volition, they would be punished, trained, and reformed by these civic institutions. The dream of the perfect city depended on the inner transformation of each resident. The ideal of stoic self-mastery takes a turn here; good citizens do not just tame the beast within themselves—they also dominate the beastliness that they perceive in others.

While certainly more heavy-handed, the aims of the prisons were surprisingly similar to those of the Doolhoven. Both structures detained people in a bounded space where they were confronted with their own failings. In each place, the labor of getting out of confinement required the effort of gaining control of the self—of recognizing and training the inner brute. The contrast to the workhouses instructively highlights what was distinctive about the Doolhof program of urban reform: it seductively used artistic innovation and aesthetic pleasure as forms of discipline. The Amsterdam mazes proudly proclaimed artistic supremacy and a classical heritage for the city while contributing to the reform of human behavior. These sites provided a controlled environment where visitors were immersed in their interactions with highly innovative works of art that were cleverly designed to trick and ensnare, to move their audience both emotively and physically. The artworks behaved like the dangerous monsters that find asylum within a labyrinth: unleashing their powers, they attacked wandering visitors along the prescribed route with “a thousand strange lows,” provoking struggles in which spectators battled to outwit the lively forms of art. Heinsius wrote that affective theater performances were a “wrestling school of the passions,”[95] and I would contend that this was the function of the Doolhoven as well. The Neostoic prescription was that people needed diligent exercise to learn how to master their instinctive urges. This type of cultural exercise—a sort of initiation rite—prepared visitors to go back into the worldly city and meet its many disturbing blows armed with dispassion and self-control.

If the preservation of contradictory forces is a necessary part of defining rational urban space, then the Doolhoven served as asylums. By releasing and containing potentially dangerous artworks, they provided a training ground to combat the fearful powers of art. As we will explore further in the next chapter, in the process of being led astray by deceptive art forms, wandering became wondering, a means of thinking through paradox.[96] After the trial of the maze, peripatetic spectators were equipped for the next part of the itinerary: the encounter with the living statues secreted within the Doolhof sculpture gallery.

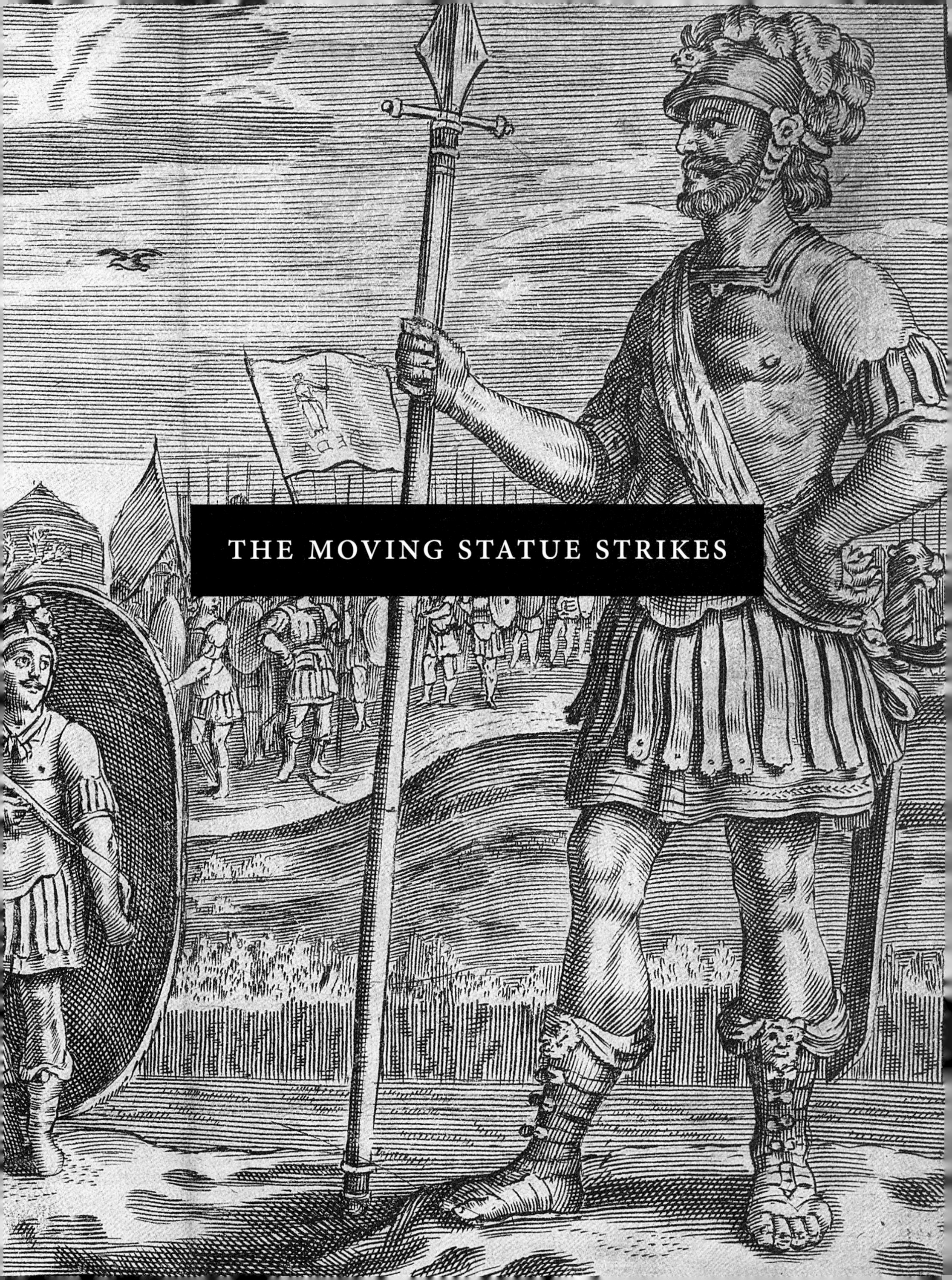

THE MOVING STATUE STRIKES

AUTOMATA

Chapter 4

Activating Human Behavior

For all ingenious and accurate mechanical inventions may be conceived as a labyrinth.
—Francis Bacon, "Daedalus, or Mechanical Skill," 1609

The maze unwinds, pulling exiting visitors along a path to a door that opens into a large building: the sculpture gallery. It is as if the sobering test of solving the maze was a rite of preparation to attain an appropriately receptive state before proceeding into the building. The admission price was one *stuiver*, roughly equivalent to the cost of two tankards of beer. Melchior Fokkens's civic history indicates that it cost a *stuiver* to enter the maze and another *stuiver* to gain access to the gallery.[1] The additional payment heightened expectations, augmenting the mood of suspenseful revelation created by moving out of the peripatetic constrictions of the labyrinth. Visitors were ushered from the open-air garden—with its noisy crowds of exuberant customers, aggressive spurting fountains, and incarcerating green hedges—into a dim interior that must have seemed calm, dark, and quiet at first. This gallery was somewhat like the innermost sanctuary of a temple precinct, where the most powerful and awesome effigies were kept.[2] The physical movements of spectators were curbed as they took their places in rows of tiered seating facing a curtained stage, much like in the theater. Accompanied by organ music, the curtains parted to reveal performances in which "each image moves and acts as if it lives, for the wonder and delight of onlookers."[3] The act of unveiling potent artworks was an awe-inspiring moment in the ritual.[4]

The shows or performances (*vertooningen*), as they were called, were wondrous indeed. They consisted of three kinds of mechanically driven works. The androids were life-sized or larger than life-sized costumed figures that looked and moved like human bodies. There were also moving-picture shows: miniature stage sets on which a series of animated figures acted out familiar classical, theatrical, or biblical narratives. On display, in addition, were complex astronomical clocks with intricate moving parts. Each work in turn was activated and introduced by a presenter. On a raised stage behind these apparatuses was a row of life-sized historical figures in wax. A house of moving pictures, the gallery was a complex pictorial and narrative space, featuring works that apparently came alive. A number of the automata made sounds as the spectacle of lifelike motion began to play. Accounts of the performances entice with claims that these lively artworks could sing and almost speak and that animal automata regaled crowds with their crowing, lowing, braying, bleating, and quacking.[5] While spectators sat still to look and listen, the artworks came to life with motion and sound in an animated performance that would move viewers with wonder and delight.

Inventions of Daedalus

Housed today in the Amsterdam Museum are the few Doolhof automata that survive: painted wooden statues of David, Goliath, and Goliath's shield bearer that were once part of the Oude Doolhof spectacles (color plate 3).[6] A guidebook print of this sculptural ensemble situates the figural group in the context of the battlefield, probably in front of a painted backdrop scene, giving an indication of the theatrical trappings of automata shows (fig. 26). Framing the staged scene is a parted curtain, opened to reveal Goliath and the invading Philistine army standing against the Israelites who line the city walls. David and the shield bearer both look up to the enormous armed figure of Goliath as the Israelite shepherd boy prepares to battle the giant from Gath. Although David stands only as high as the titan's shield, he is animated by a potent force: with mouth slightly agape, this moving statue would slowly raise its head and then roll its eyes back and forth as if in assessment of his fearful adversary. The Goliath statue responded by turning its head and rotating its eyes in glowering disdain.

These polychrome statues, especially the colossal Goliath, with a height of almost five meters (sixteen and a half feet), may have been constructed at the Amsterdam shipyards. They resemble the giant painted wooden figureheads that ornamented East and West India Company ships.[7] Ship sculptures often represented imposing monsters or warriors. As the Dutch trading companies sailed the world, their conspicuous, colorful figureheads were intended to intimidate and impress those who encountered them on the high seas or in far-off ports. The David and Goliath statues also derive from large-scale sculptural programs that had long adorned Renaissance cities. The Florentine Davids of Donatello and Michelangelo were particularly compelling, encouraging citizens to protect their cities against even the most formidable foes.[8] With the Amsterdam David, the powers of sculpture are taken to an extreme. This statue of the shepherd boy is so lively that it seemingly becomes David as it physically moves to attack the invader Goliath. Such actions resonate with the claims that visitors "were struck with wonder at the mechanisms."[9] David thus deals a double blow: as the mechanized statue moves to confront the automated giant, it simultaneously strikes viewers with wonder.

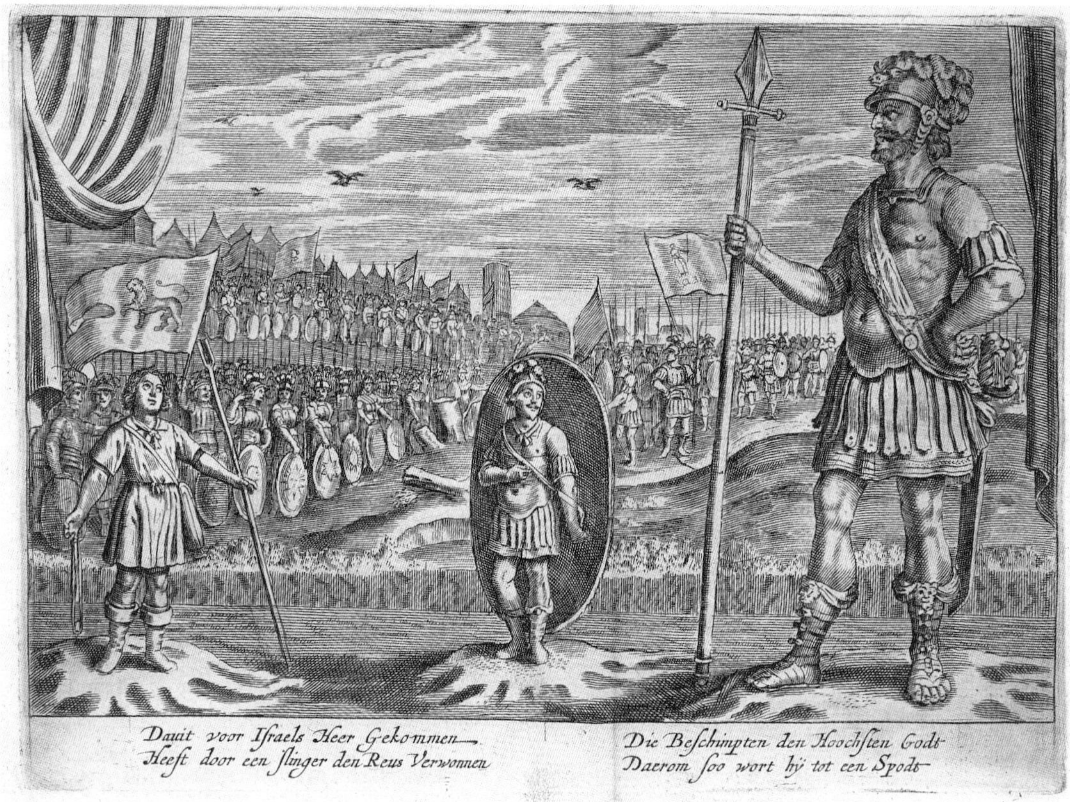

FIG. 26 *David and Goliath*, from *Verklaringe van Verscheyden Kunst-Rijcke Wercken en haer Beweginghe* (Amsterdam: Dan. Davidsz. Mansvelt, 1657). Photo: Allard Pierson, University of Amsterdam Library Special Collections, OTM O 62–2615.

A poem at the bottom of the guidebook print assures that David's defeat of Goliath with his sling was a warning that all those who mocked God would themselves become objects of ridicule. This moral message may have been included in the banter of the showman who presented the works. Androids were an especially potent type of graven image, and strategic methods of display were a critical means to integrate these ambiguous arts into society, making them familiar, understandable, and acceptable to audiences.[10] In the case of David and Goliath, an oft-represented biblical narrative was presented in theatrical fashion and framed with a religious warning to the enemies of Christianity. The novelty of potentially idolatrous moving statues was thus counterbalanced by well-known iconography and a conventional exhortation about defending the city against foreigners. Such powerful arts had civil benefits.

Like the maze, automata were intricate works designed to amaze, confound, and change those who interacted with them. This insight was articulated early on by Aristotle, who wrote that automata prompt wonder, which leads to wondering about causes, the beginnings of philosophy.[11] The metamorphic potential of automata lay in their shockingly lifelike qualities, which first startled and arrested viewers and then aroused their interest, which, in turn, could stimulate further investigation, rational thinking, and the acquisition of new knowledge. This rationale for automata was often repeated, sketched out as a temporal progression in which beholders moved from astonishment to curiosity to thinking.

Automata were considered a special kind of technology whose purposes were not practical but philosophical. The Roman architect Vitruvius classified these types of inventions as *parerga*—marginal and ornamental things. He writes that such machines are pleasurable and amusing but do not immediately serve human needs in the manner of devices like pumps or waterwheels.[12] Plato also writes about moving statues, and his pithy assertion that they set arguments in motion remains the best summation of what automata do. Plato was referring specifically to the works of the inventor Daedalus with his statement: "Nay, Socrates, I shall still say that you are the Daedalus who sets arguments in motion."[13] While previous chapters have focused on Ariadne, Bacchus, Theseus, and the Minotaur, we now turn to the role of the ingenious craftsman in this narrative, for Daedalus not only made the maze but also was credited with inventing statues that moved. The earliest textual reference to the legendary artist appears in Homer's *Iliad*, which alludes to dancing figures that Daedalus made in Knossos for Ariadne of the lovely hair.[14] This association of the Cretan maze with mechanical moving figures is key to deciphering the Amsterdam Doolhoven. The evocative combination of these Daedalic art forms is a deliberate part of the Doolhoven's program to revive—in the literal sense of bring back to life—the mythic creations of antiquity.

A guidebook to the Nieuwe Doolhof trades on the automaton's ancient lineage. Addressing its readers as "art-loving and tasteful spectators," it asserts that although we often read of great and very ingenious ancient masters who made extremely subtle works, the Greeks and Romans never saw anything like the clever artificial inventions on show here. Through art, these things can stir and move as if they have life in them. Yes, they even can sing and almost talk.[15] Another booklet boasts about bringing "an art that has long been in darkness into the daylight."[16] Such assertions legitimized the inventions by claiming to recover—and outdo—the most accomplished inventions of antiquity. They promise more than the information gained by reading about lost works of the past; instead, these were concrete material objects that could be seen and heard. Like the green labyrinth and the Bacchus fountain, they could be physically experienced in the here and now. The Doolhof's innovative works were not mere repetitions of previous ancient devices; they moved in a new direction.

Like multicursal mazes and hydraulic works, automata were already well established in Europe and had significant civic, spiritual, and political functions. Since the fourteenth century, they had enlivened the tower clocks of town halls, churches, and city gates, and they played a role in urban, religious, and courtly festivities, pageants, triumphal entries, and processions. As with fountains, such public demonstrations of ingenuity manifested the inventive powers of a court or city. In the sixteenth and early seventeenth centuries, the main centers of innovative clockwork and automata making were the German-speaking free Imperial cities, especially Augsburg, but also Strasbourg, Ulm, and Nuremberg. Besides public and urban displays, German clockmakers also made smaller-scale clocks and automata, intricate and costly prestige objects for rulers, aristocrats, and the wealthiest merchants, who proudly showed them as status symbols in their collections and on their banqueting tables.[17]

Automata have been described as the "playthings of princes."[18] Like the labyrinth, they were common features of aristocratic garden design, manifesting the ruler's power over nature itself. The earliest known garden automata were at the chateau of Hesdin, which also featured fountains, trick waterworks, and multicursal hedge mazes. Records of Hesdin's notorious *engines d'esbattement* (frolicsome machines) date to the late thirteenth century. The trend caught on, particularly in France, Germany, and Italy, where the grounds and grottos of many palaces and villas sported hydraulic moving statues and waterworks that attracted well-educated travelers in search of notable sights. The British gardener and diarist John Evelyn, for instance, wrote in 1660 of "wonderfull Automats . . . which at present so celebrate the Gardens of the greatest Princes . . . and many other famous Gardens of the most illustrious persons of the world."[19] The French Huguenot refugee and artisan-engineer Salomon de Caus created spectacular automata-filled gardens (which also featured hedge mazes and inventive fountains); the most famous was the Heidelberg *Hortus Palatinus* for the Palatine Elector Frederick V and Elisabeth Stuart.[20] As noted above, de Caus's 1615 treatise, *Les raisons des forces mouvantes avec diverses machines tant utiles que plaisantes*, probably informed Amsterdam fountain makers like Bargois. The hydraulic automata made by the brothers Tomasso and Alessandro Francini for the palace gardens of King Henry IV of France at Saint-Germaine-en-Laye were also renowned. Descartes possibly visited these gardens, and they may have provided a source of inspiration for his mechanical philosophy, an issue we return to below. Another significant philosopher, Montaigne, wrote about the various automata and hydraulic works he encountered on his travels through Europe, including those at the celebrated Italian gardens at Pratolino and the Villa d'Este in Tivoli, as well as the estate of the Fugger banking family outside Augsburg.

Montaigne's reaction to wondrous automata is revealing; after he had seen a number of comparable devices and gained an understanding of their inner workings, they began to lose the element of surprise. For instance, trick water jets such as the ones in the Doolhof gardens were quite common. Montaigne saw them everywhere. By the time he got to Tivoli, he no longer exclaimed about wonders but summarily explained that the hydraulic attack was "launched by a single spring that can be worked from far off," as "seen elsewhere on my trip."[21] Other hydraulic and pneumatic works, such as a display of flittery, twittering bird automata, evoked a comparably blasé response: "All these inventions, or similar ones, produced by these same natural causes, I have seen elsewhere."[22]

While historians of automata claim that these devices were the vanguard of early modern European technology and commanded unprecedented veneration, Montaigne's disenchantment allows us to assess another type of reaction.[23] Many of these mechanisms were designed to surprise or even assault the viewer. In parks and palaces across Europe, there were mechanical beasts that jumped out of doors, moving figures that beat their victims about the head and shoulders, chairs that grabbed their occupants, and numerous variations of waterworks that soaked the unwary, sometimes followed by devices that sprayed and coated wet bodies with flour, soot, feathers, or a combination of these.[24] Such works elicited immediate physical and emotive responses: like the David and Goliath statues, automata struck at their viewers to render them wonderstruck. However, as Aristotle points out—and Montaigne's reaction confirms—wonder was not an end in itself. It generates curiosity about causes and can prompt scientific inquiry about the rules of nature. Once the viewer becomes familiar with the workings of novel, strange, and surprising devices, these seemingly magical machines are no longer astounding, and their mysterious movements lose their powers to enchant. As Aristotle concludes, they are only wonderful to those who have not perceived their causes.[25]

What is especially significant about Montaigne's account is that it articulates how knowledge and the understanding of causes effected a change in affective behavior. Indeed, Montaigne's point seems to be that after repeated exposure to automata and their tricks, he took control of himself and was unmoved by them, reaching something like a state of Stoic *apatheia*, an absence of passion when faced with unpredictable external forces. The amazing powers of the automaton are thus short-lived. There is a distinctive temporal dimension to the process of viewing such artworks. The initial reaction is embodied, immediate, and nonrational, but sustained or repeated exposure can lead to disillusionment and a loss of interest in startling effects, as attention turns to consideration of predictable causes. These inventions accordingly had the potential to work significant emotive and behavioral changes in their viewers, moving them from astonishment to dispassion.

A comparable response to Amsterdam's hydraulic and mechanical inventions is recorded in the travel journals of Evelyn, who visited the city in 1641. Evelyn was a virtuoso, interested in the new experimental philosophy. He was most impressed by what he saw at the inn called D'Os in de Bruyloft, run by the Mennonite scholar Theunisz. Evelyn's description is one of only a handful that we have of this remarkable place:

> The next day we were entertained at a kind of tavern, called the Briloft, appertaining to a rich Anabaptist, where, in the upper rooms of the house, were divers pretty waterworks, rising 108 feet from the ground. Here were many quaint devices, fountains, artificial music, noises of beasts, and chirping of birds; but what pleased me most was a large pendant candlestick, branching into several sockets, furnished all with ordinary candles to appearance, out of the wicks spouting out streams of water, instead of flames. This seemed then and was a rarity, before the philosophy of compressed air made it intelligible.[26]

In this brief passage, Evelyn, like Montaigne, charts a transition from the pleasures of viewing an appealing array of surprising devices to dispassionate consideration of the laws of natural philosophy. The inventive spurting candlestick was only perceived as a rarity before the physics of compressed air were explained. In a manner similar to the theories of Aristotle, Evelyn tracks how wonder, the initial prediscursive response, prompted inquiry and discourse about causes.

Evelyn's experience indicates how the automata-displaying inns were sites of knowledge production, where craft know-how and creative inventions sparked discursive learning and investigation of the natural world.[27] Comparable apparatuses were shown at the Doolhof inns: a branched candlestick with spouting sockets like the one Evelyn found particularly pleasing is pictured just beneath the Bacchic chariot in Holsteyn's engraving of *The Triumph of Bacchus and Ariadne* fountain (see fig. 9). A similar automatic device was featured in the Nieuwe Doolhof courtyard. In his civic history, Domselaer states that the "most wondrous of all" of the Nieuwe Doolhof devices was a church chandelier with six candles that sprayed water.[28] Such illusionistic contraptions—which replaced the expected element of fire with surprising spurts of water—were regarded as astounding novelties and were singled out as especially marvelous.

Evelyn's experience at D'Os in de Bruyloft evidently included discussion of the causes of the illusions. His comments imply that these rare devices were made intelligible to visitors through explanation and discussion of their hidden workings. It is unclear whether instruction about the physics that drove the works was part of the staged exhibition or whether such inquiries took place more informally, perhaps in conversations over drinks at the tavern. Evelyn was wealthy, well connected, and highly educated, with a particular interest in experimental science and technology. Possibly the secrets of the maker's practice were mainly revealed to visitors of his gender and status.

Doolhof publicity materials do not disclose information about inner mechanics to the general public. These sites, after all, were designed to entice with "many fine secrets." Exposing all of the mysteries to all visitors was a risk, as it could lead to the sort of disinterest expressed by Montaigne and thus to fewer paying customers. It was common practice to guard lucrative artisanal knowledge and processes as trade secrets.[29] The advice to hide the mechanisms and keep them confidential had long been part of automata lore. One of the earliest treatises, by Hero of Alexandria (first century CE), advocates that readers keep the maker's knowledge to themselves. This caveat is repeated by de Caus: only with ignorance of causes did such devices appear marvelous.[30] While the treatises on automata reveal their workings and principles to readers, they do so in a manner that sets the learned reader apart from other spectators, making distinctions between those in the know and those who were ignorant and thus wonderstruck. In his book *Mathematical Magick* of 1648, the English Protestant clergyman and natural philosopher John Wilkins called automata "veiled arts." He advises that ancient practitioners like Daedalus had found it wise "to conceal their learning from Vulgar Apprehension or Use" in order to be able "to excite the People's Wonder and Reverence."[31] Indeed, Wilkins himself sounds somewhat wonder-weary: describing displays that sound similar to the Doolhof shows, he remarks—with the apatheia of the virtuoso—that it had become quite "common to represent persons and actions of any story by self-moving images," so that strange inventions "are derided by common opinion."[32]

What made the Amsterdam automata displays different from those in the elite gardens visited by Montaigne is that the Doolhoven were commercial establishments whose attractions were advertised to entice anyone with a few *stuivers* in their pocket to pay the admission price, luring them with promises of novelty and wonder. The Doolhof exhibits also differed from previous urban presentations of religious automata. Prior to the Reformation, the Roman Catholic Church had been a prolific patron of public automata shows. All manner of moving religious figures—God the Father, Jesus, the Holy Spirit, the Virgin Mary, apostles, saints, angels, devils—were familiar sights on clock towers and in church sanctuaries; they also enlivened civic festivals and processions.[33] Devotional figures also proliferated. A rare surviving example is a mechanical monk, today in the Smithsonian Institute (color plate 4). It was most likely made by Juanelo Turriano, who was commissioned in 1560 by King Philip II of Spain to make a votive automaton in gratitude for the recovery of the prince and heir, Don Carlos, after he suffered a serious head injury. This small figure can turn its head and eyes and open and close its mouth while its hands and arms move to beat its breast. The repetitive prayer gestures of the moving monk offer up endless thanksgiving to God for the miraculous intervention.[34]

Religious automata like the monk were apprehended as sacred things, imbued with divine spirit and agency. The spiritual vitality of seemingly miraculous or magical moving figures revealed in devotional contexts proclaimed the presence of God in the material world. Aristotle's claim that automata provoke wonder about causes still pertains in this context, but there is not the same temporal shift from wonder to knowledge of the mechanisms that was described by Montaigne, Evelyn, and Wilkins in response to courtly and commercial displays. When automata were used to animate devotion, it is unlikely that there was explanation of the technology. Their air of mystery and miracle was accounted for in religious terms and provoked veneration.[35] The repetitive gestures of mechanical devotional figures moved to inspire unending reverence for the unfathomable power of an eternal God—the ultimate cause of movement, spirit, and life in all things.

In the post-Reformation period, religious automata came under attack, particularly in Protestant regions.[36] For the Calvinists, God was understood as immaterial, invisible, and transcendent: the divinity of God therefore could not be captured in material forms. Calvin's stance hinged on his insistence that "the finite cannot contain the infinite."[37] The eternal, unending, immeasurable glory of God cannot be accessed through mere physical props, no matter how wondrous. Calvin found all image veneration misguided because vain and fleeting human-made things could not harness God's divine power. He insisted on the second commandment: "For God forbade . . . the making of any images representing him, since they would falsify his glory and change his truth into falsehood."[38] Calvin emphatically classified all imagery as the worldly work of human hands and maintained that human-made objects were not conduits to God. Needless to say, the moving statue was especially transgressive; its lifelike, spirited qualities, particularly when representing the divine, most certainly had the power to stir up idolatry—false worship misdirected at the vital object rather than given to God himself. Calvin's concerns focused on wrongful responses to devotional imagery, which he characterized as "foolish and inconsiderate passion."[39] In Calvin's view, therefore, it was the affective relation between the viewer and the work of art that stood in need of reform: improper viewing practices allowed thoughtless emotions to hold sway, and foolish passion spawned idol worship.

Post-Reformation Dutch sculptors accordingly worked "in the shadow of Daedalus," as De Jongh aptly put it.[40] They had to be careful not to transgress image prohibitions or incite idolatry. Daedalus was construed as a troubling figure in seventeenth-century Dutch art theory: while he was one of the most renowned of the legendary ancient artists, his ingenious inventions were described with a great deal of ambivalence. For instance, in his treatise *The Painting of the Ancients*, published in 1638, the Dutch Calvinist scholar Junius writes that moving mechanical figures were both "wonderfull" and "diabolicall," and he calls the automaton a "divelish deceit."[41] Of Daedalus's works specifically, he notes that they "were not very handsome to looke on, but that there was in them a certain kinde of divine majesty."[42] In other words, they were idols.

Van Mander's references to Daedalus and the Daedalic arts in his artist biographies reveal a comparable vacillation. He reserves the term "Daedalus work" for masterpieces that surpass art: wonder-inducing works in which "everything is alive."[43] For instance, when describing the unprecedented innovations of the Daedalic Ghent Altarpiece, he claims that Jan van Eyck "painted breathing figures . . . to bring the work to life."[44] However, he also lauds Cornelis Ketel's allegory of Pictura painting Daedalus and Icarus as a didactic work that served "to impress upon the artist restraint from conceit."[45] To say that Daedalic art suffers from conceit seems an understatement. After all, Daedalus's works included the wooden cow built to help Queen Pasiphaë mate with the bull, and thus the artist played an integral role in the procreation of the man-beast Minotaur. Not only this, but Daedalus secretly harbored this atrocious offspring by building the ingenious labyrinth to protect it. The labyrinth ultimately served to imprison the inventor as well as his beastly invention: Daedalus and his son, Icarus, were incarcerated in the maze by King Minos upon the discovery of Daedalus's monstrous transgressions. The ingenious craftsman, undeterred, fashioned wings of feather and wax for them to fly from confinement, resulting in the death of Icarus, who flew so close to the sun that Apollo took notice of his hubris and melted the fabricated wings. In his commentary on Ovid, Van Mander calls Daedalus a great artist and ingenious master, but he also interprets the fall of Icarus as punishment for his father's excessively harmful, unseemly, and twisted works.[46]

The moral of the tale of Daedalus was clear to Protestant artists and theorists like Van Mander and Junius: human ingenuity could go too far, and when artists' godlike creativity or lively creations attempted to vie with the power and majesty of God, they had to be curtailed. Dutch art treatises tended to reiterate Calvin's main concern: it was human behavior that needed reform, most of all when powerful passions were aroused by lively works that were equally wonderful and diabolical.[47] Van Mander urged his readers not to imitate cunning Daedalus but to become like Theseus and find the way out of the crooked labyrinth of artifice by following "the straight path of reason."[48]

Daedalic arts emphatically do not follow a straight path. Like their creator, who flew from the labyrinth, Daedalus's moving works are elusive; they too could deviate, dart away, and resist confinement. Plato writes of Daedalus's creations: "If no one ties them down, [they] run away and escape. If tied, they stay where they are put. . . . If you have one of his works untethered, it is not worth much; it gives you the slip like a runaway slave. But a tethered specimen is very valuable, for they are magnificent creations."[49] Vitruvius notes something similar about automata with his claim that they "are not subservient to our needs."[50] If the purpose of the machine is to serve humanity, then the automaton was perceived as a recalcitrant servant or a runaway slave, an issue we will revisit in the following chapter. Automata were a special class of self-moving machine that could escape human control. Echoing Plato, Wilkins writes that "unless violently detained [the inventions of Daedalus] would of themselves run away."[51] The display of such things in the post-Reformation period thus required specific strategies, as they transgressed the limits of artistic practice in Calvinist-dominated Amsterdam.

Like Van Mander, Bacon writes insightful commentary about Ovid's tale of Daedalus, and he, too, expresses ambivalence about works associated with the canny artist-inventor. In *The Wisdom of the Ancients* of 1609, Bacon importantly addresses the potential social impact of the mechanical arts. He sums up the contradictory status of the legendary inventor by calling him a "most ingenious but execrable artist."[52] Bacon disparages Daedalus for putting his extraordinary mechanical skill—his "destructive genius"—to ill uses, both in the conception of the bestial Minotaur and in the invention of cunning mechanical devices. The story of Daedalus is an important founding myth of technology, and Ovid's suggestive description of this crafty craftsman evokes what was to become a central concern in the early modern period: "And turning his mind toward unknown arts, he transformed nature."[53] Daedalus's daring experimentation allowed him to create new forms not found in nature: the labyrinth, the Minotaur, and the automaton. All were paradoxical configurations: natural and artificial, excessive and controlled, they kept irreconcilable opposites in play.

The social functions of ambivalent art forms were also perceived as contradictory. Bacon asserts that humanity is indebted to the mechanical arts, which are an "ornament of civic society," but he concurrently expresses concerns that the same arsenal could supply instruments of lust, violence, and death. Guns and engines of war, for instance, are wicked mechanical inventions that far exceed "the cruelty and barbarity of the Minotaur himself."[54] Daedalus's monstrous mechanics could unleash brutality into the world: when placed in the wrong hands, they might do as much harm as good. Bacon writes that the Daedalic arts resemble their creator, who "always found refuge and a means of escaping."[55] Like "runaway slaves," these cunning devices could dart off along unforeseen paths. Once invented, they are let loose in society where their potential uses are equivocal, elusive, and difficult to judiciously control. Bacon's discussion turns on his comparison of the mechanical arts to Daedalus's other unnatural innovations—the Minotaur and the labyrinth—and he concludes that such arts "have an ambiguous or double use, and serve as well to produce as to prevent mischief and destruction; so that their virtue almost destroys or unwinds itself."[56] These marvelous and monstrous creations proffer civil benefits even as they wend their crooked ways.

Pressing further with the paradoxical nature of the mechanical arts, Bacon notes that—like their maker—they are always banned and always maintained in the city: "Unlawful arts, and indeed frequently arts themselves, are persecuted by Minos, that is, by laws, which prohibit and forbid their use among the people; but notwithstanding this, they are hid, concealed, retained, and everywhere find reception and skulking-places."[57] If Minos represents law, structure, authority, and civic order, then Daedalus epitomizes the wily intelligence and subversive nature of craft.[58] Bacon's anxieties about mechanical technologies focus on the reception of new inventions by the people and the inability or unwillingness of governors to entirely forbid equivocal arts that were perceived as equally malicious and advantageous to civic life.

With this, we discover another monster skulking within the Doolhof urban mazes—the beastly machine.

One cannot help but wonder if Bacon's ideas helped to inspire the highly idiosyncratic Doolhof exhibition sites. His works were well known in the Netherlands, and the obvious interests of the inns' proprietors in experimental science, the mechanical arts, and the advancement of learning were likely informed by Bacon's views.[59] The manner in which Bacon compares automata to the labyrinth certainly seems relevant for understanding the logic of these establishments: "The labyrinth contains a beautiful allegory, representing the nature of the mechanic arts in general; for all ingenious and accurate mechanical inventions may be conceived as a labyrinth, which, by reason of their subtilty, intricacy, crossing and interfering with one another, and the apparent resemblances they have among themselves, scarce any power of the judgement can unravel and distinguish; so that they are only to be understood and traced by the clue of experience."[60] As with the labyrinth, the complex artifices of machines require a specific mode of investigation: they can only be deciphered by experience. As experimental trial-and-error processes, the labyrinth and the mechanical arts are both emblematic of the methods of the new science itself.

In his preface to *The Great Instauration*, Bacon, in fact, conceives of the universe as a labyrinth to be decoded:

> But the universe to the eye of the human understanding is framed like a labyrinth, presenting as it does on every side so many ambiguities of way, such deceitful resemblances of objects and signs, natures so irregular in their lines and so knotted and entangled. And then the way is still to be made by the uncertain light of the sense, sometimes shining out, sometimes clouded over. . . . In circumstances so difficult neither the natural force of man's judgement nor even any accidental felicity offers any chance of success. No excellence of wit, no repetition of chance experiments, can overcome such difficulties as these. Our steps must be guided by a clue.[61]

In the trial of solving the labyrinth, reliance on the human senses or accidents of chance are not dependable guides. While Bacon echoes Calvin's assertion that the divine countenance "is like an inexplicable labyrinth to us, unless we are directed by the line of the Word,"[62] the philosopher shifts the terms by claiming that the irregularities, ambiguities, and deceptions of a multicursal labyrinth are like nature. Its causal laws are hidden in densely interlaced manifest phenomena, which cannot be disentangled by luck or the senses alone. What is needed is scientific method, especially trial-and-error processes of experiential interrogation, which provide the clue to knowledge, "delivered as a thread to be spunne on."[63]

The gap between complex phenomenal effects and ignorance of causes prompts further inquiry. Preternatural phenomena, which appear to exceed or deviate from the laws of nature, had traditionally been understood in metaphysical terms, as was the case with religious automata like the seemingly miraculous mechanical monk (color plate 4). Bacon took issue with such nonrational explanations, advocating a new natural history that approaches marvels as articles for further examination. Experience and the process of experimentation are the means to assess the labyrinthine deceptions of the external world and the uncertain testimony of the senses.[64] Labyrinths and automata are ways to test human understanding, examine how it is best attained, and then proceed to new discoveries by analyzing nature and its processes as if they are intricate machinery or convoluted mazes. Daedalic arts prompt seekers to devise a solution and to find a way by using their creative powers of invention and investigation.[65]

Another important guiding thread, according to Bacon, is ancient wisdom. The myths of antiquity can direct human understanding, as they convey a great deal of practical learning. Knowledge gained from ancient texts has to be actively examined and tested, however, not just passively believed: "Whereas antiquity used to deliver the knowledge which the mind of man had gathered, in observations, aphorisms, or short and dispersed sentences, or small tractates of some parts that they had diligently meditated and laboured; which did invite men, both to ponder that which was invented, and to add and supply farther. But now sciences are delivered to be believed and accepted, and not to be examined and farther discovered."[66] This passage is from Bacon's fragmentary work *The Clue to the Maze* (*Filium Labyrinthii*), dated to around 1607. In it, he advocates the importance of contemporary inventors who would both continue and advance the work of the ancients so that the sciences would cease to languish and could proceed in new directions.

This congruity between old and new knowledge is essential to the advancement of learning. As at the Doolhoven, there is a guiding thread that runs through the past, unravelling its secrets in order to move to the future. Bacon stresses that the wisdom of the ancients is not easy to apprehend, as it is both obscured and illustrated by equivocal fables: "They serve to disguise and veil the meaning, and they serve also to clear and throw light upon it." Ancient parables are like "a kind of ark," a protective container "in which the most precious portions of the science were deposited."[67] There are many "secrets and mysteries" hidden and consequently safeguarded within classical texts. Fables and parables have to be deciphered and assessed, and this process of active experiential learning stimulates further inquiry.[68] In the manner of a Silenus box, which looks unimportant on the outside but secretes precious things, ancient knowledge is both arcane and accessible.

If we return to the Amsterdam maze gardens with this in mind, we recognize the revival of the labyrinth's apotropaic function as a pattern of protection. Entered with difficulty by those who had wound their way through the Daedalic maze, the exhibition gallery was a closeted space that protectively concealed the secrets of ancient wisdom and transgressive new inventions—beastly machines that could emerge as benevolent or malevolent social forces. By re-creating the mythic works of Daedalus, these experimental labyrinth gardens activated esoteric and ambivalent ancient wisdom. They gave material form to aspects of the classical fable, literally bringing it to life with wondrous artworks designed to provoke immediate responses from visitors. In this manner, the Doolhoven allowed people to partake in the wisdom of the ancients and decipher the mysteries of nature by providing the clue of direct experience. Unlawful and potentially idolatrous arts were thus given specific functions—they were presented as wonders that could prompt investigation and reform human behavior.

Beast Machines

In the writings of Bacon, Van Mander, and Junius, there is equivocation between positive and negative assessments of human-made artifacts and technologies. Daedalic invention is described as wonderful, vital, ingenious, magnificent, intricate, delightful, and virtuous but also as diabolical, deceitful, execrable, destructive, excessive, mischievous, unlawful, and unruly. This intermingling of praise and disparagement stems from fearful hopes that the very triumphs of human accomplishment could rival the works of God and nature. Transgressive ingenuity soars into the domain of the divine. But to be too successful at this game is to fail. The flight and fall of Icarus serve as a cautionary tale that testifies to the heights of human achievement as well as the plummet of its breakdown. It also indicates a middle course en route to new knowledge and invention—one that avoids extremes of godlike flight and punishing fall by not "raveling too far into God's secrets."[69] Finding this middle path was central to Bacon's plan for the advancement of learning, and it required reforming the behavior of practitioners so that they proceeded with moderation, conducting themselves in a civil and gentlemanly manner.[70] Van Mander's art theory also points in this direction, indicating the straight path of reason as the best way. As we have seen, this type of disposition toward startling automata displays was adopted by virtuosi like Montaigne, Evelyn, Wilkins, and Bacon: it involved curbing spontaneous reactions and controlling the passions in order to observe and learn from direct experience. The acquisition of knowledge was a technique of the self, as much an affective as an intellectual endeavor.

If the main purpose of the automaton was to prompt viewers to think, then it primarily generated consideration of what differentiates the human from the machine. Pondering the similarities between humans and costumed automata, Descartes observes: "Were I perchance to look out my window and observe men crossing the square, I would ordinarily say I see the men themselves. . . . But what do I see aside from hats and clothes, which could conceal automata?"[71] Notably, the philosopher was living in Amsterdam in the 1630s and 1640s, and he may well have penned this statement while looking out of his window onto the busy streets and squares of the Jordaan neighborhood, where he lived for a time within easy walking distance of the Oude and Nieuwe Doolhoven. Jessica Riskin argues that "the great bustling population of lifelike devices that enlivened the world in which Descartes lived, before arriving at his radical philosophical and scientific proposal . . . shaped how Descartes conceived the idea during the 1630s and '40s. He had the machines in mind as he wrote."[72] Aristotle's claim that automata stimulated philosophical thinking is born out in the case of Descartes, whose engagement with these apparatuses certainly prompted the philosopher to wonder about causes.

While scholars agree that Descartes's personal fascination with automata contributed significantly to his mechanical philosophy, there is only speculation about where he encountered such devices. Stephen Gaukroger suggests that Descartes's mechanistic theories were influenced by the hydraulic automata that he might have seen while living outside Paris in the 1610s if, during that time, he visited the nearby château gardens with the famed inventions of the Francini brothers at Saint-Germain-en-Laye.[73] Because the Amsterdam Doolhoven have not received much scholarly attention, the possibility that Descartes had Amsterdam's mechanized, moving exhibits in mind as he wrote has not been considered. Given Descartes's sustained interest in such devices, it is probable that the philosopher visited the well-publicized automata shows in his own neighborhood. The Doolhoven were at their zenith in the 1630s and 1640s, suggesting the likelihood that these inventive urban displays influenced Descartes's thinking about mechanical philosophy.

Notably, Descartes described the exhibition of automata in terms similar to Bacon, and both philosophers followed Aristotle's account of a response process that involved a temporal and affective shift from amazement to understanding. Descartes writes: "After having stirred wonder in you by exhibiting to you machines, extremely powerful, very strange and rare automata, visual appearances seemingly real and in postures the subtlest that artificers can devise, I will then uncover the secret devices on which they rest and these are so simple that you will no longer be tempted to feel wonder regarding any product of human devising."[74] For the philosopher, the initial passionate reaction triggered by automata displays was based on partial and deceptive sensory information. By means of informed technological explanation and observation—uncovering the secret devices—wonder would be dispelled, thus moving the viewer from feeling to thinking.

This ability to make a transition from sensation to contemplation was, for Descartes, precisely what distinguished the human from the automaton or the animal. He famously coined the term *bête machine*—beast machine—to encapsulate this idea.[75] The analogy centers on the idea that beasts and machines lack a rational soul, which is only possessed by humans. The beast machine is a configuration that moves on its own according to mechanistic principles; it is not driven by an intelligent mind or soul. For Descartes, this absence of reason was manifested in specific ways. Beast machines do not possess the human capacity to use words or signs to convey thoughts; while they can make noise and almost speak, they cannot converse or answer questions in dialogue with another. And though they can do many things with greater perfection than humankind, they do not act from knowledge. While machines can imitate movements that indicate the passions, they have no affect or sensations, and so they cannot respond to unexpected external challenges. Animals, on the other hand, are sensate and can think, but only in a limited way. While Descartes allows that animals have cognition, he asserts that they only see as we do when our minds are elsewhere. Animals' perceptions and sensations are inattentive due to the fact that we cannot demonstrate the presence of rational souls in them. The philosopher accordingly advocated that weak minds stray "from the straight path of virtue" by mistakenly imagining "that the soul of beasts is of the same nature as ours."[76] While crooked paths could create entanglements of self with beast, the straight path keeps them separate.

By pinpointing what differentiated the human from the machine and the animal, Descartes is able to extend the beast machine analogy to aspects of the human body, especially movements that are not controlled by reason. In a well-known passage, he describes the body as a machine made by God:

> I suppose the body to be just a statue or a machine made of earth, which God forms with the explicit intention of making it as much as possible like us. . . . We see clocks, artificial fountains, mills, and other similar machines which, even though they are only made by men, have the power to move of their own accord in various ways. And, as I am supposing that this machine is made by God, I think you will agree that it is capable of a greater variety

of movements than I could possibly imagine in it, and that it exhibits a greater ingenuity than I could possibly ascribe to it.[77]

Descartes revisits the biblical creation story here in order to imagine how the mechanisms that drive the body could move independently of the soul, in a machinelike or beastlike manner. This is especially the case with bodily functions, reflex movements, and habitual reactions that occur automatically and without thought. A key example is the type of unthinking, involuntary motor response that is triggered by external corporeal stimuli: for instance, a foot in the fire causes a body to move before it thinks.[78]

If we consider Descartes's observations together with some of the actual automata displays that he might have seen in Amsterdam, the philosophical interests of these peculiar objects come to the fore, and there are two important connections we can draw. First, and most obvious, self-moving mechanical figures were designed to deceive the senses by mimicking so closely the mechanistic motions of actual bodies that they appear to be alive: this is what Descartes implied with his hats and clothes comment. Second, automata were strategically exhibited to startle their viewers. As unexpected external stimuli, they made people move in ways that were automatic and unthinking. In other words, the beast machine imitates the living body in order to make the human body behave like a beast machine.

Instinctive, unreasoning bodily motion was instigated by works that surprised and amazed. In the case of the trick fountains, for instance, the sudden shock of cold water on bare flesh caused women to "hop about like newborn calves."[79] The implication is that their instinctive physical reaction is both automatic and brutish—this is exactly the response of the beast machine. Strategic exhibition tactics particularly equated female bodies with the bodies of animals and automata, all moving on impulse and without thought.

If we consider the gendered nature of other Doolhof displays, it is notable how many of the automata mechanically mimic the powers of irrational female passions run amok. Several of the moving-picture shows enacted biblical and classical stories focusing on scenes of "women on top," in which empowered female characters dominate men. The mechanical pictures were intricate assemblages in which clockwork drove the actions and gestures of numerous small costumed figures in architectural settings. Familiar stories unfolded in a series of automated sequential scenes. The moving picture shows illustrated in the Doolhof guidebooks indicate an interest in such mechanically driven narratives of domineering women: Salome dances with the head of John the Baptist; the Persian princess Granida chooses the lowly shepherd boy Daiphilo as her lover; Venus humiliates Mars by committing adultery with the inventor Vulcan; Queen Esther outsmarts her enemy, Haman, resulting in his execution; Queen Semiramis orders an assassin to murder her husband so that she can usurp the throne. To take the tale of the infamous Semiramis as an example, Dante's *Divine Comedy* consigns her to the Second Circle of Hell for her lasciviousness: "To sensual vices she was so abandoned / That lustful she made licit in her law, / To remove the blame to which she had been led."[80] Moreover, Boccaccio's *Famous Women* describes the Assyrian queen as the very embodiment of female inconstancy, who "like others of her sex . . . was constantly burning with carnal desire," which caused her to commit sexual acts "more

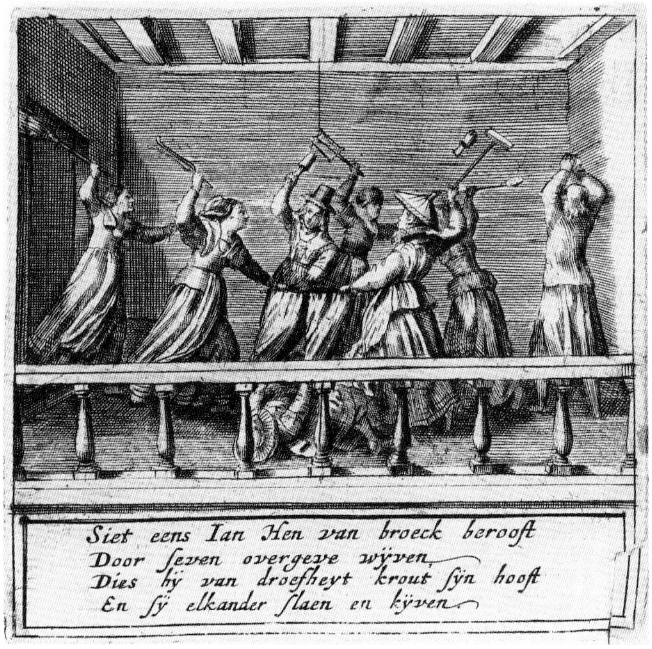

FIG. 27 Crispijn van de Passe, *Seven Wives Fighting for a Man's Pair of Pants*, from *Den Nieuwen en Vermaeckelijcken Dool-hof, Staende op de Roose-Gracht* (Amsterdam: voor David Lingelbach, 1648). Photo: Allard Pierson, University of Amsterdam Library Special Collections, OTM: O 63–7272.

beastly than human."[81] Semiramis exemplifies human beastliness.

The comic moving picture scenario in the Nieuwe Doolhof of seven women fighting for a man's pair of pants also thematizes the insatiability of violent female sexual desires (fig. 27). This scene plays on a well-established proverb tradition, originally derived from the biblical text in Isaiah 4:1: "And in that day seven women shall take hold of one man."[82] Walter Gibson has analyzed the numerous early modern Netherlandish prints and paintings that picture seven women battling for a man's breeches and concludes that the humor of this popular proverb is that the women are vying for what is in the pants—all seven desire the sexual favors of the one man.[83] This scenario would have been especially hilarious when the familiar comic battle was brought to life by moving automata. The print depicting this moving picture shows three women at center who all clutch at the waistband of the trousers while beating one another with tongs, slippers, and a brush. A fourth woman has been pushed to the ground but keeps a tight grip on a pant leg. A fifth, brandishing a broom, runs onto the scene through a doorway at left. Behind the central group, two others go at each other with distaff and ladle. At right, we see the back of the man under attack, his arms raised as he claws at his head. The printed poem under the image calls him "Jan Hen" designating a stock comic type: the henpecked man, who is unable to fight off the squabbling women.

The real joke, of course, is that none of these overwrought, unruly figures is truly out of control. In exhibit after exhibit, dramatic disorderly emotions—desire, jealousy, anger, fear, bloodlust—are expressed by measured, repetitive bodily movements. Automata can go through the motions of violence and passion, but their mechanistic behavior is, as Descartes points out, completely devoid of affect and understanding—they neither think nor feel. These figures may act and move as if driven by sensations, but in truth, they are propelled by the regular, rational workings of their interior clockwork mechanisms. Motion and emotion diverge as mechanical movement orders and regulates passionate gestures. Unlike volatile human subjects, the docile, well-run machine can behave with complete constancy. The automaton is not buffeted about by every passing event, and it wholly lacks the understanding to respond to the inevitable difficulties and contingencies of life. Animate yet lifeless, it is not vulnerable as mortals are; it is never moved by pain or pleasure. It is only moved by mechanics.

Almost every automaton exhibit plays with such distinctions between the human and the machine. The absurdity of a Nieuwe Doolhof show of android shepherds that personify the five senses, for instance, is that clockwork figures have no sense organs. They are blind, deaf, dumb, and unfeeling: completely insensate.[84] Sources also refer, albeit somewhat obliquely, to seemingly sensual mechanical figures that mimic sex acts. For instance, there were the figures of Venus and Vulcan making love in a tent at the Oude Doolhof. And at the Roode Doolhof, there was a display of copulating machines, described as two reclining naked figures with a cupid.[85] Here the jest is that the repetitive movements of these mechanical devices were not driven by physical desires or sexual urges. These apparently ardent apparatuses could neither feel pleasure nor procreate. But by simulating intercourse, they could certainly arouse their viewers. Again, the motions of the machines could trigger an instinctive, corporeal response in the audience.

The Amsterdam displays also included scores of noisy animal automata—crowing, lowing, braying, bleating, and quacking things. Brereton, who, like Evelyn, visited the inn D'Os in de Bruyloft, lists "a bull which roared, a cow and her calf which lowed, a cock stood overhead which crowed, but hoarsely," and he also highlights "an ox which pissed strongly."[86] This ox, or *os* in Dutch, may well have been the mascot of the inn. Less is known about the Roode Doolhof, which was established in 1630 near the city walls at the Reguliersspoort and demolished in 1663 as part of the ongoing urban expansion. The main attraction of this exhibition site was the display of wooden animal automata. Sources describe a lowing cow and calf, some bleating sheep, a deer, and a dog that moved its ears and eyes.[87] This Doolhof was situated very near to the *ossenmarkt*, the oxen market, where farmers and livestock sellers did business in the city.[88] The proximity to so many live animals suggests how the Roode Doolhof's beastlike machines could prompt consideration of the similarities and differences between carnal, irrational animals, thought to be driven by sensations, and unfeeling, unthinking machine-animals, driven by mechanics.[89]

Brereton's reference to the pissing ox implies that bodily functions were of especial interest. He additionally notes: "Two boys sat opposite one to the other about three yards distant, which pissed stoutly one at the other, one upon the other; a woman out of whose breast water sprung"; and "the proportion of a woman milking a cow, who seemed to move with her hand, and milked until the vessel, where into the water distilled or was milked, ran over."[90] Here, hydraulic mechanisms appear to urinate and lactate, drawing attention to the reflex discharge of bodily fluids in both animals

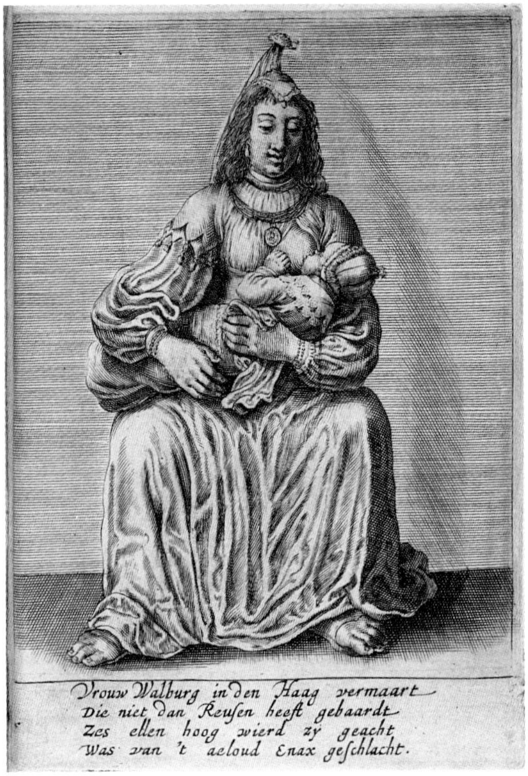

FIG. 28 *Vrouw Walburg*, from *Verklaringe van Verscheyden Kunst-Rijcke Wercken en haer Beweginghe* (Amsterdam: Dan. Davidsz. Mansvelt, 1657). Photo: Allard Pierson, University of Amsterdam Library Special Collections, OTM O 62–2615.

and humans—another instance of bodily behavior that was automatic and not directed by reason.

In keeping with the gendered nature of many exhibits, female wetness flowing as breast milk was an oft-repeated visual witticism. We have already noted the streams expressed from the breasts of Ariadne in the Bacchus fountain. And at the Oude Doolhof, the mechanical action of the giantess Vrouw Walberg was to suckle her enormous baby. The printed image in the guidebooks depicts her seated with legs spread wide, cradling the giant infant to her left nipple, which protrudes from her clothing (fig. 28). In the late eighteenth century, a French visitor was particularly shocked, writing of a giant automaton that "exposed her breast in an extremely indecent manner." (By that time, the copulating figures had no doubt been removed.) The Frenchman concluded by making some disapproving remarks about the type of audience that enjoyed this sort of indecency, stating that the showing of automata, "which used to be the only spectacle known here, is very popular with the general populace and with children."[91] These sensational spectacles may have been deemed coarse, childish, and inappropriate by later commentators like this eighteenth-century visitor and the nineteenth- and early twentieth-century folklorists whose disparaging accounts were examined in chapter 1. However, in the seventeenth century, the idiosyncratic characteristics of automata, which simulated bodily functions in order to stimulate physical responses, were an innovative means of thinking about how some of the automatic workings of the body could be separate from the workings of the mind.

Brereton wrote of D'Os in de Bruyloft that there was "no room without some rare invention for pleasure and delight; . . . and those all various to affect the outward sense, and draw on ghuests [*sic*] to apply there."[92] Indeed, the inn was sometimes called *'t Huys te Sinnelust* (the house of sensual pleasure).[93] Besides the visual displays, Brereton "heard all sorts of music upon strings, upon wind-instruments, and upon an instrument which did in a pleasant tune and harmony make the bells to sound, playing thereon as you do upon virginalls."[94] Evelyn adds, "There was likewise a cylinder that entertained the company with a variety of chimes, the hammers striking upon the brims of porcelain dishes, suited to the tones and notes, without cracking any of them. Many other waterworks were shown."[95] The implication of this and other descriptions of the various Amsterdam exhibits is that they aimed to provoke a wide range of sensory responses in their audiences. And, as Brereton notes, by pleasurably affecting the outward senses, these displays drew many guests. The visual and audible stimulus of lively devices displayed together with music, sounds, waterworks, food, and drink all directly affected the body, engaging spectators in an intense multisensorial experience.

Automata thus perceived stop short of transforming humans into affectless machines who behave automatically and with complete constancy.[96] To the contrary, these unfeeling apparatuses had the power to arouse intense sensations and sentiments—to make humans beastlike—and then, ideally, to move them to dispassionate apatheia, which is the affective quality of the machine. Ultimately, this was why the automaton mattered: it made people feel and, in so doing, prompted them to think about what it meant to be alive. The automaton moved to strike at the limits of the human, challenging viewers to consider the paradox of how they were both like and unlike the beast machine.

What was really being put on display, therefore, was human behavior. The performing automata allowed viewers to see how a person with very little reason resembles a beast, which is driven only by its senses, and how a person without passions or stirrings is like a machine, which moves but does not feel or think. Passions and sensations, as both Heinsius and Descartes point out, were neither virtues nor vices: they were essential to human life. The trick was to temper them and bring them into balance, using reason to find the middle path between extremes. Heinsius advocated dramatic theatrical and musical performances as effective ways to stimulate, purge, and refine the passions, considering them cathartic exercises that prepared people to imbibe knowledge and virtue.[97] The functions of the Amsterdam automata displays accorded with such prescriptions. In the ritual itinerary of these sites, the crooked path of the labyrinth paradoxically led to the straight path of reason by vividly provoking awareness of the importance of reason as the unique capacity that makes the human different from the machine and the beast.

The Doolhof exhibits perpetuated self-control through loss of control, and this method could serve to reinforce the relative standing of diverse individuals within the community. Since the equivocal status of Daedalic mechanical arts teetered on anxieties about misuses and misunderstandings, the model audience would ideally comprise those who were civil, dispassionate, and well educated. Much of the Doolhof publicity specifically targets gentlemen amateurs. This method endorses the preferred viewer, and the inns certainly did attract this class of person: gentlemen like Evelyn, Brereton, and possibly Descartes. Most of the historical accounts of these sites were written by learned European men and accordingly tend to focus on the reactions of this type of favored guest. A standard narrative seems to be that exposure to rare things moved them from startled wonder to learned aloofness.

However, a number of sources indicate, usually with disparagement, the presence of other social groups at the Doolhoven who were deemed less prepared to apprehend transgressive inventions: women, children, peasants, commoners, and non-European foreigners. Indeed, the diversity of the audience was the true novelty of these exhibition sites. Commercial displays like the ones at the Doolhoven brought about what Katharine Park and Lorraine Daston call a democratization of wonder. Broader accessibility to formerly elite culture often prompted mockery of disorderly or ignorant responses. This kind of ridicule was a means of making distinctions between learned wonder and vulgar wonder by reinforcing what was considered a proper reaction.[98] The ritual route of a visit to the sites was designed to bring those perceived as ill-mannered into line by inculcating an appropriate mode of receptivity, which involved moral self-examination and modification of behavior. If the well-planned exhibition of automata was intended as a civilizing process, then the words of Norbert Elias are particularly apt: "Molding by such means aims at making socially desirable behaviour automatic, a matter of self-control."[99]

The civilizing ritual was by no means an equalizing rite. In his consideration of Aristotle's concept of tragedy, Heinsius followed the ancient philosopher in his characterization of the types of people deemed most emotional and least capable of attaining knowledge: "A slave or a shepherd does not have a correct understanding of public affairs, or ordinary life, or wisdom, or learning, nor does a youth or a woman."[100] The differentiations that were made between learned and vulgar wonder thus reinforced social perceptions about groups who were thought to lack reason based on categories of class, ethnicity, race, nationality, age, and gender. Specific techniques of display were thus devised to control not just the ambiguous arts but also their audiences. Through vivid and direct performative experiences, these exhibits transmitted the wisdom of the ancients in a manner that aimed to instill ideals of conduct and convey values and beliefs about normative social hierarchies. The following chapter accordingly considers how the Doolhof displays made distinctions between those who were able to move from wonder to thinking and those who were described as being akin to the beast machine.

STRANGE THINGS FOR STRANGERS

Chapter 5

Transcultural Automata

All these machines open, within the protective enclosure, a space which is also that of marvelous communication.
—Michel Foucault, *Death and the Labyrinth*, 1963

By all accounts, seventeenth-century Amsterdam was a city teeming with immigrants and travelers. As one visitor exclaimed: "No part of Europe is so haunted with all sorts of foreigners!"[1] Like today, some felt threatened by the prospect of daily contact with all manner of people. Yet, the civic magistrates understood that heterogeneity generated creativity and prosperity, and they actively encouraged immigration, especially of skilled craftspeople. Talented émigrés included Bargois, the French hydraulic engineer who created the city's first monumental fountains for the Oude Doolhof and later fashioned more spectacular waterworks for the Nieuwe Doolhof. Also of note were David and Philips Lingelbach, Frankfurt clockmakers who brought the latest in German horology to Amsterdam, displaying marvelous mechanical devices at the Nieuwe Doolhof and the Franse Doolhof. Clearly, the wonders of Amsterdam were indebted to foreign invention.

As points of intersection in rapidly expanding networks of global transit, inns and taverns were gathering places where travelers and locals could meet over drinks and the appreciation of artworks. Van Mander highlights the art of cross-cultural conversation when he encourages "amiable and generous" tavern keepers to show "praiseworthy and important pieces by the most accomplished hands . . . to foreign gentlemen, strangers or other lovers of art."[2] A similar appeal to internationally oriented art lovers runs through publicity about the Doolhof exhibits. The targeted audience of educated gentlemen included visitors from abroad.

Commelin stresses that the Doolhoven offer "fine things for foreigners."[3] Fokkens alleges that the sights at these inns are particularly "worthwhile for all foreigners who have not been here before, strange and wondrous for them," and he reiterates a few pages later that the sights are especially pleasurable "for strangers and foreigners who do not live here in Amsterdam . . . wondrous for those who have never seen it before."[4] The larger claim was that Amsterdam was a center of learning for all peoples, and its automata-displaying inns would captivate the global influx of newcomers with rare inventions that they could only see in the Dutch city.

Jovial assertions about the cultural superiority of Amsterdam fluctuated from the welcoming cordiality advocated by Van Mander into the realm of insults and derisive sneers about the ignorance of foreigners. Doolhof publicity booklets report that international sightseers declared the exhibited works "impossible" and avowed that nothing like them could be found anywhere else in the world. Foreigners "were struck with wonder at the mechanisms of the clockwork here, and could not even understand it, but only gazed in wonder."[5] Along the ritual route through these labyrinthine sites, strange things were purported to strike strangers with wonder. Derogatory attitudes toward outsiders were made visible in the exhibits as well, especially in presentations of humanlike automata dressed in exotic costumes.

Historical evidence, not surprisingly, challenges this promotional hype and indicates that international guests—far from being dumbstruck—made significant intellectual contributions to the cultural life of the Dutch Republic, including fostering the esoteric knowledge needed to create automata, which were themselves foreign things that haunted the city with the powers of other worlds. In mediating transcultural knowledge exchange through entertainment, Amsterdam's taverns prompted the testing of identities and creative experimentation with various forms of visual and material culture. The affective dynamics of these urban exhibition sites—which can be characterized as a vacillation between hospitality and hostility toward outsiders—suggest how early modern artistic invention was activated in confrontations between the mobile spectators and moving artworks that circulated in an increasingly interconnected world.

Dumbstruck

The most strangely human of the Doolhoven's various moving statues were the androids. Predating the David and Goliath statues was a life-sized mechanical figure at the Oude Doolhof named Jochum (fig. 29).[6] This work does not survive, but a print of it appears in several of the guidebooks, advertising it as a key attraction. The print pays particular attention to the figure's pursed lips and wide staring eyes as well as its hands and fingers. An accompanying description declares that Jochum could play an innumerable variety of tunes on his *moesel* (bagpipes). While he performed, his head, eyes, and hands all moved "as if he lived." He is very artfully put together, the pamphlet states, and the Master's work is hidden inside.[7] This phrase, "Master's work," has double significance: it could refer to the Daedalic genius who made

FIG. 29 *Jochum*, from *Verklaringe van Verscheyden Kunst-Rijcke Wercken en haer Beweginghe* (Amsterdam: Dan. Davidsz. Mansvelt, 1657). Photo: Allard Pierson, University of Amsterdam Library Special Collections, OTM O 62–2615.

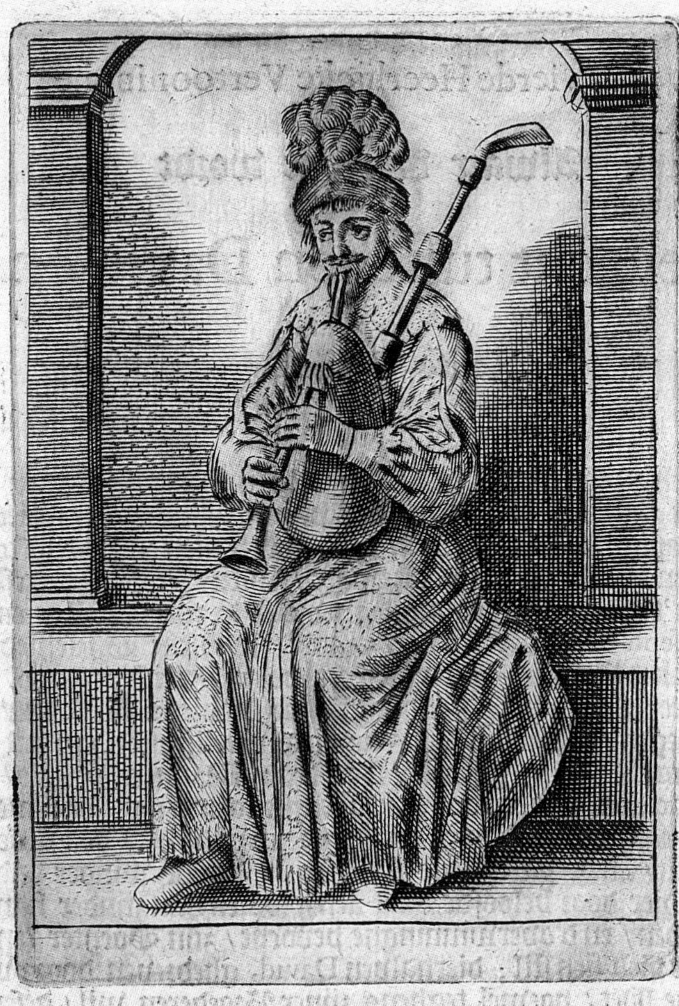

Jochemus sit en speelt hier vast
'Verscheiden Deuntjes by de tast,
Zijn handen, oogen, en zijn hooft
bewegen dat men 't nauw gelooft,
Met zulck een kunst en aardigheid,
Doch 't Meesters Werck van binnen leijd.

the android, but it also implies the work of God. It evokes a "ghost in the machine," raising questions about what activates this humanlike being whose mechanisms are well hidden within its costumed body. The illusion created by the clockwork is that the repetitive movements of Jochum's fingers, eyes, and head are powered by muscles, whereas the bagpipes externalize the functioning of unseen internal lungs. The invisible life force of breath appears to pulsate from the lips into the pipe to inflate the sack and produce various sounds. This action endows Jochum with interiority, and the naming of the automaton further thematizes the agency of personified machinery. A marvel of technology, the android performs as a living being, and thus viewers experience Jochum's vivid lifelikeness as if it (or he) were human.

The mimesis of this particular kind of artwork does not just imitate nature with deceptive illusionism in the manner of many Dutch paintings. As Stoichita observes, the art of the copy, which Plato termed the *eikastiké*, has triumphed as the main focus of Western art history.[8] This assessment certainly applies in the case of Dutch realism. The *phantastiké*, the art of the simulacrum, by contrast, remains vague, ambiguous, and unheeded. We are reminded of the *parergon* status given automata by Vitruvius's and Schlosser's assertion that automata have always lurked at the threshold of art. Pushing beyond the art of describing, automata move as if they live. As the Roman scholar Cassiodorus observed, the machine is the only thing that seeks to surpass nature: "It makes mute things sing, lifeless things live, immobile things move."[9] A lively dead thing, the automaton confronts viewers with the paradox of being animate and lifeless at the same time.[10] Observations about how art and nature interact in the mechanical arts were reiterated in the early modern period. As Wilkins explains: "Now Art may be said, either to *imitate* Nature as in Limning and Pictures; or to *help* Nature, as in Medicine; or to *overcome* and *advance* Nature, as in these Mechanical Disciplines."[11] For Bacon, experimental mechanical arts are a method of intervening in nature and thus devising its hidden workings: "The secrets of nature reveal themselves more readily under the vexations of art than when they go their own way."[12] Activated by "the Master's work," the automaton moves at the edges of artistic practice where it interferes with nature, interrogating and exposing its hidden workings. Automata stop just on the verge of realizing nature's greatest secret—and art's most transgressive aim: the creation of life.[13]

Especially intriguing is how the Doolhof exhibits presented the strangeness of the android, which appears as lively person and lifeless thing, similar to but not the same as the viewer. Descriptions of the Amsterdam automata highlight the appeal of their *uytheemsche kleedinghe* (outlandish clothing),[14] indicating that the foreign appearance of these devices was a significant part of their attraction. Almost all of the moving-picture shows were set in legendary sites in the Near East, which evoked mythic tales and biblical times but also the seventeenth-century territories of the Ottoman and Safavid empires. The Oude Doolhof moving pictures animated the tale of Semiramis, the legendary Assyrian queen in her Babylonian citadel; the Old Testament story of the Ethiopian Queen of Sheba visiting the palace of King Solomon in Jerusalem; and an amorous encounter between the Persian princess Granida and the shepherd Daiphilo, a pastoral scene derived from a 1605 play by P. C. Hooft. Among the automated pictures in the Nieuwe Doolhof were the dream vision of the Jewish prophet Daniel, exiled in Babylon, and the trials of Esther, Jewish queen of the Persian king Ahasuerus at Susa. Notably, all of the figures in these mechanized shows wore Eastern costumes comprising robes and turbans or elaborate headdresses. None of the Amsterdam automata were clothed to look like European men or women.

The printed image of Jochum shows that the figure was clothed in a long patterned and fringed robe, possibly a figured silk *entari*, and wore a curious headpiece—a turban or furred cap adorned with a large plume of feathers. A later sketch of the long-lived Jochum, done in 1861 by the antiquarian J. G. L. Rieke, emphasizes the android's dark complexion and coarse black hair, which, together with the outlandish costume, endowed the moving statue with a generically Eastern appearance (color plate 5). Other aspects of the costuming do not quite fit this image, however. For instance, the garment's cuffs and collar look like European lacework. And Jochum's musical instrument—the bagpipe or moesel—was customary in Netherlandish music. It is depicted in numerous early modern Dutch and Flemish representations of peasant dances, kermis revelries, and comic tavern scenes. The moesel player was a stock figure in long-standing local festive customs. Jochum's "countless variety of tunes" may well have been familiar folk songs, although the machine plays on tradition in an entirely novel way.

Such oscillation between familiarity and exoticism seems to characterize the staging of automata. Both indigenous and Orientalized, the android's mishmashed trappings create the impression that the humanlike machine is somewhat Dutch and somewhat foreign, both ordinary and unfamiliar. The apparent ethnic and cultural difference of the figure thematizes its alterity as a machine even as its outer appearance hides inner mechanics and makes it look like a person. Outward show thus complicates beholders' apprehension of the ways *he* was like them—and *it* was not like them.

In this manner, the strange thing appears as a stranger. As Georg Simmel observes, the social position of the stranger combines the contradictory qualities of proximity and distance. Interactions with strangers sharpen perceptions of general commonalities as well as differences, provoking a mix of friendliness and prejudice.[15] We have already encountered this type of response in Doolhof publicity, which both welcomes and derides foreign guests. It also can be discerned in how the almost-human automaton was never clothed as a European, so that its disquieting and entertaining qualities were associated with the stranger.

The interplay between nearness and remoteness in Amsterdam was intensified by the fast expansion of trade routes forged by the East and West India Companies, which brought ever more commodities and peoples of distant origin into the port city. As Claudia Swan argues, a sort of "ethnographic eclecticism" prevailed; there was much interest in the foreign, the strange, the rare, and the wondrous, but cultural and geographic precision was not of central concern. The figure of Jochum has much in common with a genre made popular by Rembrandt van Rijn and Jan Lievens in the 1630s. They produced numerous paintings and etchings of figures wearing turbans and robes that vaguely call up the dress of Ottoman Turkey or Safavid Persia without being specifically identifiable. Some of these works are portraits of Dutch sitters sporting outlandish outfits, whereas others are *tronies* or character studies.[16] The remarks of the poet and statesman Constantijn Huygens in 1630 about such a painting by Lievens illuminate how this type of imagery was perceived (color plate 6); Huygens calls it "a picture of a Turkish sort of nobleman, done from the head of some Dutchman."[17] This brief description reveals how feathered turbans and patterned tunics could function like theater costumes, a performative means of interrogating the dividing line between a Dutchman and a Turkish man. As was the case with costumed androids like Jochum, the combination of familiarity and foreignness prompted awareness of similarities and differences between self and stranger.

Jochum also resembles the costumed figures that appear in a specific subgenre of Dutch still-life painting that developed in tandem with the colonial exploitation that undergirded the Dutch Republic's rise to global wealth and power. Juriaen van Streek's version depicts a table covered with an imported rug and heaped with luminous porcelain wares from China, citrus fruits from the Mediterranean, and local oysters and breads (color plate 7). Behind this sumptuous display is a turbaned African youth wearing a silky orange-and-gold-striped robe. He looks out and holds up a porcelain platter of fruit. Things from far away have been brought near and domesticated.[18] Van Streek's painting records the European custom of dressing enslaved persons and household servants in attire that was not congruent with their actual place of origin but gave them a generically exotic appearance. The turbaned youth is like the possessable luxury goods that surround him, brought from afar to serve the Dutch. He is as much thing as person, conflated with the importable commodities of the trading empire.[19]

While the ownership of enslaved persons was not permitted in the Dutch cities, Dutch merchants certainly were enslavers in the colonies. They set up large slave-owning households in the East and West Indies. In the Americas, especially Brazil, the Dutch implemented plantation slavery. Driven by profit, the Dutch West India Company took control of the transatlantic slave trade in 1637 and was responsible for the seventeenth-century escalation of African enslavement.[20] Paintings like Van Streek's, in which African figures obligingly present viewers with costly imports, manage to give no hint of the atrocities of the slave trade while still showing how African people conveyed enormous wealth to the Dutch.[21]

When we situate a costumed automaton like Jochum in the context of the brutal practices of global capitalism, such a figure can conjure the dream of a self-moving machine that performs as a perfect slave. This particular fantasy weaves through the literature on automata.[22] Recall Plato's observation that the problem with the work of Daedalus is that, if untethered, it can give you the slip "like a runaway slave."[23] And Aristotle writes: "For if each instrument were able to accomplish its own task, either in obedience or anticipation, like the [statues of] Daidalos . . . , then in same manner . . . master builders would not need apprentices, nor masters, slaves."[24] For these philosophers, automata and enslaved people both lacked the capacity for reason, a prejudice that continued to inform seventeenth-century attitudes. As Heinsius opines in his commentary on Aristotle, a slave does not have a correct understanding of learning or wisdom.[25] And the Amsterdam philosopher Baruch Spinoza declares that "one has to regard" people who are ignorant and uncomprehending "as automata that are utterly mindless."[26]

While Heinsius and Spinoza were paraphrasing classical sources, such statements take on specific relevance in the setting of a society that was heavily invested in slavery. Spurious claims about the intellectual deficiency of the enslaved could justify violent practices of human exploitation that brought staggering financial profits. For the Dutch merchants in the Doolhof audience, an exoticized moving statue like Jochum may well have called up long-standing comparisons between automata and enslaved persons. If so, then particular qualities, especially the automaton's routine, repetitive, automatic, mechanistic activity, could merge with the desire to perceive foreigners as controllable, obliging, unthinking, and less than human.[27]

The use of non-European costumes sets up a dynamic in which European viewers are in command of the exoticized machine, a subservient thing that exists for their pleasure. Woven into the labyrinth's larger pattern of apotropaic protection was the mocking deflection of beast-machinelike behavior onto those already positioned at the margins of society. In previous chapters, we saw how distinctions between the automated subject and the rational subject were drawn along lines of gender, class, and age. Women, children, and members of the lower classes especially were mocked as being like mindless automata, lacking learning and understanding. The staged presentation of foreign-looking humanlike machines at the Doolhof exhibits drew this differentiation along ethnic, nationalistic, and racist lines as well.

Indeed, according to the Doolhof publicity, there was not much difference between the stranger and the strange thing. "The Chinese," claims one booklet, "the subtlest nation in the whole world, were struck with wonder at the mechanisms of the clockwork here, and could not even understand it, but only gazed in wonder."[28] Such an assertion should not be taken as documentary evidence of the actual responses of Chinese visitors to the Doolhof; descriptions of this type of reaction played a role in philosophical theories about automata. In fact, the precise condition of being immobilized by too much wonder is discussed by Descartes in *Passions of the Soul*.[29] In his passages on wonder, the philosopher considers both the beneficial and harmful effects of the most powerful of the passions. To lack wonder is a hindrance, for it usefully stimulates scientific curiosity; only dull and stupid folk, he writes, are not naturally disposed to wonder. The main way to induce wonder is with novel objects, particularly if displayed in a manner that is surprising or unexpected. However, the strength and power of a sudden surprise can cause excessive wonder, which the philosopher terms astonishment. Astonishment is detrimental, as it "can entirely block or pervert the use of reason" with a paralyzing outcome: "The whole body remains still as a statue."[30] In Descartes's terms, then, the affective presentation of unusual objects like automata had a strong Medusa effect: it could startle and arrest viewers in a manner similar to the guidebook's description of the instinctive reactions of the Chinese to the Doolhof's mechanical displays. This was the power of the moving statue: it could transform viewers into stationary statues.

The cure for astonishment, advises Descartes, is repeated encounters with strange things. By attaining a special state of reflection and attention, the viewer could closely examine the unfamiliar object, learning more about it and dispelling its mysterious powers.[31] The travel writings of the German gentleman scholar Zacharias von Uffenbach provide evidence of this type of disenchanted analysis. Like Descartes, von Uffenbach desired to observe and comprehend the inner workings of a rare automaton, illustrated in his published travelogues (fig. 30).[32] The android depicted here is called

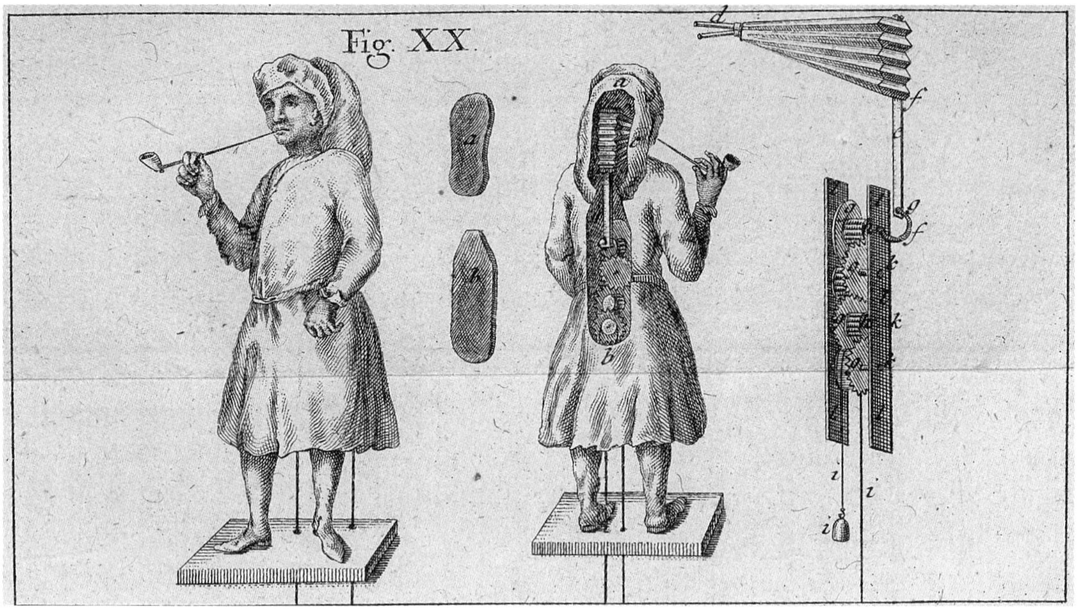

FIG. 30 Illustration of a *statuum fumantem* (smoking statue) in Zacharias Konrad von Uffenbach, *Merkwürdige Reisen durch Niedersachsen, Holland und Engelland* (Frankfurt, 1754), vol. 1, fig. XX. Photo: The Yale Collection of German Literature, Beinecke Rare Book and Manuscript Library, Yale University, New Haven.

a *statuum fumantem* (smoking statue) and was seen by von Uffenbach in a private German collection. Akin to Jochum, this mustached pipe smoker wears a turban and robe, giving it a similarly generic Eastern appearance. Here we see a practical purpose of the turban, which houses mechanically driven bellows. The illustration of the back of the android shows that it can be opened up—its secrets revealed—by removing two small panels, visible as dark oblong shapes between the front and back views of the figure. In the inner cavity are gears, pulleys, and bellows. The entire apparatus is depicted at the far right of the page as if it has been pulled out of the statue. By considering all three views together, it becomes evident that weight-driven pulleys move the interlocking toothed wheels to make the bellows expand and contract, forcing air through the lips of the costumed figure. This mechanism created the illusion that the exhalations of the android caused it to puff smoke. Similar to Jochum's bagpipes, the tobacco pipe makes the process of breathing visible, but what appears as life force is merely forced air.

Von Uffenbach anatomizes the android, prizing it open to reveal its insides. With its mysteries eviscerated, it is exposed as a human-made device that is clever but not powerfully rare and wonderful. The diagram in von Uffenbach's book seems to illustrate Descartes's confident assertion about the cure for wonder: "After having stirred wonder in you by exhibiting to you . . . very strange and rare automata, . . . I will then uncover the secret devices on which they rest and these are so simple that you will no longer be tempted to feel wonder regarding any prduct o huma dvising."[33] The exotic automaton thus understood does not hold sway over thoughtful human viewers. Its unsettling powers diminish when viewed as a work of mechanical ingenuity, a pleasurable puzzle to be solved by the intelligent beholder.

As explored in the previous chapter, Aristotle claims that the initial stirring response to startling automata displays was considered useful mainly as a stimulus for contemplation and discourse. The unthinking machine makes us think, and thinking makes us human. The denigration of Chinese viewers who were "struck with wonder at the mechanisms of the clockwork here, and could not even understand it" thus suggests that these astonishing moving machines had the power to transform foreign viewers into thoughtless automata. The reason-blocking, immobilizing force of excessive wonder prevented the progress of the Chinese. To be wonderstruck was to be rendered incapable of overcoming the automaton's paralyzing powers with investigation, discussion, and comprehension.

This slight against the Chinese was a particularly pointed insult, considering that the guidebook text also recognized China as "the subtlest nation in the whole world." The backhanded compliment alludes to the fact that China had exceeded Europe in technological innovation for centuries. There are tales dating as early as the third century BCE of Chinese inventors who constructed deceptively human-looking automata, and in the eleventh century, the inventor Su Sung designed sophisticated clockwork that incorporated moving mechanical figures.[34] Scholars note the difficulty of tracing the entangled global transmission of clockwork technologies, a multidirectional diffusion of ideas, people, and things that provoked local inventions in various places. Knowledge of early Chinese mechanical automata may have been conveyed to India and thence via travelers and texts to Europe.[35] Know-how about the workings of these types of devices was passed on by craftspeople and voyagers, as well as in treatises, especially the important Hellenistic works of Philo of Byzantium (third century BCE), Apollonius of Perga (second century BCE), and Hero of Alexandria (first century CE). These texts, which we return to below, were preserved and circulated in Arabic translations in the Middle Ages; they were taken up in Byzantium and Persia, where the knowledge probably commingled with Chinese and Indian elements.[36] Ancient ingenious inventions became legendary, and fables about the wonders of the East were conveyed to European audiences in medieval romances and travel accounts that describe courtly displays of artificial marvels encountered or imagined in faraway sites like China, Constantinople, or Persia. Automata journeyed to Europe from places beyond Latin Christendom. In the European cultural imaginary, these strange things had long been associated with distant places and mythic times.[37]

By the seventeenth century, however, European craftspeople had become accomplished automata makers, and the diplomatic gifting of mechanical devices became a widespread practice, a means to impress and influence foreign rulers with wonders of Western ingenuity.[38] The Doolhof guidebook demonstrates awareness of such transactions. In fact, its condescending comments about the wonderstruck Chinese were borrowed almost word for word from the writings of the Italian Jesuit missionary Matteo Ricci. Ricci is credited with introducing European clocks and automata to China. His presentation of these technologies at the court of Emperor Wan-li in 1601 created much interest in the mechanical marvels of the West, and there was particular fascination with automata. The Chinese called them *qiqi*, strange things.[39] In Ricci's words, the clockwork presented at the court "struck all the Chinese dumb with astonishment" for such things "had never been seen, nor heard, nor even imagined, in Chinese history."[40] In these passages, echoed at the Doolhof, the Chinese fall behind Europe, into the belated time of the Other. Ricci's point was to underline the advancement of Western knowledge by denying the extensive history of Chinese technological innovation, a claim that legitimated the Jesuit's mission of conversion.[41] Exposure to European mechanical inventions, the missionary implied, would beneficially cure the dumbfounded Chinese. Clockwork would move them forward in time, from astonishment to the enriching philosophical understandings that wonders stimulated.

Simon Schaffer argues that automata were distinctive of the milieu that produced them: "Automata were both arguments and entertainments, designed seductively to place craft skill within the setting of power."[42] While Ricci advocated the wonders of mechanical devices at the emperor's court to further his religious mission, in mercantile Amsterdam, similar rhetoric may have been mobilized mainly for the commercial conversion of the Chinese. There were many marvelous Chinese-made things that Dutch merchants were actively seeking to import in the face of significant trading obstacles. These goods are inventoried in Visscher's engraved city profile, which depicts the various peoples of the world paying tribute to the Amsterdam Maid, who is represented as "Empress of the whole world" (see figs. 5 and 6). The print describes the Chinese as *diepsinnighe*—profound, abstruse, and complex—and itemizes priceless treasures that they bestowed freely as gifts to the Maid: silks, precious stones, gold- and silverware, ivory combs, and porcelain. Contradicting this appealing fiction, the text of Visscher's print openly criticizes the restrictive trading practices of the Chinese, which were disadvantageous to the Dutch. Visscher laments the loss of an "inexhaustible richness" of potential profits on both sides. The flow of wealth into Amsterdam was obstructed by the Chinese emperor, who blocked direct trade with middle-class Dutch merchants in part because they had no royal or courtly status. The Chinese also required that Dutch traders make lavish payments of costly gifts—mechanical clocks and automata were especially desirable.[43] Visscher's imagined scenario in which Empress Amsterdam receives homage and tribute from the Chinese thus fictively reverses actual trade relations.

The guidebooks also cite a Portuguese visitor to the Nieuwe Doolhof who purportedly exclaimed that the works displayed there were "impossible" and that nothing like them could be found in all of Europe.[44] The Portuguese were significant rivals in global trade, and the Dutch East and West India companies engaged them in numerous battles over the control of strategic trading posts worldwide. The contention that both Portugal and China recognized Amsterdam's technological supremacy was obviously a means to legitimate the city's aspirations to global dominance. In this way, the migration of automata from their traditional courtly and religious settings into new venues of civic display signifies a transfer of power from older governing structures to this nascent center of merchant capitalism.

Providing moral justification for these inflated claims was the declaration that Amsterdam, as the "new Athens," offered cultural advancements to the rest of the world. This ideal informed the founding of the Athenaeum, Amsterdam's "Illustrious School," in 1632. Professor Caspar Barlaeus delivered the inaugural speech titled "On the Wise Merchant" and declared the whole city a treasure chamber. Barlaeus did not just refer to commercial imports but called Amsterdam a treasury of wisdom and knowledge, especially in the teaching of the arts and sciences. In these areas, the scholar asserted, there was much that caused foreigners to wonder, so that they were "struck dumb" by the city's many marvels.[45] This trope was often repeated by Amsterdam's leading cultural producers. For instance, Joan Blaeu, in his atlas of Dutch city maps, cites P. C. Hooft's declaration that Amsterdam had climbed to such heights in such a short time that "the wonder of it strikes people dumb" and "foreigners stand in wonder."[46]

The Doolhoven thus addressed complex urban transformations as multiple global influences were infiltrating the city. The spectacles sought to enthrall foreigners with marvelously incomprehensible technological innovations designed to strike them dumb with wonder and then, through repeated exposure, to cure their astonishment and cultivate learning and virtue. As Barlaeus summed it up, folk from all parts of the world could benefit from Amsterdam's *wetenschappelijke markt*—its knowledge market. The city had much to teach its visitors in exchange for their trade goods.[47] In this context, Amsterdam's automata attractions promoted the technological and philosophical superiority of the mercantile city as a center for the European-made curiosities that were most coveted by recalcitrant trading partners: namely, the qiqi, the strange things that facilitated global commerce. Advertised as "fine things for foreigners," automata were a means of bringing the fabled wonders of the world into Amsterdam.

Whether foreigners apprehended these sites as especially wondrous or beneficial is, of course, an open question. A French pamphlet, published in 1666, overtly ridicules the inflated claims of the Dutch civic descriptions. Titled *Description of the City of Amsterdam in Burlesque Verses*, the author of this satirical work used the pseudonym Jacques le Curieux. This parody of an *amateur* who visits Amsterdam in search of curiosities singles out the Doolhof displays for particular scorn. The author points out the fraudulent nature of the sites and declares that they were a waste of time and mainly frequented by inebriated Dutch peasants who would get trapped and walk round and round the labyrinth without looking up or down or ever becoming the least bit tired.[48] This derision turns the tables: instead of curious foreigners being captivated by exhibits of Amsterdam's marvelous technologies, it is the indigenous Dutch who are ensnared, stuck in a meaningless circuit with no self-improvement in sight.

Conversations

Running counter to descriptions of dumbstruck foreigners and drunken peasants, compelling evidence suggests that actual meetings between locals and foreigners at the city's automata-displaying taverns created opportunities for reciprocal dialogue and new understanding. For example, D'Os in de Bruyloft, run by Theunisz, was a well-known gathering place for discussion and debate, especially pertaining to matters of religion and natural philosophy. This inn was centrally situated on Oudenbrugsteeg, just off the Damrak Canal,

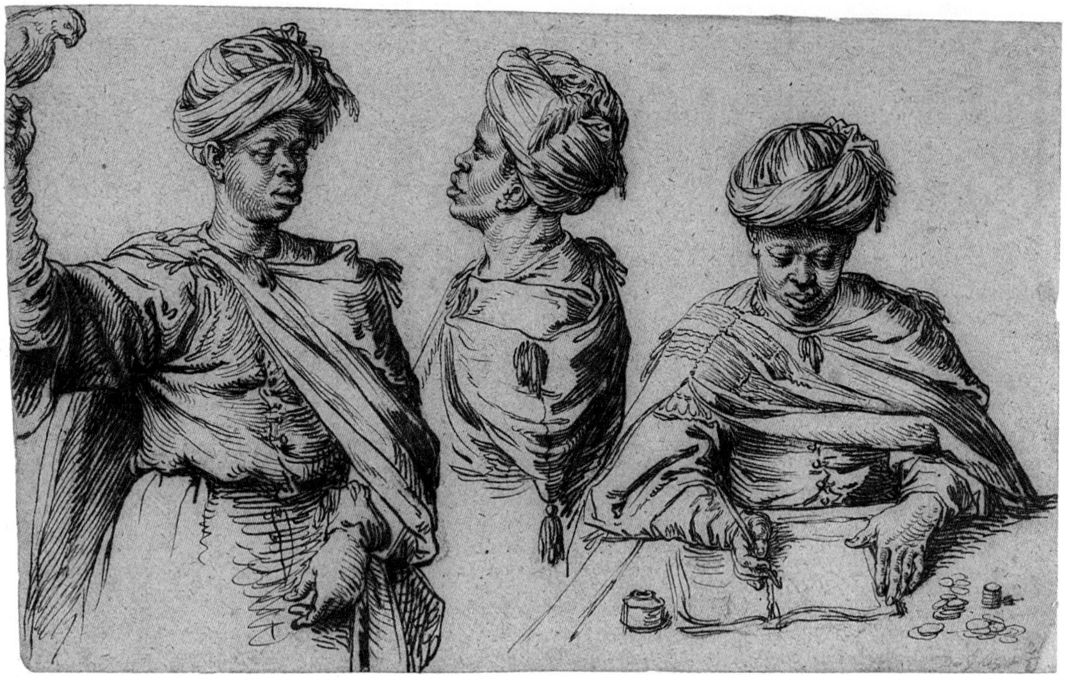

FIG. 31 Jacques de Gheyn II, *Three Studies of an African Man*, ca. 1605–10. Photo © Trustees of the British Museum, London.

the major artery that ran from the harbor to Dam Square, where the town hall, the weigh house, the New Church, the stock exchange, and numerous shops selling books, prints, and maps were located. The tavern drew learned international visitors, including the Englishmen Brereton and Evelyn, as well as a group of prominent Puritan intellectuals. Theunisz's coreligionists, such as the merchant and preacher Cornelis Claesz Anslo, one of Rembrandt's patrons, and Joost van den Vondel, the city's foremost playwright and poet, probably also gathered there.[49] The Frankfurt cartographer and publisher Caspar Merian, son of Swiss-born engraver and publisher Matthäus Merian and half-brother of the naturalist Maria Sybilla Merian, described the inn in his *Topographia Germaniae Inferioris* (Topography of lower Germany) of 1659, which gives a sense of its international renown.[50] The German lawyer Gottfried Hegenitius visited in 1627 and characterized it as an organ house or music hall that featured concerts and wonderful fountains as well as extremely unusual and hitherto unknown musical instruments. An avid recorder of epitaphs, Hegenitius transcribes a poem that laid out "the rules of the house," designating it as a paradise. He concludes that to hear and see all of the shows teaches the gentle art of conversation, an observation that is central to my analysis of this exhibition site.[51]

The colorful career of the publican Theunisz has interesting implications for understanding Amsterdam's automata. Theunisz was highly educated: he had been a student at Leiden University in the 1590s, studying Hebrew and mathematics with Professor Rudolph Snellius and Arabic with

Professor Franciscus Raphelengius.[52] He worked with Raphelengius, son-in-law of Antwerp publisher Christophe Plantin, at the Leiden branch of the Plantin Press. This print shop was a gathering place for Leiden scholars, including renowned humanist scholars Joseph Scaliger and Franciscus Junius.[53] As the first European publisher outside Rome to design Arabic types, Raphelengius made Leiden a center for Arabic publishing. He produced the first Arabic-Latin lexicon ever to be printed, and Theunisz probably assisted in its publication.[54]

Theunisz moved to Amsterdam in 1604, where he worked as a bookseller, printer, and innkeeper. It was during this time that he encountered on the streets a man who, as Theunisz later wrote, attracted his attention because of his "dark complexion" and "curious clothes."[55] He approached the stranger and—making practical use of his university training—struck up a conversation in Arabic. The man was ʿAbd al-ʿAzīz ibn Muhammad, who was visiting the city as part of a Moroccan diplomatic deputation sent to Holland by Sultan Mulay Zaydan in 1609. Such delegations were of great interest to locals. A sketch by De Gheyn II depicts three views of a turbaned African man (fig. 31). In one, the figure sits at a table and writes in a book; an inkpot and pile of coins lie adjacent. The drawing, today in the British Museum, has been dated to around 1605 to 1610, making it possible that the artist encountered this visitor to the Dutch Republic because he was a member of the sultan's delegation, if not ʿAbd al-ʿAzīz himself, who served in the role of secretary. In contrast to the caricatured turbaned figures in paintings like Van Streek's still life (color plate 7) or in prints and drawings of costumed automata (see figs. 29 and 30 and color plate 5), De Gheyn's sketch conveys the agency and intelligence of the African man.

As evidenced by the exotic figures in tronies and still-life paintings, people who appeared foreign attracted the notice of Dutch artists. Turbaned figures in Eastern garb were included in numerous city scenes, especially of Amsterdam's market squares and stock exchange but also church interiors.[56] For example, a small group of Eastern figures appears in the bottom right corner of a busy view of Dam Square by Johannes Lingelbach (1622–1674) (fig. 32 and color plate 8). Lingelbach was the son of David and brother of Philips, the German automata makers who ran the Nieuwe Doolhof. While such imagery indicates the apprehension of visible foreigners as objects of curiosity, it also points to the possibilities of everyday encounters in the city's public spaces, such as the one that occurred between ʿAbd al-ʿAzīz and Theunisz.

Winter was setting in at the time of their meeting. The Moroccan man told the innkeeper that he had decided to prolong his stay and planned to overwinter in Amsterdam and set sail in the spring when the seas were calmer. Seizing this opportunity, Theunisz invited the visitor to reside with him, offering bed and board in exchange for instruction in Arabic.[57] ʿAbd al-ʿAzīz agreed and stayed on for four months, during which time he taught Theunisz and a few other interested scholars. The group included the English intellectual Matthew Slade, a former Brownist who had converted to Calvinism and was the rector of Amsterdam's Latin school as well as the curator of Amsterdam's library, located in the Nieuwe Kerk. There was also his compatriot, the Puritan John Paget, who had studied Arabic at Cambridge University and was pastor of the English Reformed church in Amsterdam.[58] This eclectic group met regularly to converse in Arabic. Theunisz later wrote a summary of their lively cross-cultural Arabic debates about "the

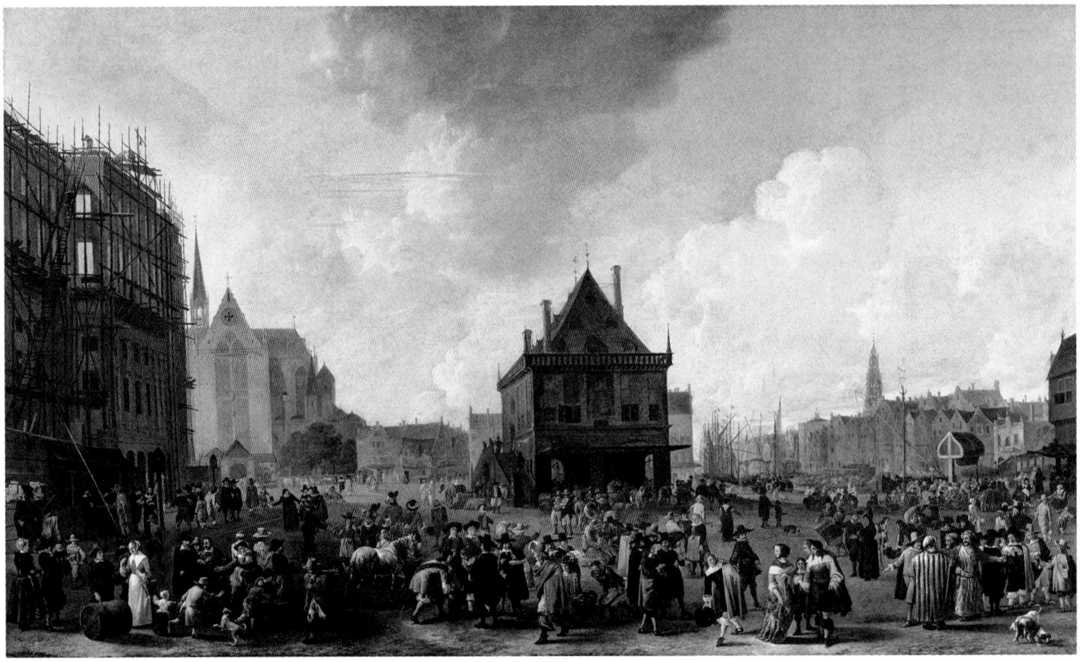

FIG. 32 Johannes Lingelbach, *The Dam*, 1656. Photo courtesy of Amsterdam Museum.

Christian religion and the Quran," highlighting areas of agreement and commonality as well as irreconcilable points of doctrine.[59] These discussions were quite frank. For instance, the publican Theunisz recorded ʿAbd al-ʿAzīz's insistence that God forbade the consumption of alcohol.[60]

The keen interest of Theunisz, Slade, and Paget to learn about Arabic and Islam was in keeping with larger academic trends. A number of scholars and theologians at the universities and academies advocated study of the Quran and the Torah as well as language training in Arabic and Hebrew. The impetus for opening up these areas of study was not disinterested: it facilitated trade and diplomacy. Moreover, knowledge of Semitic languages was regarded as a valuable means to attain deeper comprehension of the Bible. An oft-repeated justification was that the study of Islam and Judaism aided in the Calvinist mission of converting all non-Christians: once this was accomplished, the end-times would come, a belief that we will explore further in chapter 7.

The Calvinist missionary approach had much in common with that of Jesuits like Ricci who believed that informed debate based on familiarity with languages, texts, culture, and religious traditions was a more persuasive and effective technique of conversion than the violent and coercive methods of the Roman Catholic Inquisition. We do not know if conversion was the intent of Theunisz and his learned friends, but if it was, it seems that ʿAbd al-ʿAzīz's sojourn among the Christians only served to reinforce his Islamic convictions. If anything, he may have sought to convert his Protestant hosts.

Before he departed, ʿAbd al-ʿAzīz gave some precious gifts to Theunisz in thanks for his hospitality: one was an Arabic copy he made of an Islamic prayer book; another was his own manuscript of the Quran.[61] The Moroccan man then boarded a ship, bound not for Marrakesh but for Medina, having resolved to go on pilgrimage to the burial place of the prophet Muhammad.

This kind of dialogue between people from different parts of the world could have significant cultural reverberations, some of which can be traced through the subsequent career of Theunisz. The innkeeper later described his meeting with ʿAbd al-ʿAzīz as a life-changing experience that occurred "not without divine intervention."[62] The language skills he gained from a winter in study and conversation with ʿAbd al-ʿAzīz were so impressive that he was hired as a lector to teach Arabic at Leiden University in 1612. A passage from the Quran written in Arabic in Theunisz's hand is preserved in the collections of the Leiden University Library, testifying to his skills.[63] As part of his academic work, Theunisz produced a translation of the Quran given to him by ʿAbd al-ʿAzīz. He also served the States General, composing and translating diplomatic correspondence.

Theunisz's knowledge of Hebrew must have been equally impressive—he built on his university training by fostering a relationship with Amsterdam's Jewish immigrants and, in 1617, was appointed professor of Hebrew at the newly established Dutch Academy.[64] Theunisz was one of the first Amsterdam printers to use Hebrew type; he possibly designed and made it himself, a skill he had opportunity to learn during his apprenticeship with Raphelengius.[65] Notably, he published the Hebraic scholarship of the English Puritan Hugh Broughton, who was an associate of Slade, Paget, and Raphelengius. This publication led to a public debate between Broughton and David Farar, a rabbi in Amsterdam's Portuguese Jewish community; the two men may have first met at Theunisz's inn or his print shop.[66] The Puritan and the rabbi discussed the value of the Gospels, another instance of open conversation about contentious religious differences.[67]

In spite (or perhaps because) of Theunisz's significant contributions, Reformed church leaders protested his prestigious academic appointments, resulting in his dismissal first from Leiden University and later from the Dutch Academy on the grounds that he was a "dissenting Mennonite."[68] Theunisz belonged to the Amsterdam Waterlander Mennonite community, and his inn was sometimes called *Mennisten Bruyloft* (Mennonite marriage). The Mennonites or Anabaptists were a radical religious group that formed in the Reformation. They were harshly persecuted in the sixteenth century and were regarded with suspicion in the seventeenth due to extremist beliefs such as rejection of infant baptism, pacifist nonresistance, and refusal to swear oaths. The Waterlanders were a subsect, considered more tolerant than the strictly orthodox Mennonites. Like every non-Reformed confessional group, Anabaptists were not permitted to hold public office—these positions were reserved for confessing Calvinists—but they were allowed to gather for worship, as long as it was done privately, mainly in house churches.[69] They also came together in places like Theunisz's inn, which seems to have been a center for nonconformist conventicles.

Indeed, Theunisz's unconventional views may account for the fact that, unlike the Doolhof taverns, his internationally known tavern and its marvelous displays were not publicized in any of the official histories of Amsterdam. The Anabaptist innkeeper seemed to court controversy throughout his life. Jan van der Veen composed a poem in 1630 that lists the multiple sects and dissenters that could freely express their opinions in the Dutch cities. The poem makes explicit reference to Theunisz's inn. It begins with obvious denominations like the Calvinists, Catholics, Lutherans, Mennonites, Zwinglians, and Puritans, followed by the more radical Libertines, Socinians, and Sophists. Van der Veen then turns to extremists, heretics, and pagans:

> Robbert Robbertsen den flouwer.
> Den Mennisten Bruyloft-houwer,
> Ian Taurens, int suchtend' huys,
> Broeders van de Rose-cruys,
> Turken, Ioden ende Heyden . . .[70]

Robbert Robbertsen was a extremist Anabaptist schoolmaster, astronomer, and prophet—and a friend of Theunisz. "Ian Taurens" possibly refers to Jan Torrentius, an artist who was condemned for blasphemy and the practice of magic due to a secret chemical compound he used to make his paintings; his artworks were deemed so dangerous that they were publicly burned in the town square of Haarlem.[71] "Broeders van den Rose-Cruys" alludes to the Rosicrucians, a radical Protestant secret society engaged in the alchemical transformation of matter, including the quest to transmute base metals into gold and the creation of a homunculus, a small living human.[72] Rounding out this list were the "Turks, Jews, and pagans." That Theunisz's inn, "Den Mennisten Bruyloft," was grouped with such heretics gives a sense of the notoriety of the place.

Its inclusion might have had to do with a doctrinal battle within the Waterlander community in the early 1620s that erupted into a nasty pamphlet war with Theunisz, the printer, leading the charge for the more radical side. He was disciplined by the congregation, then banned outright in 1626, and only reinstated after publicly confessing his guilt and avowing his contrition in 1634. Moreover, Theunisz's inn, D'Os in de Bruyloft, itself came under censure in 1633, when the Amsterdam Reformed church council made a formal complaint to the civic government about immoral goings-on, with allusion to "blasphemous inscriptions."[73] What these inscriptions were we do not know, although it is tempting to speculate that they were Arabic or Hebrew script and may have been connected with the lifelike things that the innkeeper was creating. These creations certainly could have appeared magical, blasphemous, or alchemical to church leaders.[74]

After he had been dismissed from his academic appointments, Theunisz devoted his energies to his inn and to developing its fabulous hydraulic automata in the 1620s and 1630s. While this may seem like a retreat from scholarly pursuits, I want to explore the idea that Theunisz's experimental creations were borne out of his active engagement with Arabic. As a dealer in books, Theunisz no doubt took a keen interest in the Arabic manuscripts that were being sold in Amsterdam's bookshops. Moreover, his close associations with the most notable scholars and publishers of Arabic in the Dutch Republic would have given Theunisz access to the manuscripts that they were avidly collecting and studying. Some of these entered the collections of Leiden University and were counted among its most prized possessions. In 1609, Heinsius, who oversaw the library collection, commissioned the construction of a special closed bookcase to house the Eastern manuscripts

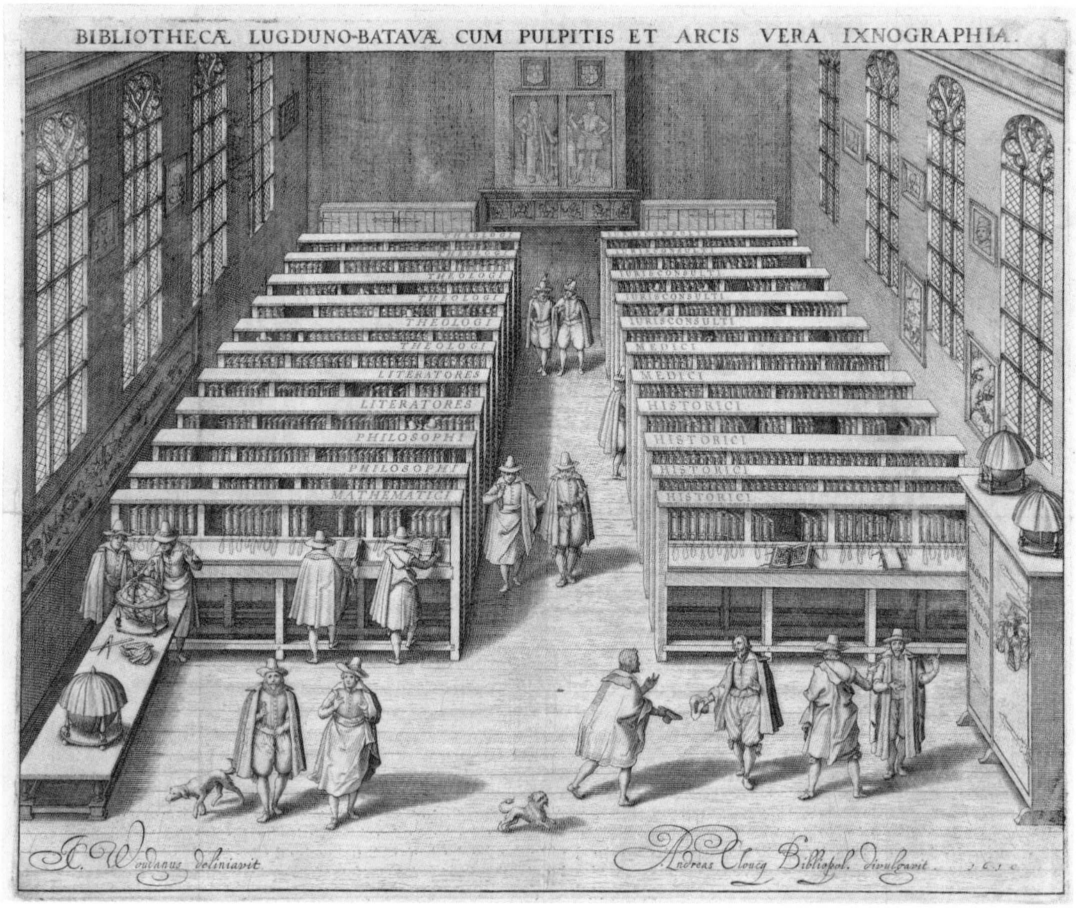

FIG. 33 Willem Isaacz van Swanenburgh, after Jan Cornelisz van 't Woudt, *University Library Leiden*, 1610. Photo: Rijksmuseum, Amsterdam.

bequeathed to the university by Theunisz's professor Raphelengius and his close colleague Scaliger. Scaliger's manuscripts served his larger endeavor, which was to create a unified linear chronology of world history. Anthony Grafton connects Scaliger's historiography to concomitant fascination with clocks and automata—all symptoms of the period's obsession with time and history, an issue we will return to in chapter 7.[75] The bookcase, called Arca Scaligerana, is depicted in the right foreground of Jan Cornelisz van 't Woudt's 1610 print of the library interior (fig. 33).[76] As lector of Arabic in 1612, Theunisz no doubt was familiar with the esoteric contents of this ark.

Theunisz's successor in Arabic studies, Thomas Erpenius, also amassed an important collection of books and manuscripts, which were auctioned after his death in 1624; many of these works eventually entered the collections of Cambridge University. In the 1620s, Erpenius's pupil and successor, Jacobus Golius, traveled to Morocco and through the Levant, including Aleppo and Constantinople, where he purchased more than two hundred manuscripts for the Leiden University Library.[77] He also gained a significant private collection, which later went to auction; many of these manuscripts ended up in Oxford's Bodleian Library.[78] Such manuscript collections were renowned and had a significant impact on European scholars. Descartes, for instance, moved to Leiden in 1630 so that he could join Golius in the study of a newly acquired and exceedingly rare Arabic manuscript of Apollonius of Perga's *Conics*.[79] And Huygens penned a poem in praise of the treasury of manuscripts brought by Golius, comparing it to Admiral Piet Hein's famed capture of a Spanish fleet loaded with silver from the Americas. These Eastern manuscripts bring more glory and prosperity to the *patria*, the fatherland, than the silver fleet, orated Huygens; they are true gold, of benefit to everyone. The spoils of the whole globe, East and West, belonged to the Dutch.[80] Like Barlaeus, Huygens expressed the sentiment that the astonishing material wealth and riches of the Dutch Republic were only exceeded by its vast treasury of knowledge.

Of importance for our study of the mechanical wonders of Amsterdam is the fact that Golius brought back manuscript copies of some of the most important Arabic books on automata making: Ismāʿīl ibn al-Razzāz al-Jazarī's *Book of Knowledge of Ingenious Mechanical Devices* as well as a fragment of the Banū Mūsà bin Shākir's *Book of Ingenious Devices* and an Arabic manuscript of Hero of Alexandria's *Mechanics*.[81] The Banū Mūsà brothers were three ninth-century scholars from Baghdad, and their *Book of Ingenious Devices* describes the construction of one hundred mechanisms, including automata, fountains, and musical instruments, some of which are elaborations and improvements on the more basic ideas conveyed in the Hellenistic texts of Hero as well as Philo of Byzantium and Apollonius of Perga.[82] The thirteenth-century *Book of Knowledge of Ingenious Mechanical Devices* by the Mesopotamian engineer al-Jazarī was also indebted to the works of Hero, Philo, and Apollonius, and the author specifically notes the improvements he had made to the previous designs of the Banū Mūsà.[83] Besides the Arabic copy of *Mechanics*, there were two Greek manuscripts of Hero's works in Leiden. One of these had belonged to Scaliger and was housed in the Scaliger ark: it contains Hero's *Pneumatica* and *Automata* as well as the intriguingly titled *Labyrinthi descriptio*.[84] Another Greek manuscript of Hero's *Pneumatica* was in the possession of Professor Gerardus Vossius. Vossius (like Theunisz) was dismissed from Leiden University due to his nonconformist religious views: he was a Remonstrant supporter of Arminius. He was then hired as one of the first professors at the Amsterdam Athenaeum in 1632. Vossius's son Dionysius had studied Hebrew and Arabic at Leiden University while still an adolescent, and he became associated with Theunisz after the Vossius family moved to Amsterdam.[85] Theunisz thus potentially had access to a number of important ancient manuscripts about the crafting of automata. Moreover, the innkeeper had the language skills to decode these esoteric works and a vibrant scholarly community with which to discuss the ideas and designs.[86]

That Theunisz might have studied al-Jazarī's text is of particular significance because it was written as a sort of how-to manual for craftspeople.[87] This treatise contains detailed instructions as well as illustrations for the manufacture, construction, and assembly of various automata, fountains, water clocks, and trick drinking vessels. The components needed to make these things include tanks, pipes, floats, siphons, axles, levers, taps, valves, and gears, and there are illustrations showing how to fit parts together in order to harness the principles of hydraulics, pneumatics, and mechanics to spectacular effect.[88] The eclectic Theunisz would have been able to manage such instructions because, in addition to everything else, he was a distiller of brandy.[89] This ability meant that he routinely handled the tanks, vessels, pipes, and siphons used in distillation processes—the very equipment required for creating fountains and hydraulic apparatuses (and for carrying out alchemical experiments).

None of the curious devices displayed at Theunisz's inn survive, nor are there any images of them. Brereton's 1634 account gives the most detailed description of the exhibits. He calls Theunisz "a man of most strange invention" and makes a point of stating that he "hath been professor in Leyden of the Arabic language," even though the innkeeper had only briefly held the position of lector (not professor) more than twenty years prior to Brereton's visit.[90] Evidently, Theunisz's knowledge of Arabic was communicated to visitors who came to see his strange inventions, perhaps because there was a connection, suggesting that D'Os in de Bruyloft's highly unusual waterworks were informed by the innkeeper's idiosyncratic talents.

The travel accounts of international visitors such as Brereton, Evelyn, and Hegenitius all describe highly innovative hydraulic works. Besides the pissing and lactating figures discussed in the previous chapter, Brereton gives special mention to figures that carried vessels from which water overflowed, as well as "a lively representation of the five wise and five foolish virgins; the one having oil plentifully, the other wanting." Some of these fountains may have been derived from al-Jazarī and the Banū Mūsà, who explain how to make all manner of trick vessels for dispensing liquids—hot and cold water as well as wine and oil—including those that overflow and refill automatically. The Englishman singles out "a most curious water-work at an infinite charge" and a work in which "the water did spring directly upwards three yards: in this work did the water issue at least in forty or fifty places."[91] These creations, too, may have come from al-Jazarī, who details the workings of six different fountains with variations of alternating devices, allowing for water to emanate for a while as a single jet, then as multiple sprays, and then again as a jet, without any diminishing of the water supply, so that the fountain appears to flow "at infinite charge," as Brereton put it.[92]

Brereton also describes "divers beasts placed upon the rails round about whereon we leaned, out of all whose mouths water strongly sprung, and was carried two or three yards."[93] Al-Jazarī's *Book of Knowledge of Ingenious Mechanical Devices* includes a number of water-spouting animals and birds. A striking example is the illuminated manuscript page currently in the collection of the Museum of Fine Arts in Boston (color plate 9). Painted in opaque watercolor with gold leaf, it depicts a peacock, neck outstretched, perched on the top of a fountain. Water spurts from the bird's beak into a basin below; underneath the basin is a reservoir where a water wheel ensures the continuous circulation of water. This illustration and accompanying instructions give a vivid sense of the types of inventions that Theunisz was making.

Evelyn, visiting Theunisz's inn in 1641, wrote of learning about the natural philosophy of compressed air while being entertained by "many quaint devices, fountains, artificial music, noises of beasts, and chirping of birds."[94] Al-Jazarī's book also includes instructions for creating chirping and whistling bird automata, which were an invention of Hero of Alexandria. These inventions became quite popular in the sixteenth and seventeenth centuries: de Caus describes them in his treatise, and Montaigne encountered them in many estate gardens throughout Europe. The trick was using compressed air and forced water to create noises that mimic the sounds of birds or animals. As with all of these devices, once the basic principles were understood, the automata maker could introduce inventive and entertaining variations, creating all manner of noisy moving beast machines.

According to al-Jazarī, who worked for the Artuqid kings of Diyār Bakr, his hydraulic assemblages were made to be used in palace entertainments and drinking games.[95] For instance, he describes how to make a goblet topped with a bird so that when wine is poured, the bird whistles and spins until it stops with its beak pointing to someone in the drinking party. That person has to drain the cup, imbibing through a spout. If any wine is left, it flows back into the closed body of the vessel, which compresses the air within so that the bird continues to whistle, and the drinker has to drink again to make the noise stop.[96] An elaboration of this type of device is depicted in an illustrated manuscript page from al-Jazarī's *Book of Knowledge of Ingenious Mechanical Devices* (color plate 10). While the top edge of the sheet has been cropped, we can still see part of a bird, which may have whistled and spun on the roof of a pavilion. Kneeling inside is an automaton of a human figure wearing a colorful turban and robes. This automated figure draws up a basin with one arm while the other arm pours liquid from a jug, the spout of which is shaped like the head of a dragon or beast.

Like most of the humanlike automata figures that adorned al-Jazarī's devices, this moving figure plays the role of a servant or slave. Many of the androids move to fill, empty, or convey various vessels, and some play musical instruments. For instance, the manuscript's instructions describe an elaborate drinking-game device with a seated figure called a "slave girl," who grasps a bottle and fills a goblet. Above her is a balcony with four more "slave girls" playing the flute, tambourine, lute, and drum, while above them is a dancer with a ball. Topping this complicated assemblage is a rider on horseback who twirls and stops to point his lance at the one in the party who must drink the goblet filled by the automaton slave. This sequence continues until the reservoir is empty, which triggers another automaton to emerge from a door and make a hand gesture that indicates "no more wine" so that the host of the party must decide whether to refill the device and continue the drinking game.[97] Enhancing the pleasures of palace entertainments, the fantasy of the automaton acting as agreeable slave served the court. Evidently, the use of senseless machines to create intense sensation and drunkenness in guests had long been an appealing form of diversion.

Al-Jazarī's simulacra of slaves and servants are dressed in turbans and robes, pointing to a likely precedent for the European practice of clothing androids in the sorts of costumes that we see on Jochum and the pipe-smoker (see figs. 29 and 30 and color plate 5). The possibility that exoticized automata came to Amsterdam via Arabic manuscripts accordingly prompts reassessment of the claims that were made about them. While the Doolhof publicity materials declared that Amsterdam's automata were wondrous inventions that had never been seen, or heard, or even thought of before, the objects themselves tell a different story. Both their technology and their iconography betray an entanglement with prior craft traditions that had developed centuries earlier in the East. The circulation of people, objects, and manuscripts stimulated complex interactions and the creation of inventions in which elements from various places were so intertwined that it became impossible to distinguish between them in terms of national accomplishments. While the Amsterdam automata may have been displayed as strange things that parodied strangers by mimicking them as beast machines that lacked understanding, their mime show could also be apprehended as a form of emulation, an attempt to imitate and understand the foreign technological accomplishments to which they were indebted.

Theunisz's pub can thus be understood as a sort of laboratory where he worked at re-creating ancient forms of knowledge. Sven Dupré notes that "experimental philosophy still lacked specific places of experimentation in the seventeenth century."[98] Taverns, especially those that engaged in the activities of brewing, assaying, and distilling, contained artisanal workshops. In the case of Theunisz, the publican's academic expertise was complemented by his practical know-how, allowing him to experiment with the invention of ingenious devices. His inn was connected to other urban locales dedicated to the production of knowledge, such as print shops, libraries, bookshops, universities, and academies.[99] Scholarly methods of learning about history by reading, writing, and studying were tested by artisan ways of knowing through making and experimenting with materials. As a scholar-craftsman, Theunisz combined book knowledge with hands-on experience in order to bring antiquity to life.

Amsterdam's automata exhibitions probably shared the pedagogical aims of the short-lived Dutch Academy where Theunisz had taught: to disseminate knowledge of the arts and sciences to a broad public "for the edification and amusement of everyone."[100] Brereton describes the Bruyloft inn as having "no room without some rare invention for pleasure and delight."[101] And Hegenitius records the loftier objective of emotive and sensory engagement with the entertainments, which was to learn the art of conversation.[102] As places where strangers met in encounters with strange things, these exhibition sites stimulated conversations between cultural traditions. The gathering of curious artifacts and diverse people tested identities, stimulating discussion about commonalities and differences and evoking reactions that ranged from hospitality to belittlement. Automata displays had the potential to provoke asymmetrical relations, where ethnic and religious dissimilarities were disparaged, as with the assertions of Amsterdam's world supremacy in contrast to foreigners' dumbfounded lack of understanding. Yet the mobile dynamics of such places and their animated attractions also could spark reciprocal conversations and creative engagement among people of different faiths and nationalities, as well as experimentation with various ways of knowing. Amsterdam's inventive pub exhibits had the potential to set arguments in motion. By bringing the faraway into dialogue with the local, they were important venues for crafting mutable transcultural understandings of an interconnected world.

PROTESTANT PAGANISM

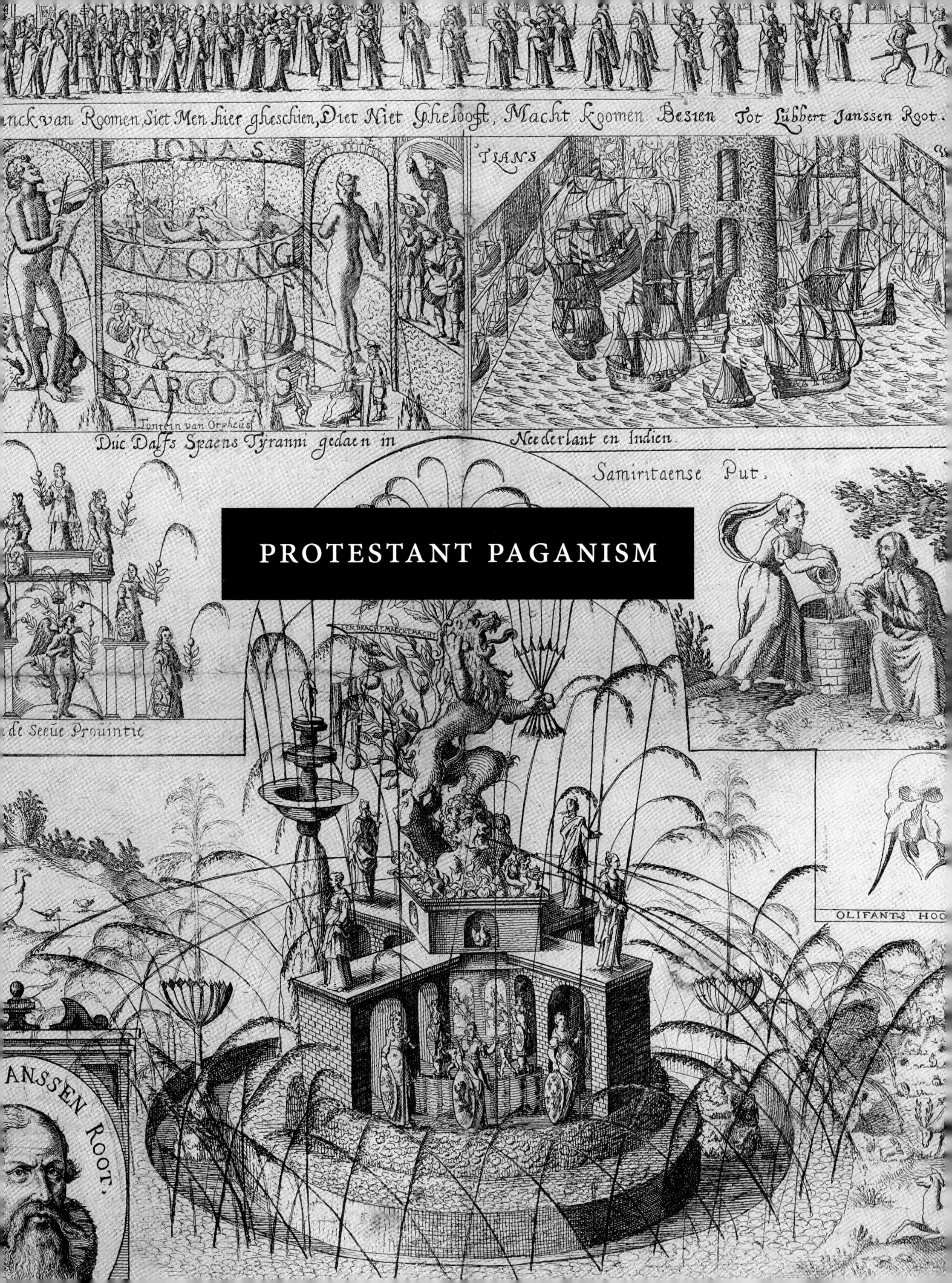

WAX PORTRAITS *Chapter 6*

Body Politics

One would be concerned with the "body politic" as a set of material elements and techniques that serve as weapons, relays, communication routes and supports of the power and knowledge relations that invest human bodies.
—Michel Foucault, *Discipline and Punish*, 1975

Presiding over the automata exhibits at the Oude Doolhof was an impressive display of large figures in wax seated in a row on a raised stage behind the moving mechanical shows. None of the figures survive. They are described briefly but not pictured in the publicity booklets, which highlight "various life-sized statues of renowned Kings and Princes of this century."[1] Listed first was "the great king of France and Navarre, Henry IV, as if he lived," followed by "the famed war hero Gustavus Adolphus, king of Sweden and Gotland and the Queen his wife as well as the young Queen Christina his daughter." The fifth figure was "the illustrious William of Orange, Count of Nassau, who was so lamentably murdered in Delft." He was set next to the then-living stadtholder "Prince William and his wife her Royal Highness the daughter of the King of Great Britain." After this came the "widely famed war hero Maurits, the Prince of Orange, Count of Nassau, who in his lifetime was the Father of the Fatherland." Following the deceased Prince Maurits, the exhibit turned to the infamous "Duke of Alba, the primary cause of the Netherland's troubles." Then came a Chinese man "from the mighty kingdom of China." And finally, another infamous character: Eva Vliegen, "whom men say did not eat for thirty-two years."[2] A later seventeenth-century addition was the figure of Oliver Cromwell, "king murderer and tyrant over England, Scotland, and Ireland."[3] At the Nieuwe Doolhof, Domselaer makes mention of "other artworks modeled in wax" and describes seeing "the gruesome atrocities of Alva, the Paris massacre, the untimely murders of Henry IV and William I."[4] Brereton writes that he went to the "Dole-hoofe" to see "the pictures made in wax most livelily." In addition to works listed in the pamphlets, he records seeing effigies of the stadtholder Frederick Hendrik and his wife, Amalia von Solms, with "their son, about seven or eight years of age"—the future William II—accompanied by "an English nurse, and new-born daughter, a little daughter dead, laid in a cradle at her feet and a page." Brereton concludes that these waxworks, or pictures, as they were called in England, were "all of these so much to the life, as they seem living as they sit."[5]

Brereton's account and the information from the publicity materials that I have just sketched out provide the only historical evidence that we have about the Doolhof waxwork displays. This chapter accordingly attempts to piece together the purpose of this strange assortment of effigies by situating them within long-standing traditions of exhibiting wax portraits of ruling families and notorious figures. Brereton's observations serve as a starting point. He indicates how the presentation of figures both lively and dead had a striking impact on viewers, who saw an arrangement of motionless bodies that appeared strangely animated. As with the automaton, the initial encounter with a wax figure triggers an intense, even visceral, response, as viewers first experience the presence of the portrayed person as if they are alive. The ability of wax to closely mimic flesh and life allowed it to channel the animated presence of historical personages into the present moment. As we will explore below, in the case of Amsterdam's exhibits, the powerful impact of the wax portrait was mainly deployed to inspire political action. These wax bodies served as a potent means to create a living image of the political structure—called the body politic—allowing people see a grouping of the Dutch Republic's past and present leaders, allies, and enemies. The public exhibition of lively wax bodies thus presented the malleable image of a stable and enduring political body that appeared constant but also was flexible enough to respond to the very changeable political situation of the Dutch Republic.

Much to the Life

Brereton's response to the Doolhof waxes was not unique. It was a commonplace to describe wax portraits as being so close to life that they lacked only spirit, breath, and voice.[6] Giorgio Vasari's influential description is representative: "I shall not conceal that modern artists have discovered how to work wax of all sorts of colours, so that to make half-relief portraits from life, they make the complexions, hair, clothing and every other detail so life-like that, in a certain way, these figures lack nothing but spirit and the power of speech."[7] As these comments indicate, materials were worked to create lively effects. Vasari describes how artists added fat and turpentine to soften and cohere beeswax, also mixing in black pitch to give it "a black hue and a certain firmness." Colors could be integrated into supple wax; the addition of red earth, red lead, or vermilion gave it a "peachy colour or some other reddish tint."[8] The pliable colored wax was then modeled in the manner of clay and left to harden. The Doolhof effigies were probably made in this manner.[9] Translucent tinted wax has a particularly lifelike appearance; glass eyes with brows and lashes were inserted into flushed faces topped with wigs of human hair; fleshlike arms and hands protruded from costumes that clothed life-sized mannequin bodies. Brereton notes that the statue of William of Orange wore "a furred and almost like alderman's gown, and a little round ruff about his neck," indicating how specific aspects of status and identity were made visible through costuming.[10] Vasari praised the artistry involved in crafting "complexions, hair, clothing and every other detail so life-like," admiring waxworks that had been "done so naturally and well that they seemed to be living men rather than wax figures."[11]

The animated qualities of the Doolhof's wax portraits set them in dialogue with the moving costumed automata displayed alongside them. In his classic study of portraiture in wax, Schlosser observes that waxworks and automata had a "close relationship": both attempted "to capture the semblance of life's most unmediated utterances—movement and ultimately speech."[12] Yet, like the automaton, the wax effigy also appears as a dead thing. Schlosser cites Arthur Schopenhauer: "The wax figure causes us to shudder since its effect is like that of a stiff corpse."[13] Fleeting, individual, and particular, the wax portrait only lacked spirit, breath, and speech, but this very notable lack made it corpselike. The Doolhof waxworks seem to play on the capacities of the genre to closely mimic bodies both alive and dead by displaying the living figures of deceased leaders like Henry IV and William of Orange next to the lively pictures of existing rulers like Frederick Hendrik and Amalia van Solms, juxtaposed with the waxen corpse of their "little daughter dead."

The deathly qualities of wax portraits were closely connected with their commemorative functions. The tradition of exhibiting wax portraits of the deceased had been particularly significant in Republican Rome.[14] Roman families, especially those involved in civic governance, kept likenesses (*imagines*) of familial ancestors in the atria of their homes. These were carried in funeral processions, "and thus, when anyone died, the entire roll of his ancestors, all who ever existed, was present," as Pliny the Elder explains. The historian also notes that this custom made visible "genealogical lines of descent," manifesting the social standing of the family.[15] *Imagines* demanded an active response, challenging viewers to emulate the virtues and accomplishments of departed citizens.[16] Polybius writes that the effect of these displays was "to be inspired by the sight of men renowned for their excellence, all together and as if alive and breathing."[17] The imagines of illustrious ancestors conferred status and dignity on living family members—lively presentations of the dead constituted the social position of the family in the present moment.[18] This potent tradition continued in the Renaissance, particularly in republican cities; Vasari observes that "a number of these lifelike portraits may be seen in every house in Florence, over chimney-pieces, doors, windows and cornices."[19]

While few survive, commemorative waxwork statues were also common features in Roman Catholic churches across medieval and Renaissance Europe. Votive portraits of donors were given in gratitude for blessings, healings, and miracles, mainly at sacred sites that contained the shrines of saints and miracle-working relics or images, so that the portrayed person was placed in the vicinity of the divine.[20] Renowned were the Florentine sanctuaries of Santissima Annunziata and Orsanmichele, which were so filled with wax figures that the Dutch Calvinist antiquarian Arnoldus Buchelius declared that visiting these interiors was "like entering a field of cadavers."[21] Buchelius would have encountered upward of six hundred full-sized wax figures (*boti*) in the Annunziata alone. Most were of prominent Florentine citizens dressed in their own clothes; they were displayed in the nave, side aisles, and galleries of the church, with many hanging from the ceiling.[22]

By the sixteenth century, waxwork displays at sites like the Annunziata and Orsanmichele had become significant attractions. Aby Warburg notes that votive-filled churches were like waxwork museums. Indeed, they drew many visitors, Buchelius among them. This form of tourism generated the practice of visitors and foreign dignitaries commissioning and donating wax portraits of themselves to commemorate visits to the city.[23] The custom may have informed at least one of the Amsterdam waxworks, described as "a picture of a Chinese from China who was here in Amsterdam and actually came to the Doolhof."[24] Warburg assesses these kinds of donated portraits as "particular objects of pride," showcasing a city's importance on the international stage. He singled out the effigy of King Christian I of Denmark in Florence as a noteworthy example.[25] Such effigies were a tangible means of preserving historic memory and promoting civic identity: they inspired reverence and emulation of the important figures represented. As Megan Holmes concludes: "Florence gained prestige through association with these illustrious, high-status votaries represented in the city sanctuary."[26] The garnering of civic prestige doubtless motivated the Amsterdam displays as well.

Less well known than the Florentine examples is the vibrant votive waxwork tradition that existed in the Low Countries prior to the Reformation. In the fifteenth century, the Dukes of Burgundy fostered a cult of miraculous images and ex-votos. As explored in previous chapters, many of the cultural forms and practices of the Burgundian court—labyrinths, automata, and trick fountains—were revived at the Doolhoven, and it is likely that the wax displays also drew from this tradition. Hugo van de Velden shows how the dukes' active promotion of a culture of votive wax effigies worked to legitimate their authority, serving as a material demonstration of how the intercessions of the saints supported territorial and military ambitions.[27] Alongside their religious functions, wax memorial portraits thus had significant civic and political purposes.

Not surprisingly, votive waxworks also had idolatrous associations. Already in the fourteenth century, the Italian writer Francesco Sacchetti notes: "Everyday votive effigies [*boti*] like this are made, which are more a form of idolatry than of Christian faith."[28] Commemorative waxworks resonate closely with Calvin's claim that idol worship originated from ancient practices of making images of the dead: "Respecting the origin of idols . . . the first authors of them were persons who paid this honor to the dead, from a superstitious reverence for their memory. I grant that this perverse custom was very ancient, and deny not that it greatly contributed to increase the rage of mankind after idolatry."[29] Wax effigies accordingly became the targets of Protestant image destruction in the sixteenth century, especially in northern Europe. Van der Velden points out that the lack of extant votives in the wake of iconoclasm has resulted in waxworks being largely absent from histories of Netherlandish art.[30] Like the automaton, the lively wax statue does not fit easily within usual narratives of post-Reformation Northern art.

One way to consider the impact of lost works is by studying the cultural practices that they generated. In spite of associations with miracles, saints, idolatry, and paganism, the venerable tradition of exhibiting waxwork figures reemerged in the Amsterdam displays. In tracing the history of portraiture in wax, Schlosser asserts that "the votive museum in churches" represented the embryo stage of the public waxwork shows that became popular in the seventeenth and eighteenth centuries.[31] The Amsterdam exhibits, established in the 1620s, were among the earliest of these commercial wax museums. These sites challenge assumptions that the peculiar formal qualities of wax portraiture run counter to Protestant aesthetics. The exhibition *Like Life: Sculpture, Color, and the Body* argues that a long-standing art-historical bias in favor of white marble is in keeping with the Protestant approach: "Polychrome was a Catholic thing, and all the more so if the figures were wax."[32] Historical evidence certainly buttresses such conclusions; yet, in spite of ongoing hostilities, iconoclasm did not signal the end of this distinctive artistic tradition in northern Europe. While the religious functions of life-sized portraits in wax were suppressed, they reemerged in urban public houses, where their political efficacy was foregrounded. Again, we are reminded of Van Mander's claim that taverns could serve as refuges for beleaguered arts that risked iconoclastic attack.

An idiosyncratic feature of wax portraiture is its persistent survival in spite of the material's ephemerality and damaging associations with idolatry and image magic. While few individual works survive, the genre as a whole has proven remarkably resilient. Georges Didi-Huberman conceptualizes wax portraiture in terms of its capacity for *revenance*, or ghostly return, which disrupts the usual temporal models of art history, especially evolutionary narratives about the progressive development of style.[33] Following Schlosser, Didi-Huberman points out that the formal properties of wax portraiture, specifically its "hyperrealism," make it resistant to stylistic change. Veristic portraiture is indexical: it conforms closely to the appearance of the subject portrayed, and as a result, the style of a likeness in wax does not change significantly over time. Because the waxwork corpus does not evolve, it has been ascribed a peculiarly inert temporality, which defies art-historical interest in artistic innovation.

Revisiting this issue, it is important to emphasize that wax portraiture does develop in interesting and perhaps unexpected ways over time. Though the style may not undergo considerable change, the venue certainly does. It is difficult to determine whether a fifteenth-century wax portrait of King Christian I of Denmark was formally similar to a seventeenth-century one of King Gustavus Adolphus of Sweden, since neither work survives, but knowledge of where and how these works were exhibited and of the responses they generated reveals that wax portraiture was a historically variable genre that afforded a range of possible religious and political meanings and practices when taken up in different contexts. While the medium shapes reception, it does so in the framework of specific rituals of presentation. As we shall see, the capability of wax effigies to bring history to life was of particular relevance in the Amsterdam displays. The wax portrait does more than preserve the likeness of a face; its animated qualities make the depicted person present, even after death. The lively waxwork insistently inserts a historical figure into the here and now so that the deceased subject continues to project a commanding living presence, forging a powerful connection with the audience.[34] The Doolhof waxworks, which were perceived as "so much to the life, as they seem living," represent a theatrical return of the dead, who were placed onstage alongside wax figures of people still living, confronting spectators with their lively presences. This display was a potent way to make manifest select aspects of the historical past and articulate their urgent relevance in the present moment.[35]

Modeling a Protestant Body Politic

A specific historical narrative was staged by the waxwork effigies of "renowned Kings and Princes of this century" at the Doolhoven. Brereton's account of the arrangement of statues that "seem living as they sit" suggests that the mode of display was informed by Polybius's description of Roman practices of placing waxworks of celebrated men and war heroes "in a row upon chairs of ivory."[36] Polybius explains the ritual functions of these seated wax figures: "It would be hard to imagine a more impressive scene for a young man who aspires to win fame and to practice virtue. For who could remain unmoved at the sight of the images of all these men who have won renown in their time, now gathered together as if alive and breathing? What spectacle could be more glorious than this?"[37] Polybius relates that this rousing assembly of renowned men was accompanied by an oration of their virtues, contributing to the successes of ongoing war efforts: "By this constant renewal of the good report of brave men, the fame of those who have performed any noble deed is made immortal, and the renown of those who have served their country well becomes a matter of common knowledge and a heritage for posterity. But the most impressive consequence of the ceremony is that it inspires young men to endure the extremes of suffering for the common good in the hope of winning the glory that waits upon the brave."[38] The lively figures of illustrious men modeled in wax inspired courageous acts of citizenship in a republican context.

The political functions of portraits of eminent persons persisted in the Dutch Republic.[39] The antiquarian Thomas de Rouck summarizes the Roman ceremonial uses of imagines of important citizens and advocates that this was an important means to keep alive "a lasting impression of their notable deeds."[40] Vossius, professor at the Amsterdam Athenaeum Illustre, writes that such works aroused the desire for virtue by passing on great deeds to posterity: "It gives not only knowledge but also will-power, and makes us not only spectators but also actors . . . so our minds are also set aflame to emulate the examples that have lived a very long time before us. . . . The same power is also clearly manifest in cold stone or in wax, in silent statues and masks."[41] The emotive charge conveyed by likenesses of deceased historical figures could motivate audiences, transforming them from spectators into actors.

By showcasing military leaders and war heroes, the Doolhof waxworks were designed specially to generate active support for the Dutch Republic's war efforts. Writing about his visit to Amsterdam, Mountague underlines the importance of stopping at the city's public houses, including the "Dooly-House, a noted Tavern," adding: "The mighty Figure this Republick has, and still does make in the World, her Wealth, Power, and Reputation, both at home and abroad; her Numerous Armies, Formidable Navies, and Foreign Alliances, are all well worth seeing and knowing."[42] Mountague's comments indicate the importance of showing the "mighty Figure" of the Dutch Republic to locals and foreign visitors. The assembly of wax figures at the Oude Doolhof presented something that was not otherwise readily visible—the Republic's political and religious alliances with other European Protestant powers. This impressive scene presented a stirring vision of Europe united under strong Protestant leadership.

The effigy of Henry IV of France and Navarre was placed first among the figures of allied leaders, and his likeness may have been the oldest waxwork in the group. The earliest extant Oude Doolhof poster includes the coat of arms of the Kingdom of France and Navarre and proclaims that the Huguenot fountain maker Bargois had previously served Monseigneur the Prince of Conti (see fig. 3). The Princes of Conti were of the ruling house of Bourbon and thus princes of the blood, relatives and supporters of Henry IV. Given his illustrious connections to this noble family, Bargois may have commissioned or acquired the waxwork of the French king to show—and generate—support for the French Protestant cause.

Placed next to the wax portrait of the French king were figures representing the Swedish royal family together with the Dutch stadtholders and their wives and children. This grouping vividly manifested a united pan-European Protestant front. The inclusion of spouses and children is significant. In the manner of Roman family portraits, a genealogy is established, so that viewers could see a continuing lineage of allied ruling families committed to the success of Protestantism. This alliance included the ruling house of Great Britain, represented by the waxworks of Queen Mary and her husband, William II, with their offspring in a tableau vivant that presented the procreative intermarriage of Stuart and Orange. The Protestant coalition was staged as a formidable force in the present moment, an enduring power bloc that built on past conquests and would continue with future generations.

The emphasis of this stirring arrangement of figures was on the efforts of Protestant armies to prevail against Catholic enemies throughout Europe. The "famed war hero" Gustavus Adolphus had led the Protestant forces in the Thirty Years' War and was revered by the Dutch as a defender of the faith. There were many Dutch prints and pamphlets celebrating his victories and heroism. The most inventive of these were by the engraver Magdalena van de Passe, whose Protestant family had immigrated to the Dutch Republic from Cologne.[43] Her brother, the printer Crispijn van de Passe the Younger, published some of the Oude and Nieuwe Doolhof pamphlets, indicating the family's connection to the Doolhoven.[44] An innovative artist, Magdalena experimented with techniques of using copperplate to print on fabric. In 1630, she went before the States General to exhibit linen nightcaps engraved with the portrait of the Swedish king. The States General granted her the privilege to produce these, which was renewed in subsequent years, indicating the popularity of these decorated sleeping caps and the high regard in which the States General as well as the populace held Gustavus Adolphus.[45] To wear a picture of the Swedish military leader on sleepwear was like the practice of keeping a protective talisman of a saint close to the body—a means of forging a close physical relationship with a revered figure. While little is known about these intriguing nightcaps, when considered in relation to the wax figure of the Swedish king at the Doolhof, we glimpse a larger pattern in which banned practices surrounding the religious image were reactivated in the Protestant context.[46]

Like Gustavus Adolphus, Henry IV also was extolled as an exemplary leader. P. C. Hooft's 1626 work *Henrik de Gróte* (Henry the Great) celebrates the virtues of the deceased monarch, who was assassinated in 1610. Hooft describes the French king as an ideal leader whose prudence and virtue brought religious peace to his domain. In highlighting the exemplary piety and wisdom of Henry, one aim of Hooft's book was to encourage other European rulers—particularly the new stadtholder, Frederick Hendrik, who was appointed in 1625—to emulate the model of the French monarch.[47] Hooft's approach was in keeping with the Neostoic philosophy articulated by Lipsius. While *De Constantia* equipped citizens with a practical philosophy to help them attain endurance and constancy, Lipsius's six-volume work *Politica* was in the instructive genre of mirrors for princes, written to aid those who governed.

Politica advocated that a strong centralized monarchy was the best form of government to achieve peaceable conditions, granting the ruler absolute power to quell disruptions.[48] Due to his great influence, the monarch had to be schooled in practical wisdom and virtue, since a strong and righteous prince was of paramount benefit to the people.[49] *Politica*'s specified goal was to lead rulers to apatheia: they should govern with constancy and lead by example, having attained a stable and unchanging state of freedom from the upheavals caused by the passions. Drawing on these ideals, Hooft put forward Henry IV as a model monarch, glossing over aspects of his reign that were less than ideal. Henry's notorious sexual exploits, for instance, were framed as positive behavior, signifying the potency of a robust masculine leader.[50] Even Henry's conversion to Roman Catholicism was portrayed as a form of constancy, an instance when the wise ruler had to rise above sectarian conflicts for the greater good. In Hooft's view, Henry remained Calvinist in his heart but publicly espoused Roman Catholicism as a practical means of making peace.[51]

Similar interpretations of Henry's qualities were probably ascribed to the Doolhof waxwork of the king. It is likely that the ruler portraits were presented with "an oration of their virtues" in a manner derived from the Roman ritual described by Polybius—an effective means to show the merits of the ruler to the public. Lipsius stressed that the visibility of the leader was inseparable from his political power: the prince was always onstage and had to demonstrate his virtue, justice, constancy, and wise government. The public image of princely righteousness persuaded people to accept authority; this required the mobilization of multiple representations to make the ruler's potency visible and tangible.[52]

When the Doolhof wax figures of "renowned Kings and Princes of this century" are considered together with the Protestant political theory of writers like Lipsius and Hooft, it suggests how the idiosyncratic characteristics of wax—especially its capability to present an unchanging yet animated and lifelike image of the prince—were politically efficacious for the Protestant cause. Wax portraiture's oft-disparaged constancy of style lends itself particularly well to portraying the quality of constancy perceived as a compelling outward sign of good leadership. Wax ruler figures thus conveyed the enduring apatheia of the prince, upholding stable leadership in times of political crisis.

Placing such an image before the people was an expression of the public persona of the prince. Lipsius reminds his learned readers that maintaining an unchanging public image requires deception.[53] To this end, the deceptive qualities of wax portraiture were particularly efficacious. Because the artistry and technical ability involved in the making of a wax portrait are suppressed by a lack of stylistic innovation, the effigy appears as a close counterfeit of the sitter. Vasari's claim that such works are "living men rather than wax figures" is echoed by Schlosser: "Nowhere has pictorial art ever striven harder to attain the mirror image of reality."[54] The indexical qualities of veristic portraiture convey the presence of the individual rather than the formal inventions of the artist. The maker disappears so that the sitter comes to the fore. While waxwork has been overlooked in the history of art because of its perceived lack of artfulness, this characteristic is precisely what made it such an effective medium for royal portraiture.

The illusory qualities of the wax ruler portrait were mobilized for the greater good. Hooft compared Henry IV's peacemaking abilities to those of Orpheus, whose enchanting playing of the lyre harmonized differences among the animals.[55] At the Nieuwe Doolhof, an Orpheus fountain made by Bargois combined music and the sound of flowing water to create an atmosphere of pacific harmony. There was also an automated show of David playing the harp. His captivating musical talents manifested his princely skill at creating conditions of peaceful harmony for his people. Sources indicate that organ music accompanied the presentation of the Doolhof exhibits. Visitors did not just see the lively figures of the wax rulers but ideally experienced a sensation of happiness and well-being from the musical accords that expressed the persuasive powers of peacemaking princes. The aim was to fashion a harmonious society whose loyal subjects were assured that their welfare was produced and protected by the enduring presence of upright Protestant leaders.

Because the representation of the monarch's material body was an analogue for the political system, the unchanging, unmoving waxwork figure manifested the ideal of a stable polity. The flexible and changeable properties of wax concurrently allowed it to respond to moments of political crisis. Wax ruler figures were mobilized with particular urgency when the body politic came under attack. For the Protestant allies, the assassinations of William of Orange in 1584 and Henry IV in 1610 dealt severe blows to the image of a society united in religious concord. The material possibilities of wax made it a particularly effective medium to repair injuries to the political body. Its malleability allowed for the relatively quick fabrication of a portrait, in contrast to the time-consuming labor of working with hard materials like stone or wood. A rose-tinted, fleshlike wax effigy could swiftly fill the gap left by the unanticipated loss of the ruler's mortal body.[56]

In fact, waxworks had long been used to represent slain political leaders. Appian of Alexandria's account of the Roman civil wars includes a description of the wax figure made and displayed to the people following the assassination of Julius Caesar. The statue of Caesar was "turned round and round by a mechanical device, showing the twenty-three wounds in all parts of the body and on the face, that had been dealt to him so brutally." The effect of this mechanically driven waxwork was to arouse and channel the strong emotions of the crowd: "The people could no longer bear the pitiful sight presented to them. They groaned, and, girding up their loins, they burned the senate-chamber where Caesar was slain."[57] The animated wax body of Caesar ensured that his ruling presence actively persisted in spite of the assassins' deed. The statue continued to exert the persuasive powers of the political leader to move the populace to action.

It was precisely this capacity of the figure of Caesar that was recalled by the Dutch Calvinist art theorist Junius, who writes that "Julius Caesars Image expressed in waxe, and hideous to looke on for the three and twenty wide gaping wounds he had received, did mightily stir up the Romans to revenge his death."[58] The veristic wax effigy conveys the living presence of the dead ruler with immediacy. Such sculptures did more than create a fiction of the postmortem life of the political leader.[59] They acted as vital substitutes for the living person by channeling the leader's unique power to unite the people, stirring them to take courageous critical action.

The traumatic loss of Henry IV accordingly prompted an idiosyncratic ceroplastic response, and a number of wax portraits of the king were made after his assassination in 1611.[60] There was a competition to choose an artist to make the wax effigy for the lying in state of the king. One of the competitors, the French artist Michel Bourdin, subsequently sued François de Bechefer, secretary to the Prince of Conti, for damage done to his effigy through the illicit taking of a mold from the original. Bechefer evidently had made wax copies of Bourdin's work and toured them at provincial fairs, capitalizing on people's desire to see the figure of the deceased monarch by charging an admission price.[61] These displays of postmortem portraits of Henry IV are the first known commercial exhibitions of waxwork, marking a turning point in the tradition: as Edward V. Gatacre and Laura Dru write, "With the effigy of Henri IV an extraordinary history of the spread of replicas begins."[62] It is possible that one of these replicas was acquired for the Oude Doolhof, likely by Bargois, who may well have been acquainted with the enterprising Bechefer through their mutual service to the Prince of Conti.

A wax portrait bust of Henry IV by Bourdin survives, which is the closest we can come to envisioning the lost Amsterdam waxwork (color plate 11). While somewhat discolored and cracked, the fleshlike face has retained its peachy hue. The network of lines and wrinkles around the king's deep-set eyes intimate the life experience of a seasoned ruler, and the wax portrait's glassy stare appears pensive here, conveying characterful qualities like wisdom and resolve. The lively wax image thus ensures that although the king had been killed, the ideals he embodied would live on.

Bechefer's commercial presentations of the wax face of Henry IV exploited the centuries-old practice of publicly exhibiting lifelike effigies and death masks of rulers, which was central to royal funeral rites, especially in France and England. Such portrayals were visual expressions of political theories about the king's two bodies.[63] The body natural of the king was mortal and vulnerable. However, the monarch's second body—the body political—represented the crowned head of the political structure. Although Henry IV's natural body had been killed, the wax effigy preserved his political body, keeping intact the authority of kingship and cohering a body politic that risked fragmenting with the loss of its head. Wax portraits—which were both flexible and unchanging—ensured the stability of the polity in times of uncertainty and transition; they asserted that the body politic endured in spite of the inevitable deaths of individual leaders. The ruler's virtues and leadership qualities were shown to be undying, effectively conveyed by the static style of a portrait in wax. Displaying a lively effigy was of particular consequence in times of transition between the death of one monarch and the accession of the next. It visually proclaimed the succession statement: the king is dead, long live the king![64]

Many royal funeral effigies continued to be displayed long after the burial and decay of mortal

FIG. 34 Queen Elizabeth's wax effigy in Westminster Abbey Museum, twentieth century. Photo: Private collection / Look and Learn / Peter Jackson Collection / Bridgeman Images.

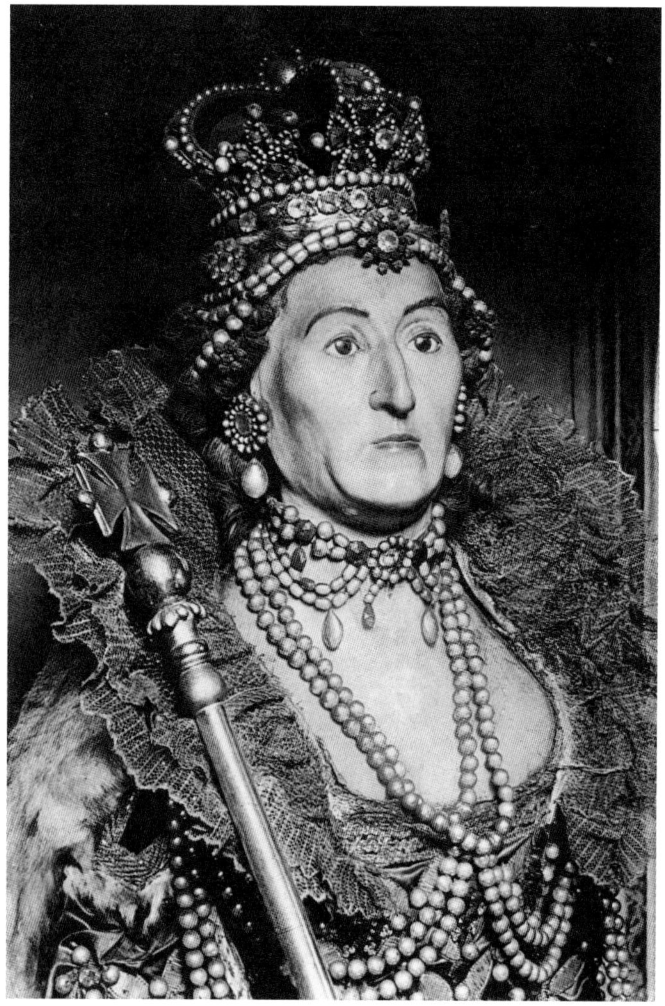

remains. Showing a living genealogy of hereditary rulers was a politically expedient means of maintaining lines of succession. Waxwork funerary figures representing the royal lineage of France were on view in the basilica of Saint-Denis until their destruction in the French Revolution. And effigies of past British monarchs are still preserved in Westminster Abbey.[65] Noteworthy is the figure of Elizabeth I (fig. 34). The current portrait is a copy of the waxwork that was paraded in the queen's funeral cortege in 1603. This costumed effigy reportedly had a profound effect on the large crowds who gathered to see the final procession of their beloved queen: "They beheld her statue and picture lying upon the coffin set forth in Royal Robes . . . there was such a general sighing and groaning and weeping, and the like hath not beene seen or knowne in the memories of man."[66] As with the ancient statue of Caesar, the wax picture of the deceased queen was agential, prompting public action and working social transformations; in this case, the emotional force of collective mourning worked to cohere a community during the volatile time following the death of a monarch who had no direct descendants and thus no clear dynastic successor.[67]

The ability of a group of waxwork figures to make visible dynastic continuity was of particular significance for the House of Orange. The stadtholderate was not a traditional hereditary position; prior to the Dutch Revolt, the stadtholders were appointed representatives of the monarch. With William of Orange, the role changed from servant of the monarch to military leader of the revolt against the Spanish crown. Even though the position of stadtholder had never been dynastic, when William was assassinated in 1584, his son Maurits succeeded him, in part because he was a skilled soldier capable of taking command of the war effort. He also stepped into the leadership gap left by his father. Without a monarchy, the Dutch Republic rallied around William of Orange and then Maurits as heroic, quasi-monarchical leaders. As if to insist on dynastic continuity as well as gifted military governance, Doolhof publicity presents Maurits's effigy as the "widely famed war hero Maurits, the Prince of Orange, Count of Nassau, who in his lifetime was the Father of the Fatherland."[68]

With the death of Maurits in 1625, the position of stadtholder passed to his half-brother Frederick Hendrik, subsequently to Frederick Hendrik's son William II, and then to his grandson William III. The visible succession of a lineage of stadtholders presented by the Doolhof wax portraits therefore emphasized the importance of dynastic continuity for political stability. The figures of William I and Maurits probably were made in the early 1620s, while the effigies of Frederick Hendrik, William II, and their wives and children were added later. On a Nieuwe Doolhof poster, the slogan "Vive Orange" (Long live the House of Orange) adorns Bargois's Orpheus fountain, emphatically linking the inn's exhibits with allegiance to the stadtholders and the Protestant cause (see fig. 44). This motto also denotes the purpose of the exhibitions: "Vive Orange" describes well the function of lively artworks that maintained the vital presence and dynastic continuity of the House of Orange as a means to achieve social harmony.

After William of Orange was assassinated in 1584, a cult of his image was fostered during the political and religious crises of the 1610s and 1620s: he was a historical figure that appealed to various dissenting factions.[69] De Keyser's 1621 monumental tomb of William of Orange, with its two effigies—of the living political body in bronze and deceased mortal body in marble—served as a means to cohere the Dutch Republic during the unstable period of the Twelve Year Truce.[70] De Keyser was the leading sculptor of his day, with significant links to the Doolhoven. As discussed in chapter 2, the sculptor's bust of Coster, the first proprietor of the Oude Doolhof, is the oldest known marble portrait bust of a citizen of the Dutch Republic (color plate 2).[71] With this work, De Keyser was reviving the ancient Roman tradition of veristic portraiture, an effectual means of manifesting the qualities of citizens who contributed to the common good.[72] This innovative endeavor was in keeping with the Doolhof's aim of bringing ancient arts back to life. A fragment of a life-sized terra-cotta bust was unearthed near the Oude Doolhof site in 1986. Frits Scholten suggests that the piece is a remnant of another portrait of Coster by De Keyser.[73] De Keyser did make works in terra-cotta, a pliable material that is modeled like wax. Notably, he fashioned veristic terra-cotta busts of William of

Orange (color plate 12). Given the sculptor's efforts to resuscitate ancient Roman portrait traditions in a range of materials, it is possible that the nonextant Doolhof waxwork of William I also was made by De Keyser or copied from one of the sculptor's lively portraits of the prince.

To sum up, the tradition of wax portraiture was revived in Amsterdam mainly because of its peculiar political efficacy. The Doolhof waxworks merged aspects of the royal funeral effigy customs of France and England with the republican portrait practices of ancient Rome and Renaissance Florence. Funerary traditions assured the populace that the political body lived on in spite of death. The republican portrait tradition was adapted to vividly present family genealogies and show that the virtues of the dead were maintained by their descendants. Such imagery was designed to inspire viewers to emulate the excellence of the deceased, and the Amsterdam waxworks especially encouraged involvement in Protestant military efforts. The mix of royal and republican forms thus effectively served the distinctive political and religious culture of the Dutch Republic, where monarchical reverence for the Calvinist House of Orange coexisted with republican respect for the deeds of notable citizens.

In spite of prohibitions against graven images, a range of sculptural practices that paid honor to the dead came to characterize the Dutch Republic. As the British visitor John Ray explained in 1663: "When any famous deed or exploit is done by any of their nation, it is represented to the people with all insinuating circumstances, to make them proud of the honour of being subjects in a state where such mighty deeds are done. Besides, for encouragement, there is, for the most part, a monument or statue erected to the memory of them who do great things." This artistic practice, he declares, was of great use "for those . . . expressions of joy do make glad the people and give them better thoughts of their governors, and dispose them to their service."[74] Ray could have been paraphrasing Lipsius. The sight of the Doolhof effigies of heroic Protestant leaders, which were "represented to the people with all insinuating circumstances," such as oratory and music, were designed to generate intense positive reactions like joy, national pride, and social harmony. To thus move the people was to transform them, disposing them to the service of their governors. The transmission of emotion via the ritual display of historical wax portraits accordingly constituted the present political moment as a critical age, motivating people to actively fight for the Protestant cause.

Infamy: Fragmentation of the Body Politic

Wax portraits had long been used to manipulate emotions and work changes in spectators; perhaps the malleability of the material lent itself to this transformative purpose. In many of the historical examples, waxwork figures of leaders elicited strong responses of grief, anger, gladness, pride, or wellbeing, which had the potential to bind people together, conducting shared passions toward a political goal. At the Doolhoven, the unifying purpose was to achieve the victory of Protestantism in Europe. Besides the statues of famed rulers, the Doolhof exhibits also included wax images of political enemies and infamous figures. In these cases, we see how the ephemeral qualities of wax—its friability and vulnerability—reveal something that its deceptive qualities seek to conceal: the continuous degradation of the body politic.

Dutch hatred of their Roman Catholic foes was fixated on one figure—the Duke of Alba. Fernando Alvarez de Toledo, third Duke of Alba (or Alva), had been governor of the Spanish Netherlands from 1567 until 1573 and leader of Spanish military action against the Dutch Revolt. Even though Alba died in 1582, his image continued to circulate as a recognizable symbol of tyranny and became a staple in Dutch patriotic propaganda even into the eighteenth and nineteenth centuries.[75] Wax effigies of Alba were exhibited at both the Oude Doolhof and the Nieuwe Doolhof. A 1630 Nieuwe Doolhof poster makes overt reference to the duke with a line of text that runs beneath the image of the Orpheus statue, as if to underline the need for harmony: "Duc Dalfs Spaens Tyranni gedaen in Nederlant en Indien" (The Duke of Alba's Spanish tyrannies carried out in the Netherlands and Indies) (see fig. 44).[76] The notoriety of Alba was part of the "Black Legend": polemic that vilified Spanish cruelty and intolerance in both Europe and the Americas.

As Ben Schmidt shows, this reference to Alba at the Doolhoven was part of larger efforts to create historical consciousness. Chroniclers identified Dutch victims of Spanish brutality with the inhabitants of the Americas, who also suffered under the harsh yoke of Spain. For example, one commentator exhorted readers to "detect, as in a mirror, the character, nature, cunning, deceit, falseness, faithlessness, ambition, cruelty, tyranny and dominion of the Spaniards . . . committed in the New World. It behooves us always to remember, to recollect in lively and plain terms and never to permit to lapse into the house of forgetfulness these memories; but rather to remain assiduously alert, diligently prepared, and always on our guard . . . that we shall never again fall under the dominion, tyranny, and violence of Spain."[77] The waxworks of Alba evidently served this function and likely were presented with similar rhetoric, keeping alive hatred of this most despised figure as a tangible locus of memory so that people would remain alert to the need to fight off aggressors. The wax portrait of a loathed Spaniard was a means "to recollect in lively and plain terms . . . the dominion, tyranny, and violence of Spain" in the present moment. Each time the animated statue was presented to the people, it could trigger an emotional recall of every Spanish atrocity. In this way, the traumas of distant times and places were reignited to keep people on guard and ensure loyalty to their current political and military protectors, the Princes of Orange.

Wax portraits of Alba did more than malign a long-dead former governor. The reference to Alba's Spanish tyrannies carried out in the West Indies was a way to project Dutch history into a faraway place and onto distant peoples.[78] By identifying with the Spanish-inflicted sufferings of the Indigenous populations of the Americas, the Dutch engaged in crafting and disseminating a negative image of their political enemy as uniquely inhuman and sadistic. Taking up the role of perpetually innocent victims also allowed them to deny their own brutally oppressive practices. As the Dutch West India Company increased colonization efforts in the Americas and took over the transatlantic slave trade, the Dutch, too, were described as particularly ruthless in their treatment of Indigenous peoples and enslaved African peoples. Indeed, the mirror that reflected "the character, nature, cunning, deceit, falseness, faithlessness, ambition, cruelty, tyranny and dominion of the Spaniards committed in the New World" could have been held up to the Dutch.[79] If the ethics of erecting statues of famous leaders was to make people glad and "proud of the honour of being subjects in a state where such mighty deeds are done," then the integrity of such an image was precarious and apparently needed to be propped up by other types of statues such those made to defame unpopular governors like Alba. In this manner, disruptions to social harmony could be attributed to the immoral actions of external enemies and political adversaries. Continuous denouncement of Alba was also a means to overlook all of the malignant practices routinely carried out by Protestant princes, soldiers, merchants, and colonizers from the Dutch Republic and other nations. By keeping alive increasingly distant evils committed by a foreign foe, the statue of Alba prompted a complicated response. It reactivated painful trauma memories and inflamed hatred of perceived enemies of the Dutch, while simultaneously denying ongoing colonial and military atrocities committed by the Dutch themselves and upholding an image of Protestant righteousness.

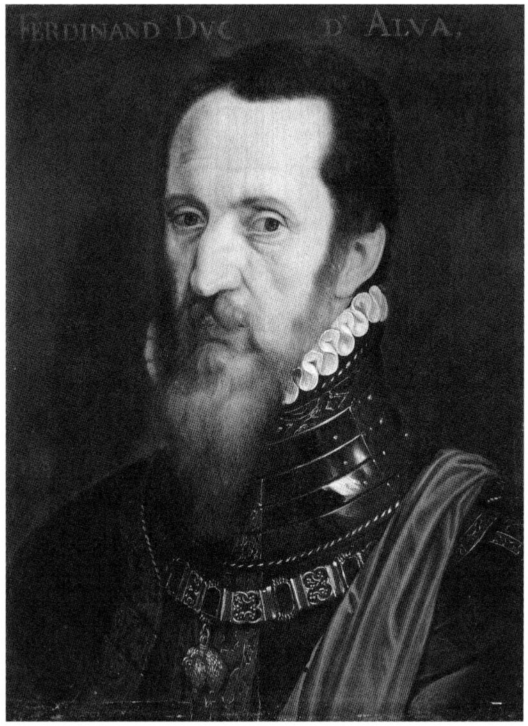

FIG. 35 Willem Key, *Portrait of Fernando Álvarez de Toledo, Duke of Alba*, ca. 1620 (copy). Photo: Rijksmuseum, Amsterdam.

The strange fixation on Alba as the most reviled nemesis of the Dutch Republic meant that lifelike likenesses of the former governor were invested with uncanny powers. In his artist biographies, Van Mander includes an anecdote about the Antwerp painter Willem Key, who had been commissioned to make a portrait of the duke (fig. 35). Van Mander describes Key as a "noble spirit gifted by Nature . . . a very good portraitist after life, achieving naturalness in everything very precisely."[80] This description drew from the earlier artist biographies written by Vasari and the Flemish poet Dominicus Lampsonius: both lauded Key as a "spirited and lively" artist who made art that was "precisely like the real thing."[81] Key's ingenious talent for re-creating his model made the artist vulnerable, however. While Van Mander expresses doubts about the truth of the story, he nonetheless incorporates an account of the artist's final illness and death as a direct result of attentiveness to the physiognomy of his sitter: "Others say that he was so appalled by the cruel appearance of the Duke of Alva that he became ill from it and died; but I think that that is invention."[82] Van Mander follows this account of the artist's death-by-portraiture with Lampsonius's verses: "Key's trained hand rendered human faces so / That one would think one was really seeing them."[83] The close resemblance of the veristic portrait to the sitter's actual face made the depicted person present. While this was a long-standing means to perpetuate the virtues of the wise man, Van Mander suggests that it could likewise convey the vices of the evil man.

Assumptions that veristic portraiture lends itself to image magic trouble the literature on waxworks, although scholars have shied away from overtly supernatural explanations. While magic is invoked, it remains under-theorized.[84] The writings of early modern art theorists such as Van Mander, however, do explain how the animated qualities of the artwork can convey the living presence of the sitter. "Spirited and lively" artists like Key were thought to have received a large measure of God-given spirit at birth. As explored in chapter 2, spirit was considered the source of artistic talent, endowing artists with the ability to create lively works. Animated artworks were thought to contain the spirit of the subject depicted, which could enter into viewers, especially through their eyes.[85] This potent process of spirit transmission between artists, sitters, artworks, and beholders could be dangerous. In Van Mander's anecdote about Key, the spirit of the murderous Alba affects the susceptible artist and kills him. To reproduce the living face of Alba was to engage with maleficent forces, and Key's ability to make art that was "precisely like the real thing" brought about his demise. This anecdote presents an extreme case of the oft-repeated criticism of veristic portraiture: it is not art because it follows life so closely that the artist disappears.

To create a portrait of a political leader was a means of producing a living image of the political body, and Van Mander's biography indicates how this body comes apart under tyrannical leadership. The biographer includes two narratives of Key's demise. The "invention" that explains Key's death as a result of looking at the cruel countenance of Alba follows another account of his final illness, which likewise begins in the studio. While working on the portrait, Key "pretended not to understand a foreign language" and consequently overheard a conversation in which Alba plotted the deaths of Lamoral (Count of Egmont) and Philip de Montmorency (Count of Hoorn). These Flemish noblemen were Roman Catholic servants of the Spanish crown. Together with William of Orange, they had attempted to mediate escalating conflicts between the subjects of the Spanish Netherlands and their king, Philip II. In spite of these efforts and their proclaimed loyalty to the Spanish king, Egmont and Hoorn were accused of treason by Alba, mainly for their lack of military action against the iconoclasm of 1566. They were sentenced to death and beheaded on June 5, 1568. This double decapitation irrevocably severed the political body of the Spanish Netherlands; it led directly to the Dutch Revolt and the eventual establishment of an independent Dutch Republic. According to Van Mander, upon hearing Alba's plot, the artist Key immediately became ill and died on the same day as the counts.

The point of this undoubtedly apocryphal story turns on the close relationship between art, body politics, and violence. The anecdote resonates with a tale told about the head of the body politic, King Philip II, whose distress upon hearing about the iconoclasm in his territories triggered a serious illness.[86] The destruction of art was an attack on the political body of the king, an injury that resulted in a sickened body. In the case of Key's portrait, the *creation* of art could also harm the political body. An entirely truthful portrayal of Alba could only convey the tyrant's destructive energy, thus wreaking damage to the body politic. There is a focus on body parts in Van Mander's narrative. Key's trained hand, his portrait of the living head of Alba, and the beheadings of Egmont and Hoorn all signal traumatic violence. While the making of art could be a potent means of creating a unified and harmonious political body, the precise likeness of a tyrant could bring about degradation and discord.

Artistic representations of Alba came to signify the breakdown of the political body. In 1571, not long after the executions of Egmont and Hoorn, the Flemish sculptor Jacques Jonghelinck was commissioned to make a bronze statue of the duke to be displayed at the citadel of Antwerp (fig. 36). A print of this sculpture shows the armored figure of Alba trampling a two-headed, six-armed hydra representing Netherlandish heresy, rebellion, and iconoclasm. An iconoclast's broken hammer is grasped in one of the hydra's many hands—Dutch violence against art is thus being crushed by a forceful statue glorifying Spain. The bronze political body of the military victor subdues the monstrous body of the rebellious Netherlands, with its proliferation of heads and art-destroying hands.

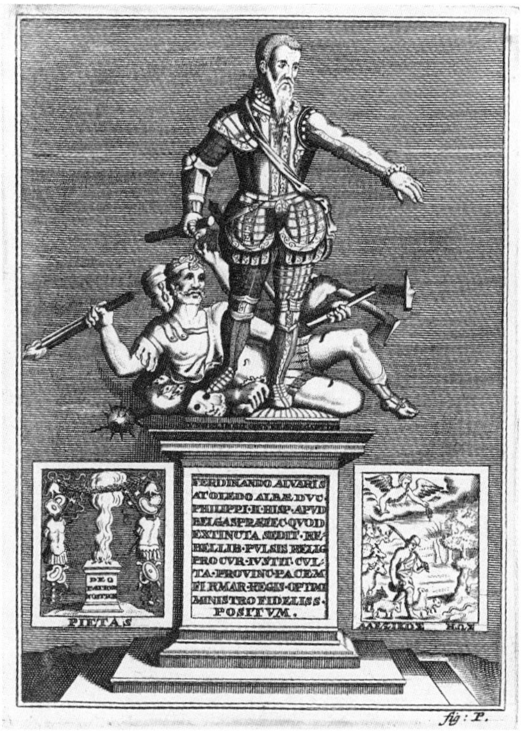

FIG. 36 Anonymous copy after Philips Galle's engraving of Jacques Jonghelinck's sculpture, *Monument to the Duke of Alva of 1571*, ca. 1650. Photo: Rijksmuseum, Amsterdam.

This statue elicited much criticism. Since sculptures of the ruler's body stood for the body politic as a whole, Spanish critics accused Alba of setting himself up in the rightful place of the true head, King Philip II.[87] And Protestant opponents condemned the statue of Alba as an idol erected by the duke to make himself an object of worship. In a letter to Philip II, William of Orange called Alba a "Nebuchadnezzar" who compelled others to worship falsely. The reference to the Babylonian king was a pointed insult. The Old Testament book of Daniel recounts how the despotic Nebuchadnezzar forced the captive Israelites to bow down and worship a golden statue (Daniel 3:1–29). The tyrant had the noncompliant Jews Shadrach, Meshach, and Abednego thrown into a burning furnace, but they were miraculously saved by God. Praising their resistance to idol worship, Calvin writes: "It seemed easier to allow their bodies to be cruelly consumed by the flames of a fiery furnace than to please the king's eye, by bending their thigh for a little before his statue!"[88]

Idolatrous and vainglorious, the statue of Alba also spurred critiques of monarchical authority, focusing especially on how absolute power relied on imagery of the king's political body, as implied by the comparison of Alba to Nebuchadnezzar. A treatise about the conflict in the Netherlands blatantly links kingship and idolatry, drawing on biblical precedents to remind readers that "God became very angry when His people of Israel, unsatisfied with a legal government, wanted to have Kings, just like the pagans had, who would be worshipped by the community like idols."[89]

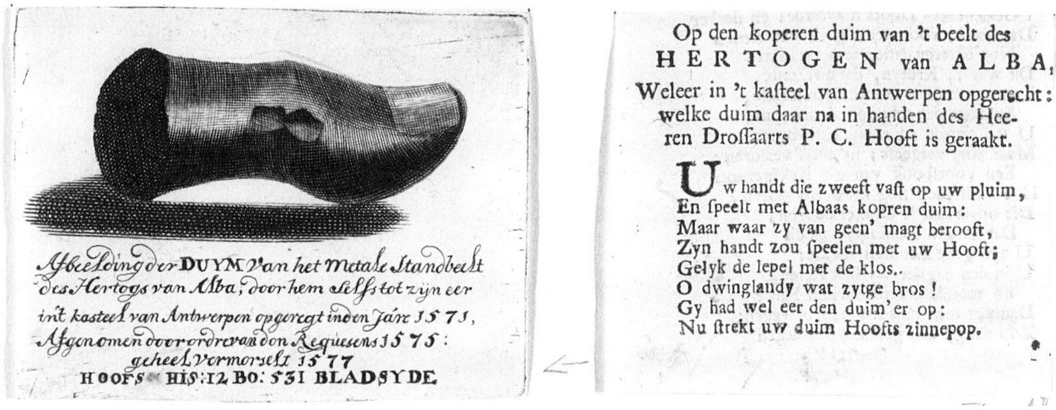

FIG. 37 Anonymous, *Alva's Thumb and a Poem by Joost van den Vondel*, ca. 1719. Photo: Rijksmuseum, Amsterdam.

In a similar vein, the controversial statue of Alba became a rallying point for opposition to the Spanish king, a means to undermine his public image of authority.[90] As a result, it had a short life. Philip II relieved Alba of the governorship of the Netherlands in 1573. The next Spanish governor ordered the removal of the statue, which was smelted down. In this turn of events, the contentious image of a political body crushing iconoclasm was itself destroyed.

Mockery of the notorious sculpture continued long after it was demolished.[91] A small scrap—the figure's bronze thumb—somehow survived and came into the possession of P. C. Hooft.[92] This broken digit points to the disintegration of the tyrant's political body. In 1641, the poet Vondel wrote a poem about the thumb for Hooft, playing with the fact that *hooft* means "head" in Dutch:

> Your hand now plays with Alva's thumb
> But were he still in power,
> His hand would have played with your head [*hooft*]
> As with a ball in a game
> How brittle and fragile tyranny is!
> Once you had the nation under your thumb
> Now your thumb only inspires this poem.[93]

An engraving was later made to illustrate the poem (fig. 37). The salvaged bit of an infamous statue represents how the Dutch escaped being under the thumb of Spanish oppression. A small sculptural fragment thus encapsulates an extended series of historical events. As in Van Mander's account of Key's portrait of Alba, there is a play with body parts in the poem. The severed hand of Alba, which had played so cruelly with the heads of his victims, shows just "how brittle and fragile tyranny is." An intact image of Alba represented the fragmentation of the Netherlands, while dismemberment of the effigy symbolized freedom from oppression.

The display of wax figures of Alba thus promulgated odium for this oppressor and perpetuated the abhorrence of his likeness. The use of veristic effigies to shame an enemy of the state had a lengthy history. Schlosser discusses rites of punishing wax figures in ancient Rome as a means to disgrace criminals.[94] Samuel Edgerton has explored the

Florentine tradition of *pitture infamanti*—defaming or insulting portrayals that were used as instruments of official state punishments.[95] Such imagery had a "fearsome potency," wreaking real damage by denigrating the person depicted. The resemblance of the likeness to the criminal was critical not for purposes of sympathetic magic but simply "so everyone would know that it is he."[96] Displayed in busy parts of the city, such imagery served as a forceful means to subject a convicted individual to public derision and disdain.[97]

Set up next to the Amsterdam image of Alba was a waxwork of the English political leader Oliver Cromwell, likewise denigrated as a despot who had violently dismembered the political body. In 1662, Fokkens describes seeing the "Duc d'Alba and his brother Oliver Cromwell" at the Oude Doolhof, and a 1674 guidebook describes the wax figure of Cromwell as "king murderer and tyrant," holding him responsible for the execution of the Stuart king Charles I, who was beheaded in 1649.[98] A wax effigy modeled from a death mask was made for Cromwell's funeral in 1658 and displayed in Westminster Abbey.[99] Possibly the Oude Doolhof portrait was a replica of this prototype. There are a number of extant copies of Cromwell's death mask. A wax impression, thought to be the face of Cromwell, is currently in the British Museum (color plate 13). It shows how closely wax can capture the texture of porous skin and preserve the contours of a postmortem face.

As with Alba, effigies of Cromwell were widely defamed. In 1660, at the time of the restoration of the monarchy under Charles II, two images were put on trial in Dorset: a figure of Cromwell and one of John Bradshaw, the judge who had ordered the regicide. An account of this mock trial reports, "The judges asked the senseless images [of Cromwell and Bradshaw] whether they would submit themselves to the judgment of the court; and when these did not reply, they were accused of High Treason because of the murder of Charles I. The surrounding crowd shouted Justice! Justice! . . . whereupon they were condemned to hang on two gallows."[100] The effigies aroused the passions of onlookers, who attacked the dummies with pikes, daggers, halberds, and other weapons, finally tearing them to pieces. The restoration of the monarchy was attended by the destruction in effigy of those held responsible for beheading the king. As with the demolition of the bronze statue of Alba, mutilation of the figures of Cromwell and Bradshaw was a means of restoring the body politic to wholeness.

Like the king, an enemy of the state had two bodies, one private and the other symbolic. While the political body of the monarch must live forever, the criminal body has to be eliminated for the good of the totality. As Michel Foucault writes: "In the darkest region of the political field the condemned man represents the symmetrical, inverted figure of the king."[101] The above examples demonstrate the complex relationship between establishing the image of a stable body politic and destroying or deriding the figure of its enemy. An infamous effigy was an effective mechanism to manage the injuries suffered by the body politic by vilifying a specific individual as the cause of troubles. However, by keeping alive memories of the damage done by a hostile adversary, the effigy also exposes the vulnerability of the political body, which in fact shares many of the material properties of the wax figure: it is friable, brittle, ephemeral, and easily broken apart. The tractability of wax and its ability to expeditiously replace a broken body with an intact body double indicates how the deceptive image of an undivided body politic required constant creative manipulation.

Infamy: Starvation of the Body Politic

Perhaps the strangest of the Doolhof waxworks is of the "lying" Eva Vliegen, also called Besje van Meurs, "whom men say did not eat for 32 years" (fig. 38). This work, too, was an image of infamy, made to shame the sitter for a seemingly small deception that attracted a large amount of publicity.

Concerning a historical figure who has been cruelly ridiculed for centuries, the case of Eva Vliegen is both curious and tragic. Born of Roman Catholic parents in 1575, she lived in Vliegenhof, just outside the German town of Meurs (present-day Mörs) in Westphalia. In 1586, the Spanish seized and occupied the town. Sources relate that Eva, who was by then an unprotected orphan, found the countryside unsafe because of the presence of soldiers, which hints at the possibility that—like many women in war-torn regions—she suffered abuse. Seeking a more secure environment, she moved to Meurs in 1592. Impoverished and alone, the young women prayed fervently to God to deliver her from misery and hunger. Her faith in a time of hardship was rewarded, but in an unanticipated manner.[102] Her desire to eat began to diminish so that she eventually required so little food and drink it was a wonder she could stay alive. While we can never know the facts of this case, lack of nutrition as well as trauma and sexual abuse can trigger eating disorders and other forms of emotional and physical distress, and this type of illness is plausibly what befell Eva.

Cultural understandings of eating disorders are historically variable. There was a venerable tradition in the Low Countries of holy women who renounced food and apparently lived for years without eating.[103] In the medieval and early modern periods, the willing denial of food and drink was apprehended as an extreme form of spiritual renunciation and a sign of holiness.[104] Eva's seemingly blessed state accordingly attracted the attention of the townspeople. The mayor of Meurs investigated and was so convinced of her sincerity that he wrote a testimony validating her claims. The Countess of Meurs also tested Eva. She invited the young woman to her castle and offered her food, which Eva refused, saying she had no need of it. The noblewoman then took her into the orchard and offered her a cherry. Pressed by her aristocratic hostess, Eva complied and ate the fruit but immediately became so violently ill that the countess feared she would die.[105] This event only increased Eva's notoriety. The Reformed preacher Conradus Velthuysen, disinclined to believe in anything deemed miraculous, was determined to expose her. He took Eva into his home so that he and his servants could watch her day and night. She was not left alone for even an *oogenblick*, a blink of an eye. Even so, their findings were consistent with those of the mayor and the countess: the minister saw with his own eyes that Eva showed no evidence of hunger or thirst.[106]

For years many people came to Meurs out of reverence for the maid, who was esteemed as a living saint. Eventually, however, the miracle of Meurs was debunked with a great deal of publicity. In 1628, Eva fell ill, and the preacher Theodorus Scriba sent three trustworthy women to care for her and also to espy if she was secretly eating. After two days, Eva asked for milk. A thorough search of the house ensued, and she was discovered to be in the possession of butter, cheese, honey, and conserves.[107] Although Eva claimed that this nourishment was for the poor, she was tried and sentenced to a fine and banishment from the town.

This news was publicized in Amsterdam's courant, normally dedicated to rather dry accounts of items related to international trade and warfare: "In Meurs men have captured a daughter named Eva Vlieghe, who for thirty years let it be known among the folk that she had not eaten / and even though the Countess as well as the Preachers and others paid sharp attention, they could not divine the deceit."[108] The notice also relates that "as many as one thousand people had been deceived." The story was further recounted in historical chronicles of the wars in the Low Countries and disseminated by prints and visual portrayals of the infamous Maid of Meurs.[109] One of these printed portraits depicts a slight and hollow-eyed figure, probably influenced by textual sources that describe Eva as weak and frail (fig. 38).[110] She is modestly attired with a simple head covering and a high-necked, long-sleeved dress. In her hands is a small bouquet of flowers: people said that she survived like the bees, taking nourishment from the scent of flowers and heaven's dew. But do not be misled by the sincere appearance of this timid orphan, the caption warns. Just as the first Eve deceived Adam, the first man, with foodstuff, so this Eva, who seems an innocent dove, is wearing a cunning disguise and deceiving everyone by claiming not to eat.

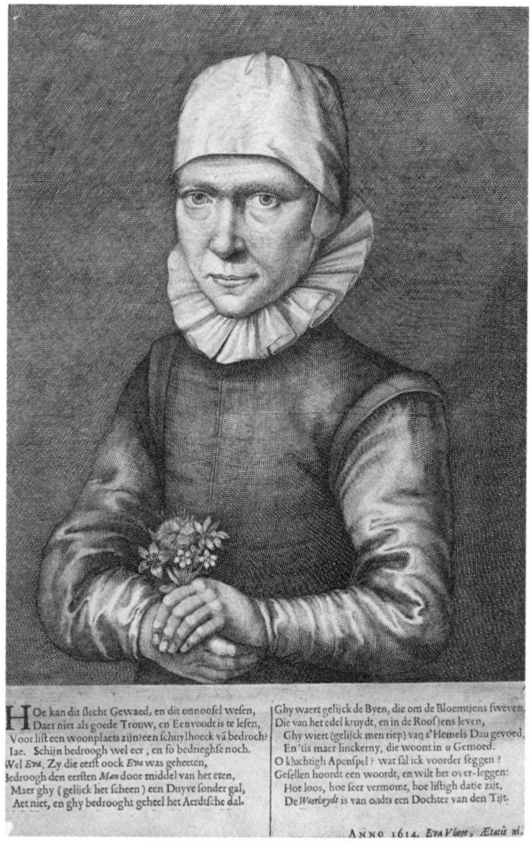

FIG. 38 Andries Jacobsz Stock, after Balthasar Flessiers, *Portrait of Eva Vliegen*, 1614. Photo: Rijksmuseum, Amsterdam.

The amount of publicity and the fact that the incident was reported in the war chronicles implies that the case of Eva Vliegen was of political import. Meurs was a town that changed hands during the wars. Invaded by the Spanish in 1586, it was liberated in a siege led by Prince Maurits and his troops in 1597. Maurits became the new governor of this territory, so Eva's crime was against him. If the body of the criminal was the inverted double of the ruler, then the miraculous powers of the Maid of Meurs ran counter to the political and religious regime established by the Prince of Orange. Obviously, the ongoing presence of female saints and miracles did not fit within the ideals of a Calvinist polity. The discovery that an apparent wonder was actually a sinful deception accorded with Calvin's views on miracles. Citing Augustine, the reformer writes: "Our Lord hath cautioned us against these miracle-mongers by his prediction, that there should arise false prophets, who, by various signs and lying wonders, 'Should deceive (if possible) the very elect.' And Paul has told us, that the kingdom of Antichrist would be 'with all power, and signs, and lying wonders.'"[111] For the Calvinist victors, the Maid of Meurs would seem to fit this description perfectly: she was a speciously innocent deceiver whose lies had fooled even the pastors. She was ridiculed and made an example of because her pseudo-saintly actions caused harm to a body headed by the Calvinist Prince Maurits. It is as though Eva's unnatural appetite starved the body politic itself, jeopardizing its health and wholeness. To mock the effigy of her malnourished body was a ritual means to counter this threat; possibly the rite symbolically renourished the body politic that had suffered depravation during the war.

Wax was a particularly evocative medium for defaming a deceitful wonderworker. Like Eva, the wax portrait appears so true that it is false.[112] In fact the word "sincere" is derived from the French *sans cire*—without wax, the vehicle of deception.[113] The defamatory waxwork thus emphasizes dishonesty. Indeed, only a wax body can remain "most lively" without ever eating or drinking. Like the bees, Eva could supposedly survive on nectar and dew. As we have seen, wax was also a preferred medium for representing the art of mastering the passions, and refusing nourishment was considered an excessive form of self-control. As Coornhert notes: "There are human actions which are diametrically opposed to natural human inclinations—we see this in people who willingly starve themselves to death."[114] By denying the natural rhythms of the body, the waxwork presents an extreme example of a body that is completely constant, appearing alive while in a deathlike state. If the staging of a persona is channeled into being through the medium and technology of the image, then the wax body double is particularly effective for the false hunger artist, as it appears completely true to her lie.[115]

In conclusion, for Schlosser, wax portraiture was an ancient and significant cultural form, embedded in meaningful social, religious, and political rituals until the seventeenth century, when through commercial practices it entered into the "lower social strata as a 'survival.'" With this transition, waxworks became "childish and passé," mere "toys for the amusement of children," and therefore no longer of interest to the scholar.[116] In Schlosser's narrative, the democratization of wax portraiture led to its decline in capitalist societies. As with the folkloric studies of the Doolhoven, there is a perceived declassing through exhibition for a commercial audience, which abases the art. However, in keeping with my approach in *Moving Statues*, this chapter has shown how the Amsterdam waxworks were not just derivative forms displayed for profit and casual amusement. Instead, I have argued that they represent an important transitional moment in the history of wax portraiture. The wax effigy in this context is not an unthinking survival; it develops out of sustained historical engagement with prior traditions and demonstrates a profound understanding of how the various possibilities of the wax portrait could be adapted inventively to serve specific social and political agendas of the day. Trading on the powers of the wax figure and its previous ceremonial practices, the Amsterdam exhibition parks created a new kind of public ritual space for this highly efficacious form of political imagery.

TIME MACHINES IN THE GOLDEN AGE

Chapter 7

The *Kairos* of Clockwork

Awareness of a *kairos* is a matter of vision. . . .
It is not a matter of detached observation but of involved experience.
—Paul Tillich, *Systematic Theology*, vol. 3, 1963

This concluding chapter takes up the representation and experience of time in future-oriented exhibits of elaborate clockwork. The main focus is the thematic program at the Nieuwe Doolhof and its particular revelations about prophecy and the unfolding of world history. A telling way to begin our consideration of these complex temporalities is by taking another turn through the labyrinth. To walk the multicursal maze is a spatial practice of time travel. When lost in its strictures, wanderers have to double back, reentering corridors already trodden to make sense of them anew. Manifold pasts are revisited, and new paths wind in unanticipated directions, weaving unexpected connections between future and past. Jorge Luis Borges describes the labyrinth as an analogy for the problem of time itself. Its branching paths are forks in time, and at each juncture, the wanderer is confronted with alternatives; choosing one way means foregoing other possibilities. In this manner, writes Borges, we create the future.[1]

The moment of hesitation at each intersection is a pregnant pause where action is required: opportunities are presented, and decisions have to be made. The chance that offers itself in the moment must be seized. The Greek term for the opportune time is *kairos*. A moment that necessitates action and transformation, kairos is a turning point; it compels a decisive change of direction, a reorientation that determines the future course of events.[2] In the Nieuwe Doolhof labyrinth garden, visitors found themselves within a kairotic space that demanded critical action in revolutionary times.

Mechanical Visions "For Such a Time as This"

The proprietors of the Nieuwe Doolhof were Lutheran immigrants from Frankfurt, the clockmakers David Lingelbach and his son Philips. They purchased the business on Looiersgracht from the widow of Bargois in 1639 and later relocated to a larger property on the Rozengracht in 1648. Clearly inspired by the idiosyncratic layout of the Oude Doolhof, their publicity pamphlets suggest a proposed itinerary that moved from a central courtyard with an impressive fountain into a labyrinth where maze treaders had to find the "right path." After prevailing in the trial of the labyrinth, participants would "see a door opening." Upon entering, they stepped directly into an automated performance of the "magnificent dream vision of the prophet Daniel."[3]

This mechanized biblical prophecy presented the interpretive key to the Nieuwe Doolhof exhibits. A Jew exiled in Babylon, Daniel was brought to the court of King Nebuchadnezzar, who had selected, from among the captured Israelites, "children in whom was no blemish, but well favored, and skillful in all wisdom, and cunning in knowledge, and understanding science, and such as had ability in them to stand in the king's palace" (Daniel 1:4). As it had for Daedalus and King Minos, cunning knowledge equipped Daniel to serve the ruler. Daniel is a far less ambivalent figure than the crafty Daedalus, however. Daniel resolutely condemned the idol worship instigated by Nebuchadnezzar. His wisdom and skill unmistakably came from God and were used in God's service, making him a fitting biblical figure to preside over an exhibition site that displayed moving statues fashioned with the understanding of science. Indeed, the Lingelbachs' curatorial decision to focus their Nieuwe Doolhof on Daniel may reveal unease with the Bacchus-inspired paganism of the Daedalic Oude Doolhof. The moving-picture shows at the Nieuwe Doolhof mostly presented biblical scenes. Combining practical knowledge of horology with Lutheran theology, the Lingelbachs designed a distinctive exhibition program that conveyed Protestant interpretations of mechanical clockwork.

Notably, Daniel possessed the gift of "understanding in all visions and dreams" (Daniel 1:17), and his prophecies were performed by more than one clockwork invention at the Nieuwe Doolhof. The opening exhibit implemented the prophet's prediction of the rise and fall of the four monarchies of the world. This moving-picture show began with four winds that raged against one another, followed by four terrible beasts that emerged in turn from the sea to engage in combat. The pamphlet explains that the first to appear was a lion with eagle's wings; it represented the king of Persia. Next came a bear with long teeth, figuring the ruler of Greece. Then emerged a leopard with four heads and four wings on its back, embodying the Roman empire. The final and most terrible creature had fangs and ten horns: this monster is described as the "Great Turk," the present-day "monarch of Mahomet."[4] A printed image accompanies this explanation (fig. 39). It shows a stage set with the robed figure of the prophet standing in the left foreground. He looks toward two angels positioned on either side of a curtained archway. As the curtains parted, spectators beheld, along

FIG. 39 Crispijn van de Passe, *The Dream of Daniel*, from *Dool-hof, Staende op de Roose-Gracht* (Amsterdam: voor Philips Lingelbach, 1660). Photo: Allard Pierson, University of Amsterdam Library Special Collections, OTM O 60–2758.

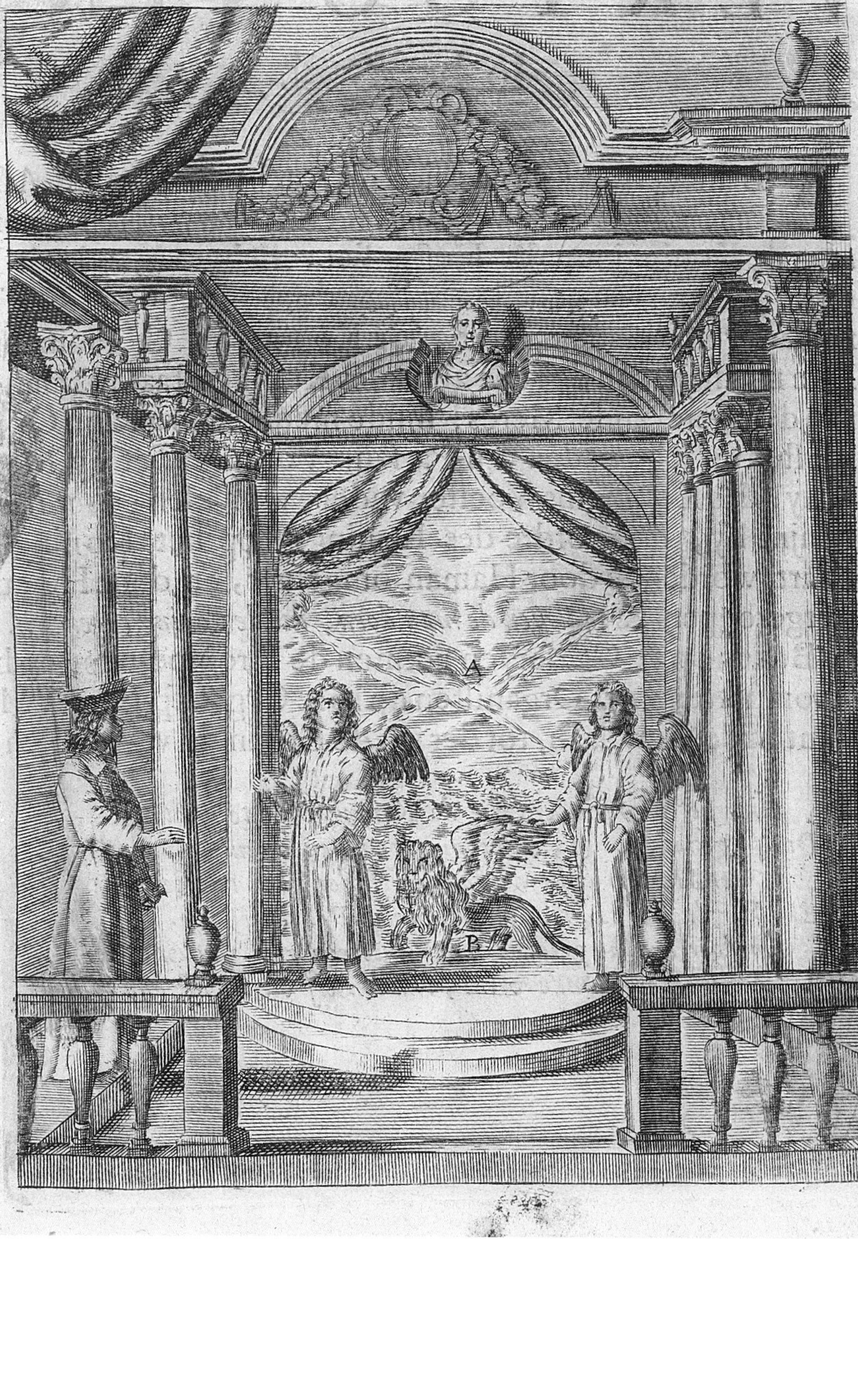
A
B

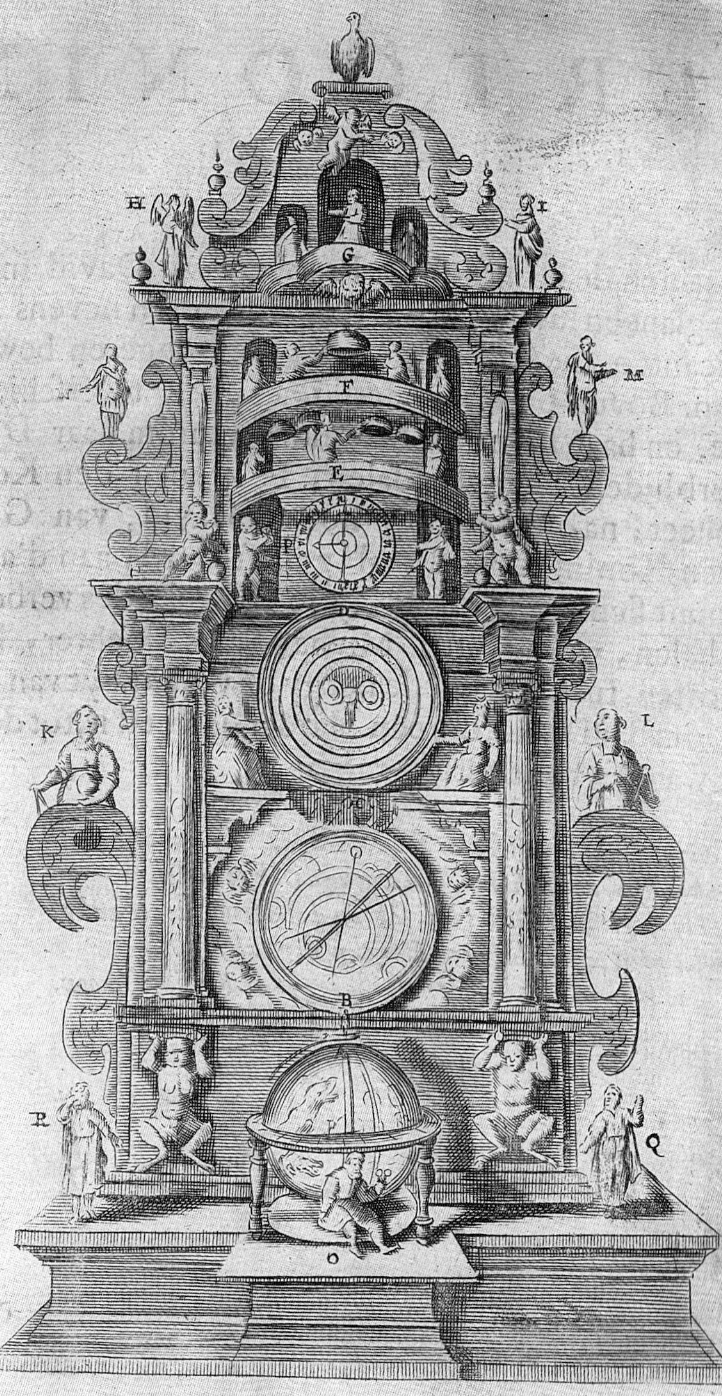

G
H
I
M
F
E
P
D
K
L
B
R
Q
O

with Daniel, the animated vision that the angels revealed. Through the arch, they witnessed the four winds coming together while a mechanically driven winged lion clambered out of the automated rolling waves. The prophecy of Daniel was presented in a series of sequential scenes; as the four beasts followed, one after the other, to arise from the depths, spectators experienced successive revelations of the vision in terms of actual movement through time.[5] The pamphlet explains: "They climb out of the sea and are exhibited to you exactly as if Daniel was telling his vision."[6] An Old Testament prophecy, activated by mechanical arts, played out in the present moment. The serial movements of the artwork and the sequential time of Daniel's prophecy unfolded in concert as a moving mechanical vision.

Besides the moving-picture show of Daniel's dream, the gallery also exhibited "an extremely artful clockwork" that likewise set the prophecy in motion. This astronomical clock tracked the course of the heavens and marked the passing of years, months, days, hours, and minutes (fig. 40). It also traced the movements of the sun and moon and could predict solar and lunar eclipses. The guidebook illustrates different parts of the clock with a point-by-point description of its main features and how they relate to Daniel's prophecy. At point *D*, the pamphlet explains, "the world's transformation is here represented through the dream that Daniel had while he was in Babylon at the court of the King of Persia." At point *E*, Daniel's vision of the four monarchs of the world played out again through an automaton of each beastly ruler appearing every fifteen minutes to strike the quarter hour. At point *F* are Satan and the world's most notorious tyrants, who came out to chime the half hour. The text notes that this sounding of the clock bells marked the inevitable downfall of every earthly despot.[7] Point *G* designates Daniel's vision of "the Ancient of Days," a figure dressed in white that emerged from an arch at the top of the clock to be crowned by a descending angel. A gesturing automaton of Daniel himself appears at point *I*. The letter *H* marks the arrival of the angel Gabriel to announce the birth of Christ, heralding the fulfilment of time. There were also automated musicians and children who play musical instruments, keep time, and turn hourglasses. Female personifications of the Liberal Arts, holding a globe, a set of compasses, and a triangle, appear at the sides of the clock, as does a writer of history at point *M*. At the bottom is Atlas with the world on his back. Flanking him are the ancient philosophers Democritus and Heraclitus, who laugh and weep at the state of the world. At the very top, a mechanical crowing rooster crowns the show, proclaiming the dawn of a new day.

As Grafton and others show, this type of astronomical clock was considered the vanguard of European technology in the seventeenth century. While the ancients may have devised timekeepers such as sundials, water clocks, and hourglasses, the mechanical clock was vaunted as a truly modern invention.[8] The main function of the astronomical clock was not accurate timekeeping: such works could be erratic, and they included many extraneous capabilities like cosmological predictions, mechanical music, and automatized theater.[9] They also were not truly new: knowledge of the mechanisms had been transmitted by craft practices and

FIG. 40 Crispijn van de Passe, *Astronomical Clock*, from *Dool-hof, Staende op de Roose-Gracht* (Amsterdam: voor Philips Lingelbach, 1660). Photo: Allard Pierson, University of Amsterdam Library Special Collections, OTM O 60–2758.

manuscripts from China, Egypt, Byzantium, Persia, India, and throughout the Arab world before being taken up in late medieval Europe.[10] Even though they were problematic timekeepers and indebted to foreign ingenuity, these kinds of clocks were chiefly valued as wonders of European craftsmanship. Astronomical clocks decorated the towers of town halls and churches in many European cities, most famously the cathedral clock in Strasbourg. They were a form of public technology seen by large numbers of people, an essential means to demonstrate the ingenuity of early modern European cities.

Indeed, the towers of so many of Amsterdam's city gates and public buildings were adorned with chiming clocks that the Englishman Edward Brown, who visited in 1668, noted in his travel journal that at every hour, one heard "a fine harmony of clockwork throughout the city."[11] Brown followed this observation with a description of the Doolhoven as music houses where you could pay a *stuiver* to see many beautiful waterworks and different movements by clockwork, paintings, and other divertissements and pleasures.[12] The Amsterdam civic histories singled out mechanical clocks as the masterworks of these eclectic exhibits.[13] Besides the astronomical clock in the gallery, there was also a clockwork tower in the main courtyard of the Nieuwe Doolhof (fig. 41). It demonstrated movements of the heavens and observed the incremental passage of time, marked by the striking of bells and the crowing of a mechanical rooster. The pamphlet includes a poem about this work, which proclaims, "Here you see that Time brings forth the course of the heavens, and that the clock strikes months, years, and days / and on the hour and half hour, a wonder in Europe / That each part in its time, does its work."[14] Amsterdam's numerous and various urban clocks made visible and audible the innovative status of the city.

Astronomical clocks were conceptual objects. The examples in the Nieuwe Doolhof overtly reference the intellectual arts by representing the quadrivium of arithmetic, geometry, music, and astronomy—all fields of learning that inform the making of a clock. Like automata, elaborate clocks were intellectual artifacts designed to challenge viewers with the labyrinthine difficulty of their intricate workings. Time takes shape in the mechanical clock as a visualizable object of knowledge, introducing audiences to complicated technological and temporal understandings of the world.[15] Historians of technology like David Landes argue that early modern clocks brought about a "revolution in time." By ordering the punctual and self-disciplined lives of mercantile Protestant city dwellers, they ultimately ushered in the modern industrial age of standardization and automation. Landes associates post-Reformation innovations in clockmaking mainly with Protestant cities. The sixteenth-century centers of invention were Strasbourg, Augsburg, Ulm, and Nuremberg, German city republics dependent on trade and craft production.[16] The Lingelbachs brought their know-how of this advanced craft tradition with them when they immigrated to Amsterdam.

The astronomical clocks at the Nieuwe Doolhof thus fit within an influential historical narrative that traces the increasing rationalization of temporality in early modern Protestant societies. The aim of this chapter, however, is to explore how

FIG. 41 Crispijn van de Passe, *Courtyard of the Nieuwe Doolhof*, from *Dool-hof, Staende op de Roose-Gracht* (Amsterdam: voor Philips Lingelbach, 1660). Photo: Allard Pierson, University of Amsterdam Library Special Collections, OTM O 60–2758.

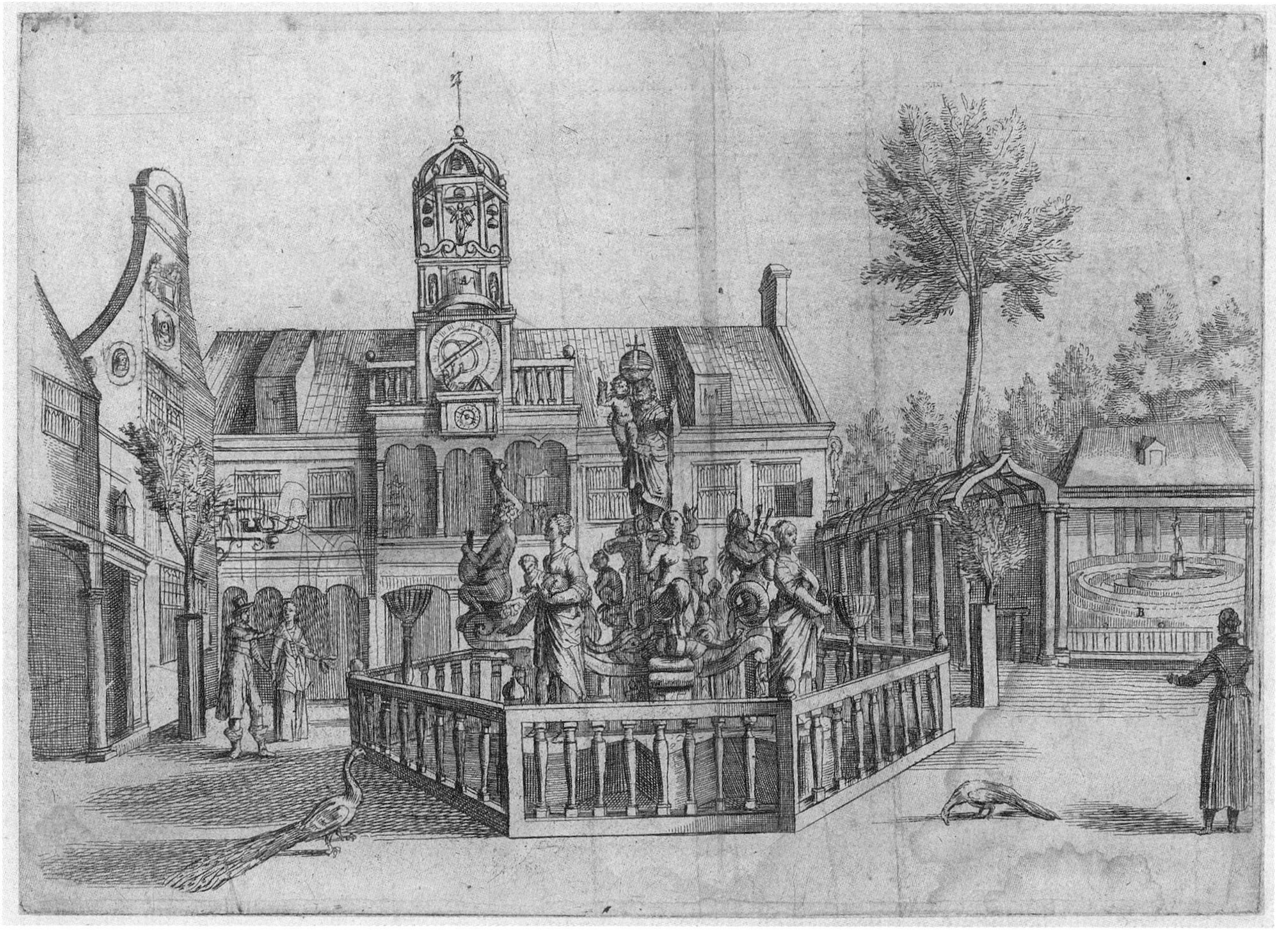

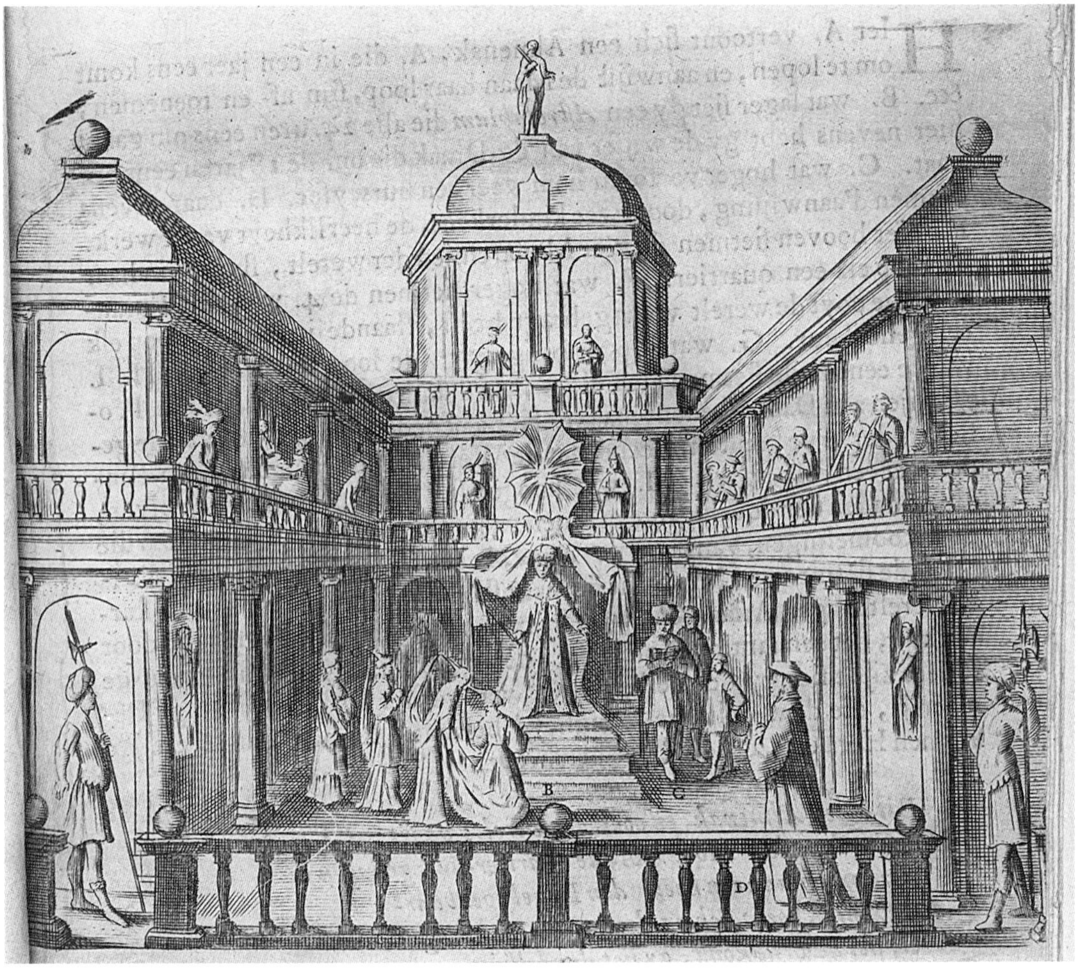

FIG. 42 Crispijn van de Passe, *Queen Esther*, from *Dool-hof, Staende op de Roose-Gracht* (Amsterdam: voor Philips Lingelbach, 1660). Photo: Allard Pierson, University of Amsterdam Library Special Collections, OTM O 60–2758.

these devices concurrently complicate linear and progressive understandings of time. Early modern clockwork cannot just be assessed as modern, secular, and forward moving. Although it does manifest a particular Protestant perception of time, this temporal mode was not conceived in terms of a straight and regular passage but rather one more akin to the winding ways of the labyrinth—multicursal, multitemporal, and kairotic.

Writing about the complicated workings and iconography of astronomical clocks, Otto Mayr raises an important point: “How meaningful these astronomical indications were to the casual viewer is hard to judge.”[17] The Nieuwe Doolhof examples and guidebook explanations provide compelling evidence of the interpretations that were conveyed to viewers via this form of clockwork. Of significance is how the Doolhof clocks enfolded the tempo of *chronos*—the undeviating, measured passing of days, hours, minutes, and the predictable movements of the sun and moon—within the mysteries of eternal, sacred, and cosmic temporalities. While these clocks insistently kept time, they concurrently transported viewers to a time out of time by calling up the hiatuses and ruptures of dreams, prophecies, and apocalyptic apparitions that cannot be contained by the regular temporal measures of earthly clocks. Kairos emerges from within these openings in time.

The Nieuwe Doolhof's prophetic clockwork presented a vision of world history that tracked the downfall of various powerful empires through the efforts of the faithful who seized the opportune moment for critical action. The first beast to rise and fall in the automated vision of Daniel—the winged lion—represents the Persian empire against which Daniel himself had prevailed. The exhibit immediately following the dream of Daniel was a moving-picture show of the story of Queen Esther, also a member of the Jewish diaspora exiled in Persia who challenged the authority of the empire (fig. 42). The exhibition of this Old Testament story in proximity to the display of Daniel's prophecy demonstrates the importance of kairos in eschatological understandings of time.

The mechanical figures in this moving performance acted out the kairotic turning point in the life of the queen, as Esther risked death to enter unbidden into the presence of her husband, the Persian king Ahasuerus. Esther was compelled to speak on behalf of the Jewish exiles who were threatened with extermination by the king's high official Haman. This act required great courage and meant that the queen had to reveal her undisclosed Jewish identity. As Esther readied herself, she declared, "So will I go in unto the king, which is not according to the law: and if I perish, I perish" (Esther 4:16). The heroine recognized that the right moment had come for her to take action. As her kinsman and guardian Mordecai exhorted her, "And who knoweth whether thou art come to the kingdom for such a time as this?" (Esther 4:14). Emboldened by such an understanding of the higher purpose of her life, Queen Esther moved decisively to mobilize the possibilities of kairos, the opportune time in which she could act to save her people and change the course of history.

The guidebook's description and image of the scene give a sense of how this narrative was presented as a moving mechanical picture show (fig. 42). The palace is depicted as a miniature stage, with two levels and a number of arched and curtained doorways. At the center is the king, flanked by his high officials and servants. In front of them is the queen. Accompanied by her ladies in waiting, Esther moves unfalteringly to enter unbidden into the presence of the king. Probably her small costumed figure glided along a track and then hinged at the waist to perform a jerky, mechanical curtsy or bow. Upon her unanticipated entrance, the king turns his head from side to side—the suspenseful moment in which he decides how to react to this audacious breach of protocol. He then rises from the throne and finally holds up his scepter as a sign of favor. The figure of Haman enters next, sliding in to join the king and queen in a feast that she has had prepared for them. Playing out in front of this scene, audiences saw the conclusion of the story, in which Esther's kinsman Mordecai is honored by the king: his small figure, robed in royal apparel, rides by on one of the king's own horses. Haman, meanwhile, is hanged on the gallows he had built for Mordechai. In the gallery above, various figures emerge from the curtained archways accompanied by music. The guidebook describes trumpeters, organists, and musicians who "strike the beat until the time of the monarch is ended."[18] The cadence of this musical performance moves toward end-times; its steady tempo signals the inevitable fall of the Persian empire.

This mechanized performance of the story of Esther provides a vivid example of what Dutch art theorists called an *oogenblikkige beweeging*, the transformative power of an artwork that intervenes in the present moment by changing the ethical outlook of viewers. Moving pictures present several moments of a persuasively charged narrative in a compressed time frame. Different from the visual representation of a single moment, this type of artwork literally moves in time. The movement, or beweeglijkheid, of the moving-picture show has kairotic potential. It generates emotional involvement to prompt timely action. Indeed, Queen Esther's courage and foresight had particular significance for the struggle against Roman Catholic Spain, in which the Dutch defined themselves as God's chosen people, the new children of Israel.[19] The story of Esther was often represented in paintings and theater plays, encouraging emulation of Esther's moral courage in standing up to the forces of a mighty empire.

Notably, kairos, seizing the right moment in time, also describes a practice of classical rhetoric, which involves using techniques of persuasion to galvanize an audience.[20] Like the persuasive Queen Esther, who convinced the king to save her people, an effective performance or work of art had to make the most of its opportunity to change the outlook of its interlocutors. The affecting force of the animated imagery that implemented the queen's actions conveyed the imperative that viewers, too, had to act with courage and conviction—for the time was at hand.

If Esther could alter the path of history, then so could Doolhof audiences. The urgency of the present moment is emphasized in a didactic poem printed in a Nieuwe Doolhof guidebook. Probably written by one of the Lingelbachs, the poem warns that an ongoing battle against the fourth and final beast, a terrible and alien monster, was still being fought. The book of Daniel describes the fourth ruler as a dreadful and exceedingly strong creature that would devour the whole earth and break it in pieces. This beastly monarch would "change times and laws: and they shall be given into his hand until a time and times and the dividing of time" (Daniel 7:25). The visionary revelation resonates with the structure of labyrinthine exhibition sites whose forking paths divided time and sheltered a hybrid beast. Daniel predicted that the last of the terrible tyrants would wage war against God's elect and would seem to overpower them: "Until the Ancient of Days came, and judgment was given to the saints of the most High; and the time came that the saints possessed the kingdom" (Daniel 7:22). As Daniel's vision of the arrival of the Ancient of Days (God himself) was enacted by moving automata on the astronomical clock (see fig. 40), the timed striking of bells measured chronos even as it signaled the demise of tyrants and thus sounded the kairotic opening toward end-times, when God and the Christian saints would prevail. The final line of the guidebook poem relates this message directly to the audience, bringing them into visionary times: "Blessed are those who will see God / He will appear on the earth at the culmination of long centuries."[21]

Significantly, the Nieuwe Doolhof interpretation of Daniel's dream departs from the Old Testament account by emphatically identifying the world's last and most powerful tyrant as the "Great Turk," the Ottoman sultan. The Lingelbachs were no doubt drawing on the influential interpretation of Martin Luther, who turned to the seventh chapter of Daniel to make sense of the expansion of Muslim forces into Europe. After the invasion of Hungary in 1520 and the siege of Vienna in 1529, Luther issued this warning: the Turk "has good fortune in waging war against Christians and usually gains the upper hand and obtains the victory . . . [but] here in Daniel it is announced beforehand that Christians are punished here on earth on account of their sin and the innocent are made into martyrs."[22] In the reformer's view, the victories of Islam were God's punishment against abuses within the Church. The ascent of the Turk signaled that the Apocalypse was impending: "The world is running faster and faster, hastening towards its end, so that I often have the strong impression that the Last Days may break before we have turned the holy Scriptures into German. For this is sure: there are no more temporal events to wait for according to the Scriptures. It has all happened, all has been fulfilled—the Roman Empire is finished, the Turk has come to the peak of his power, the power of the Pope is about to crash—and the world is cracking to pieces as though it would tumble down."[23] The rise of the "Great Turk" was interpreted as an urgent call to repentance and prayer; the imminence of the end of time demanded action and change, especially the purification of the church. The decisive time of kairos had arrived, and it required nothing less than a Reformation.

Millenarian expectations about the victory of the Reformation were prevalent in the Dutch Republic.[24] According to this interpretation, history would come to fulfilment when Protestantism had triumphed over all other faiths—Roman Catholicism, Judaism, and Islam especially. As noted in chapter 5, the intensive study of world languages, cultures, and beliefs had this aim in view: an increase in knowledge would support the mission of conversion and conquest. When this was accomplished, the day of the Lord would arrive as prophesized by Daniel.

The specific interpretation of the dream of Daniel given at the Nieuwe Doolhof was thus an exhortation to the elect not to flag in the fight against the seemingly dominant empires of the Ottomans and Habsburgs. The automata performances vividly presented assurance of the prophet's vision: spectators could watch the rise and fall of Islam, actually seeing it pass away before their eyes just as they could witness the Persian, Greek, and Roman empires ascend and descend in turn, accompanied by the striking of clock bells. The impressive workings of advanced horology together with apocalyptic imagery persuaded audiences of the certainty of the mechanical prophecy. The Doolhof's intricate clockwork artifacts presented this series of events in a programmed sequence, condensing time so that the entire passage of history advanced before the spectators' eyes. Participating in these multisensory shows, they could perceive that the moment approaching them surely was the advent of the Last Days.

If kairos is "a point in time filled with significance, charged with a meaning derived from its relation to the end,"[25] then the striking of mechanical chimes signaled the critical moment, challenging visitors to act on the knowledge that time and the prophecy were coming to fulfilment and the end of time was nigh. The message conveyed by these types of clocks, therefore, is that even the most powerful empires and the most forward-moving and sophisticated human-devised technologies were destined to pass away in the final convulsions of the world. The ticking of the mechanical clock, the wonderwork of modern ingenuity, triumphantly marked its own inevitable demise.

Protestant Pansophia

The Doolhof clockwork did not just represent or remind audiences of the end of time—their advanced technologies actively ushered it in. In the critical age before the end, vigilant efforts by the faithful were required in all fields of endeavor; for Reformed practitioners, the aim of all learning and experimentation was to establish the kingdom of God on earth. The practices of the new philosophy were understood as part of a Protestant revolution that would ultimately perfect the world, bringing it to completion. The two most important prophets of this millenarian view of science and knowledge were Comenius and Bacon.[26] Notably, both thinkers used the image of the labyrinth to articulate this vision. Comenius's widely read utopian book, *Labyrinth of the World and the Paradise of the Heart*, presented the ideal of a Reformed society perfected by an increase of knowledge and education.[27] Bacon's call for the advancement of learning was likewise taken up by Protestant intellectuals across Europe. As Charles Webster puts it, Bacon detected "a rational purpose behind the serpentine windings and intricate mazes of history."[28] For both Bacon and Comenius, the confusing labyrinth of this world, whose forking paths divided time and history, ultimately signified God's complex and inscrutable plan. While world history was experienced as an indecipherable maze by those walking in it, from God's viewpoint, it formed a flawless pattern, providential and preordained.

The conviction that the Reformed religion should preside over all aspects of human activity, including scientific experimentation, was visually conveyed at the Doolhoven by the life-sized wax effigies of Protestant rulers arrayed on a high ledge of the exhibition room, as if overseeing the mechanical marvels.[29] These stable Protestant princes stood in opposition to the transient tyrants who rose and fell in the clockwork displays. In fact, many of the Protestant leaders depicted were significant supporters of the new science. Henry IV of France, for instance, employed the engineers Tommaso and Alessandro Francini to create the famed hydraulic works and automata displays for the palace gardens at Saint-Germain-en-Laye. Queen Christina of Sweden was another influential patron, famously bringing Descartes to work for her at the Swedish court. And the Doolhof exhibits especially highlighted the wise leadership and guidance of the House of Orange in relation to their own experimental endeavors. The House of Stuart, too, was connected to innovative scientific activities. Bacon's *Great Instauration* was dedicated to King James I, proclaiming, "and surely to the times of the wisest and most learned of kings belongs of right the regeneration and restoration of the sciences."[30] Bacon advocated that the revival of knowledge had to be in service to the king: an astute ruler could best seize opportunities to deploy the potential of the sciences for social and moral improvement.[31] To buttress this view, he cites Proverbs 25:2: "The glory of God is to conceal a thing; the glory of the king to search it out."[32]

Bacon especially esteemed the Old Testament king Solomon as a model for contemporary rulers.

Solomon was renowned for his wisdom and for ambitious works like the building of the temple in Jerusalem. *New Atlantis* describes a utopian Christian society that had at its center Solomon's House, an institution for scientific experimentation.[33] Inspired by the Hebrew king, Bacon describes the aim of this edifice: "The end of our Foundation is the Knowledge of Causes, and Secret Motions of things; and the Enlarging of the bounds of Humane Empire, to the Effecting of all Things possible."[34] Bacon exhorted King James to consider King Solomon as his exemplar, urging:

> That you who resemble Solomon in so many things—in the gravity of your judgments, in the peacefulness of your reign, in the largeness of your heart, in the noble variety of the books which you have composed—would further follow his example in taking order for the collecting and perfecting of a Natural and Experimental History . . . , that so at length, after the lapse of so many ages, philosophy and the sciences may no longer float in air, but rest on the solid foundation of experience of every kind, and the same well examined and weighed.[35]

It was the weighty responsibility of the king, guided by the example of Solomon, to oversee and improve human knowledge.

The legendary wisdom of Solomon was foregrounded by the Doolhof displays. The *Derde Heerelijcke Vertoonige* (Third delightful show) at the Oude Doolhof presented Solomon's palace (fig. 43). In an automated sequence of events, the Queen of Sheba enters, accompanied by an entourage. As she arrives, four musicians in the gallery strike up their instruments. In her own manner, we are told, the queen does reverence to the king's palace and the folk. The king then looks in amazement at her, turns his head to look around, rises from the throne, and shows her the scepter. The Queen of Sheba accordingly pays homage to Solomon and then discourses with his majesty. Point *F* on the print designates the start of another series of actions. Arriving with the queen are twelve youths—six young men and six young women—all dressed as maids. The queen presents them to Solomon as a riddle to test his legendary wisdom. How will he be able to discern which are men and which are women when all are dressed in costly women's clothing? The king calls for a container of water and orders the young people to wash their hands. The men promptly plunge their hands into the basin, but the women are more demure and just touch the water with the tips of their mechanical fingers.[36] Duly impressed, the Queen of Sheba pays homage to Solomon and leaves the palace followed by her retinue, who bow to the king as they depart. The booklet sums up the entertainment by saying that everything was in motion and accompanied with artful organ playing and music.[37] In watching this sequence of events, spectators could discern the perspicuity of the ruler. Possibly a connection was drawn between the mechanized Solomon and the effigies of Protestant leaders arrayed above, whose waxy faces and glassy eyes seemingly surveyed the exemplary actions of the wise king and the advancement of mechanical arts.[38]

The prophet Daniel was another inspiring figure for Bacon. The 1620 frontispiece of Bacon's *Great Instauration* depicts a ship sailing through the Pillars of Hercules, conceived as the ancient boundary of the known world. The caption references Daniel 12:4: "But Thou, O Daniel, shut up the words, and seal the book, even to the time of the end: many shall run to and fro, and knowledge shall be increased." In Bacon's interpretation, Reformed scholars were unsealing the book of Daniel in the

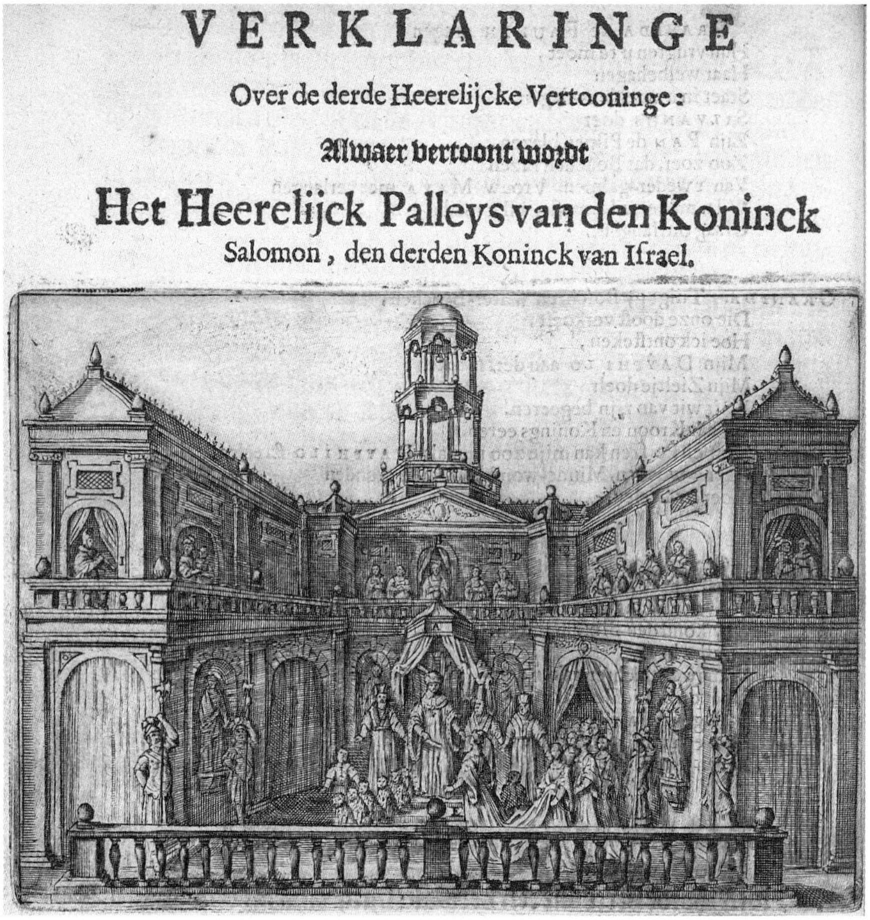

FIG. 43 *The Princely Palace of King Solomon*, from *Verklaringe van Verscheyden Kunst-Rijcke Wercken en haer Beweginghe* (Amsterdam: Dan. Davidsz. Mansvelt, 1657). Photo: Allard Pierson, University of Amsterdam Library Special Collections, OTM O 62–2615.

final age before the Apocalypse. The end-times would be blessed by a burgeoning of knowledge. Voyages of discovery and trade ("Many shall run to and fro") especially created conditions for an unprecedented influx of new information and ways of knowing the world.[39] Invoking this verse, Bacon writes that it is "as if the opennesse and through-passage of the world, and the encrease of knowledge were appointed to be in the same ages."[40] As the seventeenth-century expansion of navigation extended the known world, the sciences flourished. This opening up of space was also an opportune opening in time: the kairotic final age held "wonderful potentialities offered by God to man."[41]

The Nieuwe Doolhof thus took people out of routine tempos to reveal to them, as in a dream, the immediacy of Daniel's prophecy in the current age, as evidenced by the great increase of knowledge and craft on display in Amsterdam, presented as the world's foremost seafaring city. In keeping with the precepts of Baconian science, the Doolhof inns as well as Theunisz's tavern D'Os in de Bruyloft contained the workshops of highly skilled Protestant artisans who actively engaged in pushing beyond inherited wisdom through the practical arts of experimentation.[42] The Nieuwe Doolhof booklets claim that although people could read about the great ingenuity of ancient masters who made extremely refined works, "we do not notice anything comparable made by the Greek and Roman masters to these artful inventions."[43] The Lingelbachs had brought "an art that has long been in darkness into the daylight."[44] "Antiquity has left us many wonders, which through texts have been committed to our memory," claimed the Lingelbachs, and the aim of their establishment was to use art to re-create these ancient wonders so that they could be known by an eager public of burghers and foreigners.[45]

These exhibitions can thus be situated within a global Protestant reform movement whose objective was "to teach all things to all people in all ways."[46] This utopian objective of bringing human knowledge to completion and making it universally available was described by Comenius as "pansophia." Using mechanical experimentation to engage in the advancement of learning guided by wise Protestant leaders, the objective of the Nieuwe Doolhof was in keeping with the Reformation ideal of taking the world to its final state of perfection by seizing the wondrous potential of the moment to bring to fruition time itself.

The designation of a targeted audience of eager burghers and foreigners implies the two-pronged approach of these exhibits. Mechanical wonders were designed to make Dutch burghers see that they lived in an exigent age requiring urgent engagement with the wider goals of the Reformation. When directed at foreigners, these works showed them the technological superiority of Amsterdam and the inevitable downfall of every non-Protestant power. Chiming clocks, which purportedly struck foreigners with wonder, were set up to reveal to them the error of their ways as the striking of bells accompanied the demise of automated foreign rulers. In this manner, the Doolhof clockwork, as well as the clocks and automata that were sent around the world as diplomatic gifts and tribute, were likely intended as conversion machines. They were an advanced technology of persuasion meant to bring the Reformation to fulfillment in all the earth.[47] Like Trojan horses, the impressive exteriors of these gifts secreted the potential to bring about the downfall of all opponents of Protestantism.

Machines in the Golden Age

A pastoral or Edenic atmosphere was evoked in the city's Doolhof gardens. Sources describe the quiet restfulness of sitting in the courtyard underneath green linden trees, hearing the splashing of fountains accompanied by music, twittering birds, and bleating sheep and goats. The earliest surviving Nieuwe Doolhof poster suggests this idyllic environment (fig. 44).[48] At the center of its crowded imagery is an elaborate fountain most likely created by the Huguenot inventor Bargois; we can recognize trademark features such as water jets that spout from basins to push aloft balls through which the water sprays. On top of a rocky mount in the middle of the fountain are a rampant lion and an orange tree, symbolizing the glory of the Dutch Republic led by the House of Orange. To either side of this patriotic work are rural settings: on treed hills, deer and other animals graze, and various birds, including a peacock, roam in a meadow while rabbits pop out of holes. Probably these creatures were mechanical works that moved and made sounds. Further enhancing the tranquil mood was the "fountain of Orpheus," another innovative work made by Bargois, whose name is featured prominently together with the nationalistic slogan "Vive Orange" (Long live the House of Orange). The poster thus publicizes how Protestant rulers and accomplished craftspeople achieved both pastoral serenity and technological innovation.

The ideal of bringing about a new golden age of peace, prosperity, and perfect knowledge motivated the work of Reformed scholars and artisans across Europe.[49] The concept of the Golden Age was significant for Amsterdam, where civic chroniclers drew on Ovid's *Metamorphoses* to describe their own epoch in terms of wealth, learning, and social harmony. Ovid denotes the Golden Age as the first era, a time when "a carefree people enjoyed a life of soft ease." It was perpetually springtime and the earth bore fruit on its own so that people did not have to labor. The pristine world at the dawn of time was without arts, tools, or machinery, an age before navigation and exploration when "mortal men knew no shores but their own."[50] The Doolhof exhibits thus implemented a peculiar strategy: they employed innovative mechanical arts to re-create a Golden Age so idyllic that such inventions were neither needed nor yet invented.

The Orpheus fountain illustrates how this approach operated (fig. 44). The statue may have been made by De Keyser, as it is strikingly similar to a small-scale bronze by the artist (fig. 45).[51] The work represents Orpheus playing a violin to tame Cerberus, the three-headed dog that guards the entrance to Hades. The fountain depicted in the poster appears to rotate on a circular platform that revolves around a central cylinder. The back of a nude woman probably represents Eurydice, whom Orpheus will follow into the underworld as the platform turns. At the center of the cylinder, we can just make out Bacchus riding in on a wine barrel. Orpheus was a follower of the wine god, and the perfect harmonies of his Bacchic hymns legendarily had the power to charm even the wildest beasts. Bacon claims that learning and the arts were like the music of Orpheus, bringing harmony to society and human behavior. He describes the cohering powers of Orpheus: "All beasts and birds assembled: and forgetting their several appetites; . . . stood all sociably together listening unto the ayres and accords of the Harpe." So too could Orpheus tame men "who are full of savage and unreclaymed desires; of profite, of lust, of revenge."[52] The manner in which modern inventiveness, ancient learning, and music came together in the spellbinding fountain of Orpheus endowed it with the capacity to change its audience from savage beasts into a sociable and civil assembly. The persuasive sounds, movements, and advanced artistry of the work were designed to create social harmony.

An adjacent scenario titled *'Tlans Welvaeren* (the land's welfare) relates the Orpheus fountain directly to the Dutch Republic. This unusual hydraulic work comprises several ships, which appear to rotate round a cylinder. Around the edges of the central scene are many more ships; it looks as though there were mirrored walls that reflected the main work, creating the illusion that innumerable moving boats were sailing out to increase the good fortune of the United Provinces. Sophisticated modern technologies crafted

FIG. 44 Hessel Gerritsz, *Nieuwe Doolhof on Looiersgracht*, ca. 1632. Photo: Amsterdam City Archives (010097011866).

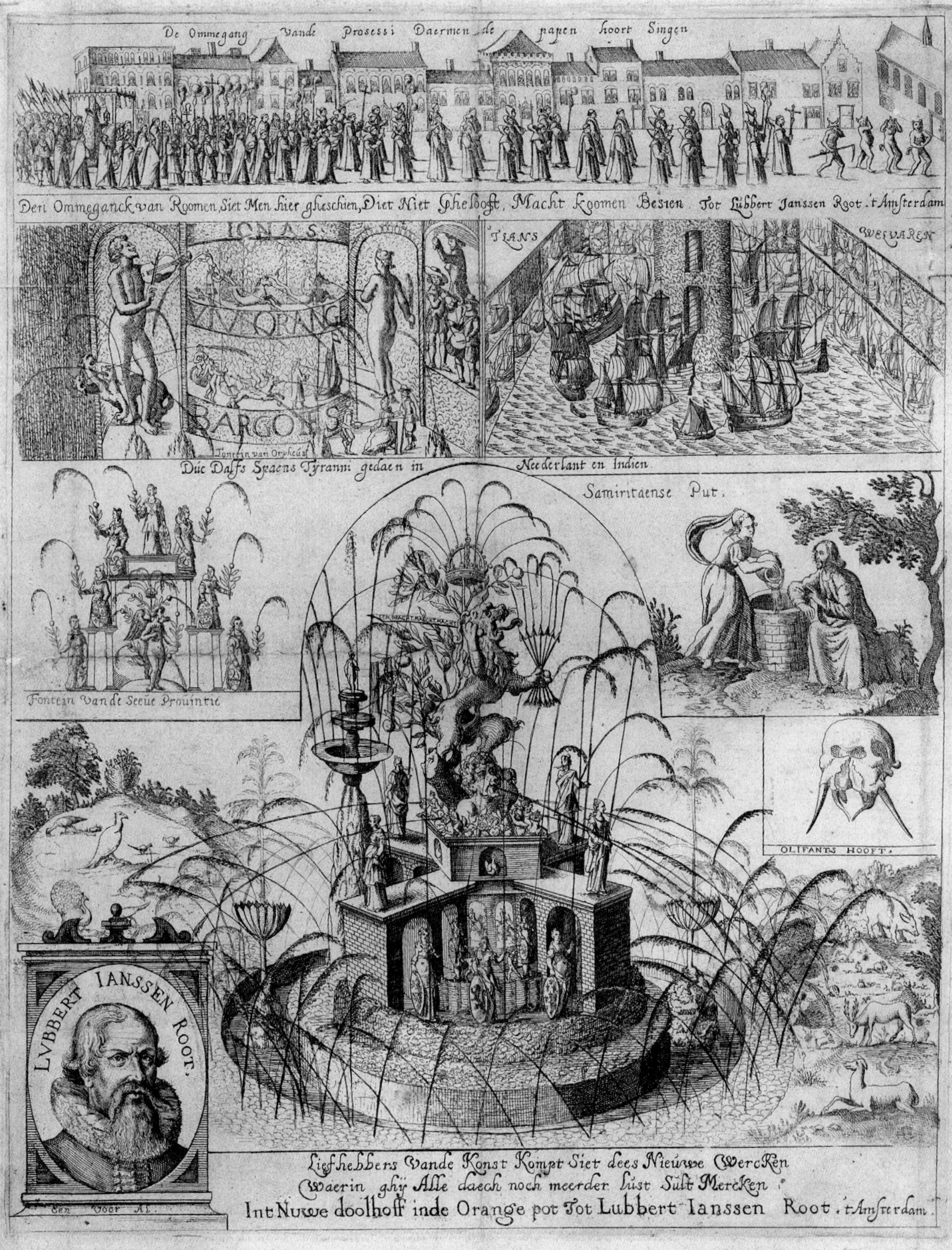
De Ommegang vande Prosessi Daermen de papen hoort Singen
Den Ommeganck van Roomen, Siet Men hier gheschien, Diet Niet Ghelooft, Macht Koomen Besien Tot Lubbert Janssen Root 't Amsterdam
VIVE ORANGE
Fontein van Orpheus
WESVAREN
Duc Dalfs Spaens Tyranni gedaen in Neederlant en Indien.
Samiritaense Put.
Fontein vande Seeue Prouintie
OLIFANTS HOOFT.
LVBBERT IANSSEN ROOT.
Een Voor Al
Liefhebbers Vande Konst Kompt Siet dees Nieuwe WercKen
Waerin ghy Alle daech noch meerder lust Sult MercKen
Int Nuwe doolhoff inde Orange pot Tot Lubbert Ianssen Root 't Amsterdam

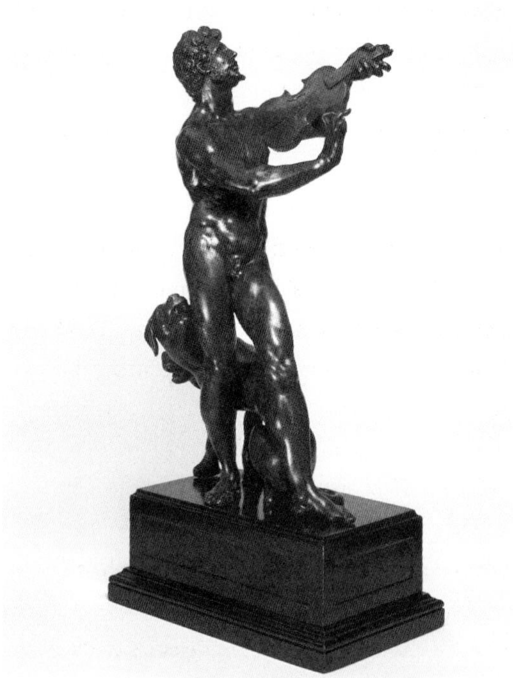

FIG. 45 Hendrik de Keyser, *Orpheus and Cerberus*, ca. 1610–15. Photo © Victoria and Albert Museum, London.

an atmosphere where music, fountain work, and moving statues together fashioned an immersive experience in which visitors experienced a return to a Golden Age that held great promise for the future of a prosperous seafaring Protestant Republic led by the House of Orange.

Almost every show at the Nieuwe Doolhof had pastoral resonance. A musical display of an automated King David playing his harp included a dance of shepherds and shepherdesses because, the pamphlet reminds us, "David was a shepherd."[53] The theme of melodic concord was developed further in the seventh show, another musical performance presented by shepherds and shepherdesses who played the flute and sitar. The booklet promises that their hands and eyes moved in a lifelike fashion. These music-making figures remained highlights of the Nieuwe Doolhof for at least a century. The Marquis de Courtanvaux, who visited in 1767, singled out two automata of a shepherd and shepherdess that played various airs for the flute with moving fingers and turning eyes.[54]

The pastoral also was evoked through references to popular theater traditions. At the Oude Doolhof, automata performed scenes from P. C. Hooft's play *Granida* in an elaborate moving-picture show.[55] The shepherd Daifilo is searching for his lover, Dorilea, when he happens upon the Persian princess Granida, who had lost her way while hunting. Daifilo invites Granida to drink from a fountain and brings the water to her in a seashell. The princess drinks and finds the water better than the costliest wine served at her father's royal banquets. She falls in love with the shepherd; however, they are banned from the court and so return to the forest to live the uncomplicated life of a shepherd and shepherdess. The moving picture is described in detail in the pamphlet, which highlights diverse points of interest: an artful fountain where the lovers meet, the god Pan playing his pipes, a rustic background with bleating sheep and scurrying rabbits, and a series of comic masquerades, such as a hunchback playing the *rommelpot* (friction drum) and an ape strumming a lyre.[56] The show ended with the arrival of John the Baptist, who preached his message to a listening crowd.

The success of Hooft's *Granida* created a vogue for the pastoral mode. Numerous artists, including Dirck van Baburen and Gerrit von Honthorst, produced paintings derived from the play. Allison Kettering explores the demand for imagery of *Granida* and scenes featuring shepherds and shepherdesses in Arcadian settings. She concludes that while there were similar pictorial traditions in other countries, such imagery was most prolific in the Dutch Republic. Traditionally a courtly and aristocratic genre favored by the European elite, the pastoral was adopted by Amsterdam's wealthy merchants, especially in the 1620s and 1630s. Kettering suggests that representations of an untroubled relationship to the land gained popularity as Amsterdam's merchant class invested in land-reclamation projects, immensely ambitious technological undertakings to drain inland lakes and marshes in order to create new agricultural land, thus transforming the rural environs of the city on an unprecedented scale.[57] The urban expansion of Amsterdam, as we have touched on above, was another enormous enterprise; the lengthy reconstruction brought much disruption to everyday life in the city and converted surrounding pastures into new urban neighborhoods. The pastoral mode allowed for nostalgic withdrawal from the drastic changes that contemporary technologies were wreaking on urban and rural spaces. The popular genre afforded city dwellers an imaginary green space in which to contemplate the humble ways of shepherds and shepherdesses dwelling in the tranquil, unchanging countryside of an unspoiled earlier age.

Pastoral time is regressive; it runs counter to progress narratives that celebrate technology and development. The ancient ideal at play is *otium*: stepping back from the complexities, demands, and artificiality of civilization. Such dreams of seeking refuge from the corrupted world were a form of social critique.[58] In representing a space and a time free from the pressures of urban life and its mechanized developments, evocations of a carefree age implied that humanity was happier living in a society untroubled by technology, navigation, and artificial inventions. As Leo Marx argues, machines disrupt the peace of the garden. The interruption of new innovations in pastoral settings introduces change and progress, which signal the loss of a golden age defined by the repudiation of human ambitions and inventions.[59]

Paradoxically, the machinery that set the idyllic Doolhof exhibits in motion simultaneously caused the Golden Age to dissipate. Of importance for understanding this contradiction is consideration of how new technologies were integrated into society. Rustic subject matter creates the illusion that comprehension of iconographic codes (shepherds and shepherdesses, grazing animals, birds, leafy trees, green pastures, fountains, and musical instruments, etc.) or recognition of well-known narratives and characters (the stories of Granida or Orpheus) entailed an understanding of these appealing clockwork displays. Machines were presented using theatrical devices, appearing like actors on the curtained stage to persuasively divert their audiences.[60] By drawing on historical genres of representation and shaping spectators' comprehension with familiar plots that were explained in oral commentaries and printed guidebooks, these highly scripted shows expedited the acceptance of new technologies.[61] In the vacillation between nostalgia and novelty, machines were assimilated into social life. Pleasurable evocation of the classical ideal of retreat into an idyllic age distracts from the forward-moving workings of the mechanisms. As audiences looked to the bucolic past, future-oriented, machine-driven devices disguised in historical costume found a kairotic opening in time and slipped into the social world. By inventively bringing history to life, these time-traveling motion machines traversed a mazelike route, turning back through the fabled Golden Age as a diverting way to enter and alter the future.

The importance of *otium* and enjoyment of the mythic past in these exhibitions prompts reassessment of well-established historical narratives about the functions of automata and mechanical clockwork. For historians of science and technology, the seventeenth century marked a turning point in the journey of the automaton from antiquity to modernity. In a pioneering article, Derek de Solla Price places the historical development of automata on a "road of evolution," which runs "straight and steady." On this upwardly ascending path, Descartes "stands on a height scaled and begins the ascent to the next plateau"—mechanistic philosophy.[62] Subsequent studies adopted a similar teleology of linear time travel. Mayr explicitly links the display of clocks and automata with mechanistic philosophy and argues that such exhibits changed the mental habits of seventeenth-century viewers, teaching Europeans how to think mechanically as they reimagined the workings of the cosmos, the state, and the human body in terms of centralized clock mechanisms.[63] For Schaffer, automata are technologies of rationalism; as "symbols of modernity," they were analogies for controlled human behavior, part of the civilizing process that aimed to produce the regulated and disciplined bodies required by Enlightenment engineering and industrialization.[64] As noted, Landes connects urban clockwork to the punctuality and discipline of Protestant societies in particular. Peter Dear expands on this view, importantly discussing the automaton in terms of the early modern ideal of cultivating the self, linking automata exhibits to the impact of the philosophies of both Lipsius and Descartes in the Netherlands. The aim of mechanical exhibits, Dear concludes, was to direct human behavior, especially along lines of reason, consistency, efficiency, discipline, and self-control.[65]

These scholars are certainly correct to stress how automata entered into early modern discourses about rational subjectivity and mechanistic philosophy. In keeping with these insights, we have seen how the Doolhof exhibitions were established with the utopian aim of creating subjects who were in step with modern civic developments and pan-European Protestant concerns about the utopian advancement of learning and the conversion of all peoples to the Reformed religion. There is scarce documentation of seventeenth-century automata displays, which makes the Amsterdam Doolhoven a particularly informative case for study, as they complicate the notion that the mechanical arts clicked along a straight path of progress to modernity. Instead, a variety of diverting clockwork artifacts were designed to bring forward the end-times. The pastoral automata show of *Granida* closed with the appearance of a mechanized John the Baptist, raising his arm to address an attentive crowd. This combination of scenes may seem random until we recall the prophetic words of the Baptist. His is the voice of one crying in the wilderness "and saying, Repent ye: for the kingdom of heaven is at hand. . . . Prepare ye the way of the Lord" (Matthew 3:1–3).

The predominant Reformed understanding of the kingdom of heaven was that the end of time was a return to the start of time. In his treatise *A History of the Constancy of Nature*, the Polish Calvinist scholar John Jonston, who had studied at Leiden University, expresses the belief that although the earth had lost something after the Fall of Adam and Eve, this event did not mean that the world "doth perpetually run to worse and worse."[66] As Jonston advocates, "We may be able to do more than any age before us. . . . We shall honour our Ancestors the more by endeavouring to go beyond them."[67] Drawing from the works of Bacon, Jonston articulates the idea that modern learning was moving toward completion, restoring the earth to its prelapsarian state. Man would then reattain the complete understanding and dominion over nature that Adam possessed before he fell into sin and the earth went into decline.[68] Jonston cites a passage from Seneca that articulates this view of time:

> All things but for time adjourn,
> And whats past, joyes to return,
> Nor is there made any thing,
> But the end and beginning,
> Touch each other in a ring.[69]

The poem evokes the paradox expressed by the form of a labyrinth: *in my beginning is my end.* The unspoiled past and the completed future touch each other because both are the everlasting kingdom of God that exists beyond earthly time and human effort.

Comenius articulates this ideal in his *Labyrinth of the World and Paradise of the Heart.* The laborious journey through the labyrinth of this world takes place in secular time, but the ultimate aim of any mortal endeavor should be to find a kairotic opening to the paradise of the heart, eternal and perfect truth, which is not contained within worldly space and time. Pastoral automata accordingly moved toward a future in which the perfect past was always already there: in these works, the ancient and the promised Golden Age came together in a ring. With hindsight, the Doolhof shows appear to progress from antiquity to modernity according to the logic of clock time, but they were, in fact, designed to achieve the fulfilment of prophecy, aiming to bring about their own obsolescence by moving history toward eternity.

For Protestant audiences and clockmakers, the ideal of a perfected earth—a new Golden Age—could only come to fruition with the downfall of every non-Reformed religious group worldwide. The seventeenth-century approach was reform: the increase of knowledge and education was directed toward the realization of spiritual and political uniformity within the Dutch Republic, which it would extend universally through its global dominion over the whole earth. Today's critiques of the celebrated Dutch Golden Age are thus well founded.[70] A number of museums no longer use the term to designate the art and culture of the seventeenth century. Scholars, too, are interrogating the nationalistic concept of a Golden Age. The main objection is that the term celebrates ideals of prosperity and peace while denying how the wealthy and powerful in Dutch society profited through practices of global warfare and colonial violence; the expropriation of land and resources; and the murder, oppression, and enslavement of peoples worldwide.[71] "The Golden Age" has become a highly contentious term, with many arguing that to designate the period as a "Dutch Golden Age" excludes diverse peoples and perspectives from the history of the Dutch Republic. These accounts are justified and could go even further. The Dutch Golden Age does more than exclude or overlook; historically, the concept served as a means to actively devalue and eradicate plural worldviews. As we have seen, the machinations of the Golden Age aimed to completely eliminate non-Reformed religions and diverse cultures in order to achieve a perfected Protestant world at the apocalyptic close of time.

EPILOGUE

Obsolescence

In conclusion, instead of imagining Amsterdam's exhibits of artistic and technological devices ascending steadily on the "road of evolution," we can perhaps better liken their trajectory to the brash escape of Daedalus and Icarus from the twisting coils of the labyrinth: the arc of technologically enhanced birdlike flight tracks the dazzling potential of human inventiveness and abruptly bifurcates in the sudden plunge of its breakdown. Bacon concludes his interpretation of Ovid's tale of Daedalus by predicting the failure of early modern innovations, writing that all mechanical arts were undone by their own vain pretensions; they "lose their reputation in tract of time; grow contemptible and perish, through their over-confidence, like Icarus; being commonly unable to perform what they boasted."[1] Acknowledgment of the limitations of human work was integral to seventeenth-century discourse about experimental arts and sciences. The labyrinth, as Bacon notes, was an accurate metaphor for the mechanical arts, as it indicated the ambiguous double uses of human-made creations, which "serve as well to produce as to prevent mischief and destruction; so that their virtue almost destroys or unwinds itself."[2] The predominant understanding was that all earthly inventions—even the most sophisticated—were limited, fleeting, and imperfect; they were never an end in themselves but only played a small part in the unfolding of God's perfect plan.[3]

Needless to say, the ambitious aims of the Doolhof exhibits did not come to fruition. With hindsight, the truest prophecy was Bacon's astute prediction that inventive devices would surely "lose their reputation in tract of time; grow contemptible and perish."[4] One after the other, the Doolhoven were shut down. The last to close its door was the Oude Doolhof, which was sold in 1862. It was completely demolished, its contents vended at public auction.

Regrets about this closure were publicized in a printed lament about the Oude Doolhof's final performance titled "A Show of Despair."[5] This poem addresses the various personages exhibited at the Doolhof. Starting with Goliath, the text warns him that his enormous wooden body would be fed to the furnace, a final fate "even worse than that day when you were slain by David!" But there was some consolation: Goliath could rejoice in the fact that David would meet a similar end or possibly be displayed at the fair dressed as "a Muhammadan" before being sent with ridicule to Amsterdam's Jewish neighborhood.[6] The racist sentiment is clear but so is the fact that the millenarian conquest of Judaism and Islam had not been accomplished. The outlandishly costumed android Jochum also faces ignominy and is advised to take to the streets with his bagpipes and become a beggar. Haman, the Persian persecutor of the Jews, is told that he will no longer be hanged on the gallows so he could take his revenge on his Jewish enemies Mordechai and Esther before they were all thrown together into the fire.

In the case of the infamous waxworks of Eva Vliegen, Oliver Cromwell, and the Duke of Alva, their shame would continue. Eva is mocked as

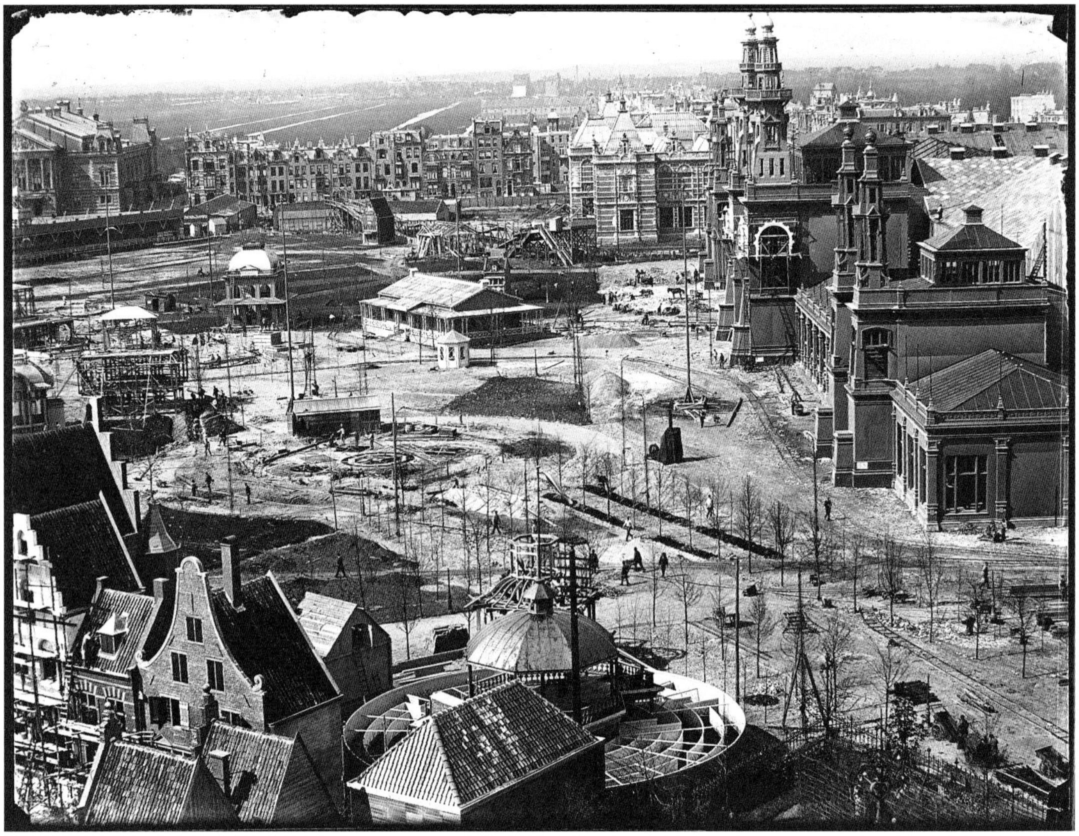

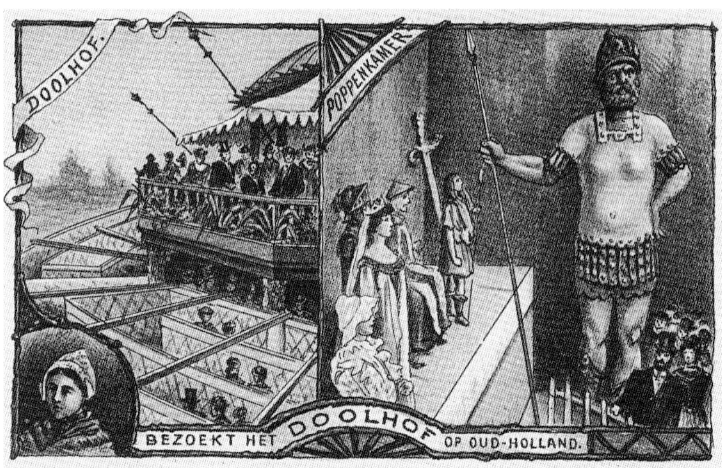

FIG. 46 Jacob Olie Jbz, *Construction of the World's Fair for Hostelry and Travel, Seen from the Tower of the Rijksmuseum*, April 13, 1895. Photo: Amsterdam City Archives (10019A000879).

FIG. 47 *Visit the Doolhof at Oud-Holland*, 1895. Photo: Amsterdam City Archives (ANWE00125000001).

being so ugly that her scrawny figure could be used as a scarecrow, whereas Cromwell's body was only fit to be hung in effigy in a public place and scorned with continuous hatred. As for Alva, who would give even a half-penny for his harmful face, since anyone who saw it would think on the lessons of olden times? The Doolhof's pastoral prophecy of new times was equally passé:

> Grieve, shepherds, shepherdesses,
> The prospect that you await is sorrowful for your spirit!
> For after my courtyard is sold, what then can you begin?[7]

There is little hope of a return to the pastoral golden age here. Insisting on the centuries-old fame and popularity of the Doolhof, this elegy ends with a plea to the city to purchase and maintain a venerable civic attraction with support from an adoring public.

Civic authorities failed to seize this kairotic opportunity, and the public did not take decisive action to save the site. In an ironic turn of events, however, the Oude Doolhof was rebuilt only a few decades after its destruction. Amsterdam hosted the World's Fair for Hostelry and Travel in 1895. In this celebration of tourism, foreign and local exhibition goers were invited to visit the Dutch past. Temporary structures were erected on Museumplein behind the Rijksmuseum. The showpiece was a village called "Oud Holland," made up of facsimiles of historic buildings from various Dutch cities. This row of replicas appears in the left foreground of a photograph of the site under construction, where laborers on scaffolding are working on the facade of a typically Dutch step-gabled house (fig. 46). Nestled between Old Holland's houses, we perceive another sentimental evocation of the past: a wooden version of the Doolhof labyrinth is being re-created.[8] A souvenir lithograph of the exhibition portrays two highlights of a visit to the "Doolhof at Oud-Holland," where visitors could experience the labyrinth and also witness the reunion of some familiar figures (fig. 47). In the reconstructed sculpture gallery—called the *Poppenkamer* (dolls' room), we can make out the surviving wax portraits of Eva Vliegen, Christina of Sweden, and the Chinese man, together with the automata of David and Goliath.

The Doolhof thus lingered in time, reappearing as a mock-up of itself. Its glory days were definitively over, however. The World's Fair village of "Oud Holland" and similarly nostalgic reconstructions of the historical past were disparaged in the press: "For an instant you get an entertaining illusion, but it's not vivid, fulsome life. Ultimately you realize it is only comedy and exhibitionism in the middle of modern society. . . . The new generation has become aware of the contrived, the unreal, the deceitful, the fake without a soul."[9] After the World's Fair site was dismantled, only David and Goliath were maintained by the city. They were first housed in city hall and then at the Amsterdam Historical Museum, established in 1926 to exhibit what detractors called "a collection of castoffs from the Town Hall and Stedelijk Museum."[10] The Doolhof's lone survivors were hailed as the dubious masterworks of this museum of rejects: "While in the Rijksmuseum, the *Nightwatch* very rightly stands in the place of honor, here pride of place goes to an astonishing apparition remembered from youth: David and Goliath."[11] This evaluation of seventeenth-century Dutch culture has had a long afterlife. Rembrandt's painting of a group of Dutch citizen militiamen taking up arms in defense of Amsterdam retains its pride of place. Although the David and Goliath statues conveyed a similar political message about fighting against foreign invaders, for decades they stood in the museum café for the amusement of children and tourists.[12]

The derogatory reassessment of the Doolhof displays returns us to the beginning of this book and its consideration of nineteenth- and early twentieth-century dismissals of the Doolhoven as mere amusement parks for children. As Erasmus notes, the ancient Athenians described superficial cultural forms as having "nothing to do with Dionysus."[13] Without the god of paradox, the Doolhof's experimental attempts to reconcile Protestantism with classical paganism no longer made any sense, appearing as "an entertaining illusion" that is "fake without a soul." Once the spirit of ancient ritual was vanquished, the exhibition of living statues was reduced to an empty puppet show.

NOTES

All translations are my own unless otherwise indicated.

Chapter 1

1. On museums as ritual sites, see the pioneering work by Duncan, "Art Museums and the Ritual," 88–103.
2. More than a hundred works by Rieke are preserved in the Amsterdam City Archives and can be searched at https://archief.amsterdam/beeldbank.
3. Ter Gouw and Van Lennep, *Uithangteekens*.
4. On seventeenth-century street posters, see San Juan, *Rome*, esp. chapter 1.
5. Colie, *Paradoxia Epidemica*, 276; Bryson, *Looking at the Overlooked*, esp. chapter 2: "Rhopography."
6. Clippel, "Rough Bacchus," 99. The painting is in the collection of the State Hermitage Museum in St. Petersburg: https://www.arthermitage.org/Pieter-Paul-Rubens/Bacchus.html.
7. See Weemans, "Herri met de Bles's *Way*," 307–31, and Clippel, "Rough Bacchus."
8. Plato, *Symposium* 215a–b (ed. Waterfield, 60).
9. Plato, *Symposium* 221d–222a (ed. Waterfield, 69).
10. There are two more drawings of the inscription by J. G. L. Rieke in the Amsterdam City Archives.
11. Later in the seventeenth century, the city of Leiden had a public maze at the top of the Burcht at the city center. See Mieris, *Beschrijvinge der Stad Leyden*, 396–98.
12. Vriese, "Nieuwe Doolhof," 354–57, and Vriese, "Nog Enkele Amsterdamse Doolhoven," 82–86. There are no surviving publicity materials for the Franse Doolhof and Roode Doolhof, and it is unclear whether these sites included labyrinths. See Reneman, "'Oude Doolhof,'" 18–20, and Spies, "Amsterdamse Doolhoven," 71.
13. Huizinga, *Autumn of the Middle Ages*, 296.
14. A derogatory tone runs through much of the literature. For instance, De Roever, *Uit Onze Oude Amstelstad*, 2:1–37, and Waal, "Oude Doolhoven," 568–72.
15. Vanhaelen, *Comic Print*, 2–11.
16. Meijer, "Oude Doolhof," esp. 33, 132.
17. The Dutch text is "algemeene vroolijkheid, echt Hollandsch, een beetje ruw misschien, maar ongekunsteld en oneerbaar." De Roever, *Uit Onze Oude Amstelstad*, 2:1–2.
18. Erasmus, *Praise of Folly*, 36.
19. Ten different editions of the various Doolhof pamphlet guidebooks survive, of which there are only fifteen extant copies. They were regularly updated with information and woodblock images of new attractions. For a full inventory, see Reneman, "'Oude Doolhof,'" 75. Many of the travel accounts are catalogued in Jensen, *Reizigers te Amsterdam*.
20. On the connection between the Oude Doolhof and De Keyser, see Scholten, "Beheaded Bust," 838–41; on Vinckenbrink, see Eisma, "Bewegende en Geklede Beelden," 76–88.
21. Passe, *Dool-hof*, n.p.: "Dinghen, di noyt te voor en zijn also bedacht."
22. Eisma, "Albert Jansz Vinckenbrinck," 33–43. The hydraulics are described as "an exclusive and modern technology" by De Jong, *Nature and Art*, 126.
23. See Meijer, "Oude Doolhof," 34–36; De Roever, "Nieuwe Doolhof," 105–6.
24. These phrases are from a Nieuwe Doolhof poster, *De Nieuwe Italiaense Fonteyn*, advertising a new fountain (Amsterdam City Archives, inv. no. 010097004373); the caption at the bottom of the 1632 poster by Hessel Gerritsz, *Nieuwe Doolhof on Looiersgracht*, solicits with this claim: "Lovers of art, come see these new works, and every day you'll notice your pleasure increasing" (see fig. 44).
25. See Chartier, *Cultural Uses of Print*, 3–12.
26. Spies, "Amsterdamse Doolhoven," 72.
27. J. Elias, *Vroedschap van Amsterdam*, 1:20–24.
28. This phrase is from a document pertaining to the founding of the Dutch Academy, promising a third of the profits to the city orphanage: "Agreement between Dr. Samuel Coster and Governors of the Orphanage at Amsterdam, 1617" (Amsterdam City Archives, P.A. 367, nr. 72), transcribed and translated in Brandt and Hogendoorn, *German and Dutch Theatre*, 349.

29. See Vanhaelen, "Local Sites, Foreign Sights." The public functions of early modern taverns are explored by Kümin, *Drinking Matters*.
30. De Jongh, *Tot Lering en Vermaak*, 27.
31. Cats, *Silenus Alcibiadis*. For commentary, see Luijten, Bloemendal, and Felix, "Ad Nova Semper Hians," 292–305, and Somos, *Secularization*, 192. The first Amsterdam edition of Cats's emblem book, with an epigram by Daniel Heinsius, was published in 1620 by Willem Blaeu.
32. Rabelais, *Histories of Gargantua*, 37.
33. Cicero, *Treatise on the Best Style of Orators* 4:528 (trans. Yonge, 1).
34. This observation was made by the British visitor Sir William Brereton (*Travels in Holland*, 57).
35. There is a growing body of scholarship on the lively image. Key studies are Freedberg, *Power of Images*, and Gell, *Art and Agency*. On moving statues, see Gross, *Dream of the Moving Statue*, esp. xii–xiii, 170–72, and Stoichita, *Pygmalion Effect*.
36. Schlosser, "History of Portraiture," 296.
37. Alpers, *Art of Describing*, xxvii.
38. Ibid., xxv.
39. Stoichita, *Self-Aware Image*, and Vanhaelen, "Public and Private Intercourse."
40. Van Hoogstraten insists on the notion of art as an honorable form of deception in *Visible World*, cited and translated with commentary in Brusati, *Artifice and Illusion*, 9.
41. Crary, *Techniques of the Observer*, 39–40.
42. For an overview of some of the theoretical debates, see Holly, "Back of the Painted Beyond"; B. Brown, "Thing Theory"; Bennet, *Vibrant Matter*.
43. Van Eck, *Art, Agency and Living Presence*; Weststeijn, *Visible World*; Jacobs, *Living Image*.
44. Dickey and Roodenburg, "Passions in the Arts"; Van Oostveldt and Bussels, "Sublime and Seventeenth-Century Netherlandish Art."
45. Loh, "Introduction"; Vanhaelen and Wilson, "Erotics of Looking"; Van Eck and Bussels, "Theatricality in Early Modern Visual Art."
46. On automata, see Keating, *Animating Empire*; on wax, see Grootenboer, "Theorizing Wax"; on labyrinths, see Brunon, *Jardin comme labyrinthe*.
47. Weststeijn, *Art and Antiquity*, 1–28.
48. Weststeijn, *Visible World*, esp. 18, 185–88; Roodenburg, "*Beweeglijkheid* Embodied."
49. Willem Goeree, *Natuurlyk en schilderkonstig ontwerp der menschkunde*, cited and translated in Roodenburg, "*Beweeglijkheid* Embodied," 314.
50. Francesco Bocchi, 1592, cited in Weststeijn, *Visible World*, 76.
51. Cicero, *Treatise on the Best Style of Orators* 4:528 (trans. Yonge, 1); see also Weststeijn, *Visible World*, 185–88.
52. The classic study on paradox is Colie, *Paradoxia Epidemica*.
53. Weststeijn, *Visible World*, 154; see also Grootenboer, "Sublime Still Life."
54. Weststeijn, "'Painting's Enchanting Poison.'"
55. On religious controversies about sculpture, see Scholten, *Sumptuous Memories*, 9–22. On the changed status of art after iconoclasm, see Vanhaelen, *Wake of Iconoclasm*.
56. Marcus van Vaernewijck, cited and translated in Heuer, *City Rehearsed*, 14.
57. Scholten, *Sumptuous Memories*, 10.
58. Brandt and Hogendoorn, *German and Dutch Theatre*, 348.
59. Somos, *Secularization*, 190–92.
60. For digital facsimiles and transcriptions of many of Cats's emblem books, see the Emblem Project Utrecht: http://emblems.let.uu.nl/c1627_introduction.html.
61. De Jongh, "Schaduwen van Daedalus," 10–15.
62. S. Morris, *Daidalos*.
63. Bacon, "Daedalus, or Mechanical Skill," 237.
64. Abrahamse, *Grote Uitleg*, 353.
65. Bakker, "Imago van de Stad," 64–65. The Visscher print states, "Ghemeyne Schole van alle Menschen: met welcke name de Stadt Athenae eertijts is ghenoemt."
66. Commelin, *Beschryvinge van Amsterdam*, 742.
67. Passe, *Nieuwen en Vermaeckelijken Dool-hof*, n.p.: "verroering doende elck Beelt, zijn actie of het leefde, tot verwonderingh en veramaeck der aenschouwers."
68. Passe, *Dool-hof*, n.p.
69. Oestreich, *Neostoicism*, 8.
70. Lipsius, *On Constancy*. Lipsius's friend and publisher Christophe Plantin took *Labore et Constantia* as the motto for his press. Voet, *Golden Compasses*, 1:31. Dear links automata to Neostoicism ("Mechanical Microcosm," 51–82).

71. On Neostoicism in art theory, see Weststeijn, *Visible World*, 39, and Dickey and Roodenburg, "Introduction," 10–11.
72. Lipsius, *On Constancy*, 33–36.
73. Montaigne, *Complete Works*, 593.
74. Heinsius, *On Plot*, xiii, 11; see also Bull, *Mirror of the Gods*, 225.
75. Oestreich, *Neostoicism*, 13; Gorski links the tenets of Neostoicism to disciplinary procedures (*Disciplinary Revolution*).

Chapter 2

1. Abrahamse, *Grote Uitleg*, 263–70.
2. Witkamp, *Amsterdam in Schetsen*, 83, 90–91.
3. This line was likely an oral street cry before it was carved in stone. Vriese, "Oude Doolhof," 270.
4. *Verklaringe van Verscheyden Kunst-rijcke Wercken* (Mansvelt, 1657), n.p.
5. Van Mander, *Lives*, 1:21.
6. Ibid., 1:46–49. Pseudo-Bacchus designates the god of beer, a name used in regions where viticulture did not thrive.
7. Duncan, "Art Museums and the Ritual," 88–103.
8. Pliny, *Natural History* 36.19 (trans. Bostock, 6:341).
9. On the Museum of Modern Art in New York City as a labyrinth, see Duncan and Wallach, "Museum of Modern Art," 28–51; see also Bann, "Retour de l'art au jardin," 199–216.
10. *Verklaringe van Treffelijcke Konstige Wercken*, n.p.
11. Fokkens, *Beschrijvinge der Wijdt-vermaarde Koop-Stadt*, 300.
12. Geyl, *Nieuw Nederlands Biografisch Woordenboek*, 641–43.
13. Another printed portrait of Coster was engraved by Jacob Matham in 1602; it was based on a now-lost painting by Cornelis Ketel, one of Amsterdam's most successful portraitists and a friend of Coster. Search: http://research.frick.org/montias.
14. Scholten, "Beheaded Bust," 838–41.
15. An inventory of Coster's art collection can be viewed in the Montias database of seventeenth-century Dutch art inventories. Search: http://research.frick.org/montias.
16. This idea was not new; Pieter Bruegel's patron, Nicolaes Jonghelinck, was the collector of Antwerp's excise taxes on wine, and his garden villa sported a large Bacchus fountain. Porras, *Pieter Bruegel's Historical Imagination*, 82.
17. Eisma, "Albert Jansz Vinckenbrinck," 33–43, and Franken, "Albert Jansz Vinckenbrinck," 72–92.
18. Mountague, *Delights of Holland*, 140.
19. On the vogue for collecting and displaying patterned shells, see Bass, Goldgar, Grootenboer, and Swan, *Conchophilia*.
20. Tolomei, letter dated July 26, 1543, cited and translated in MacDougall, *Fons Sapientiae*, 6, 12–14.
21. *De Nieuwe Italiaense Fonteyn*, Amsterdam City Archives, inv. no. 010097004373. On the Doolhof fountains as innovative works, see De Jong, *Nature and Art*, 126.
22. Reneman, "'Oude Doolhof,'" 12; Scholten, "Beheaded Bust," 840; Eisma, "Fonteinmaker Jonas Bargois," 58, 62. Bargois called the fountains "het Jonas" and "het Bolletje," and he rented them to the Oude Doolhof for 100 guilders a year. To give a sense of the expense, the average annual income of an adult male working in an urban area was around 250 guilders.
23. The treatise contains designs for garden labyrinths. For commentary, see the edited facsimile edition, Van der Groen, *Den Nederlandtsen Hovenier*, and also De Jong, *Nature and Art*, 3–9.
24. On the importance of the source and quality of water for garden fountains, see Tchikine, "'Anima del Giardino,'" 129–33.
25. Unger, *History of Brewing*, 167.
26. De Jong, *Nature and Art*, 125. Given the difficulties of collecting fresh water, it may be that the fountains were only activated at certain times of day, as indicated on some of the Doolhof posters.
27. Berkel, *Isaac Beeckman*, 30, 68, 72, 168. Berkel notes that Beeckman may have known the pneumatic theories of Hero of Alexandria via the book by Renaissance polymath Girolamo Cardano titled *De Subtilitate* of 1550. The first fountain in Amsterdam was probably the one that P. C. Hooft saw in the courtyard garden of the home of the wealthy Jewish merchant and art collector Sion Luz. Hooft wrote a poem about it that has been dated to the first decade of the seventeenth century (Van der Stroom, *P. C. Hooft*, 9–12).
28. Morgan, *Nature as Model*, 31.

29. S. de Caus, *Raisons des forces mouvantes*, 5.
30. Printed translations of Hero's important technical writings appeared in Europe in the late sixteenth century. Bernardino Baldi's vernacular translation, *De Gli Automati* of 1589, was especially important. See Boas, "Hero's *Pneumatica*," 38–48; Marr, "Understanding Automata," 209; Bedini, "Role of Automata," 26.
31. Bargois, a member of Amsterdam's Walloon congregation, was an immigrant either from France or from the southern Netherlands. Eisma, "Fonteinmaker Jonas Bargois," 55.
32. Morgan, *Nature as Model*, 21; Boas, "Hero's *Pneumatica*," 45.
33. S. de Caus, *Raisons des forces mouvantes*, book 2, p. 2, view 51, and book 2, p. 7, view 56.
34. MacDougall, "Ingegnoso artifizio," 61–83. In the north of Europe, the garden fountains in the Munich residence of Duke Wilhelm V of Bavaria employed Ovidian themes to highlight the virtues and powers of the ruler. See Maxwell, "Pursuit of Art," 420.
35. MacDougall, *Fons Sapientae*, 3.
36. Sluijter, "Rembrandt's Portrayal," 285–306.
37. Junius, *Painting of the Ancients*, 341; see also Weststeijn, "Between Mind and Body," 268.
38. Ovid, *Heroides and Amores* 10.69–75 (trans. Showerman, 127).
39. Catullus, *Poems* 64.180–81 (trans. Lee, *Catullus: The Complete Poems*, 91).
40. *Verklaringe van Verscheyden Kunst-Rijcke Wercken* (Gort, 1668), n.p.: "Haer mondt die lachte wel, maer 't oogh dat scheen te huylen," The guidebook attributes this line to Vos, and Spies identifies it as a line from a poem by Vos that was printed together with P. C. Hooft's play *Theseus and Ariadne* ("Amsterdamse Doolhoven," 72).
41. Catullus, *Poems* 64.60–62 (trans. Lee, *Catullus: The Complete Poems*, 85).
42. Heinsius, *Nederduytsche Poemata*. For commentary, see Somos, *Secularization*, 171–92, and Bull, *Mirror of the Gods*, 243–48, 255.
43. Ovid, *Heroides and Amores* 10.148–49 (trans. Showerman, 133).
44. Heinsius's colleague Scaliger owned a Catullus manuscript, which may have gone to Heinsius after Scaliger's death in 1609. De Jonge, *Auction Catalogue*, 3–5.
45. Catullus, *Poems* 64.251–64 (trans. Lee, *Catullus: The Complete Poems*, 95).
46. Heinsius, *Nederduytsche Poemata*, 16: "Ick ben de groote Godt, die droefheyt ende pijn, Doet over al vergaen, de vader van de wijn."
47. Weststeijn, *Visible World*, 185; Massey, "Reflections on Temporality," 187.
48. Even so, the extremely costly fountain was later removed; it does not appear in any of the nineteenth-century images or descriptions of the site.
49. Somos, *Secularization*, 186.
50. Heinsius entered into a correspondence and then a heated debate about paganism with the French scholar Jean Louis Guez de Balzac in 1632. Heinsius's defense of his position was published the same year as *Epistola qua dissertation D. Balzaci*, cited in Bloemendal, "Mythology," 341; see also Spies, *Rhetoric, Rhetoricians and Poets*, 74–76.
51. Cave, "Ronsard's Bacchic Fresco," 63.
52. Pers, *Bacchus Wonder-wercken*. Pers lived in close proximity to Jan Theunisz's famous inn.
53. Mountague, *Delights of Holland*, 39–40.
54. Heinsius, *Nederduytsche Poemata*, 16–17.
55. Erasmus, *All the Familiar Colloquies*, 74–75. Heinsius refers to Erasmus in his preface (*Nederduytsche Poemata*, x).
56. Brereton, *Travels in Holland*, 58–59.
57. Domselaer, *Beschryvinge van Amsterdam*, 309: "Die de gapended Vrouw-luyden veeltijts met haar uytspruytende koele water-stralen van onderen als nuchtere Kalveren doen hippelen, dat de ommestaanders vermakelijk is."
58. MacDougall, *Fons Sapientiae*, 7.
59. See Sherwood, "Magic and Mechanics," 588–89; Riskin, *Restless Clock*, 28; Bedini, "Role of Automata," 20–21.
60. Montaigne, *Complete Works*, 1097–98; Riskin, *Restless Clock*, 29–30.
61. Montaigne, *Complete Works*, 1175.
62. Heinsius, *On Plot*, 13.
63. Ibid., 103.
64. Grotius, *Introduction to the Dutch Jurisprudence*, cited in Filipczak, *Hot Dry Men*, 68.
65. Shakespeare, *Taming of the Shrew*, 302.
66. Heinsius, *Nederduytsche Poemata*, 3: "Lyaeus heet ghy meest, om dat ghy kont ontbinden, Ons sinnen en verstant, de sorgen komt verslinden." See also Clippel, "Rough Bacchus," 106.

67. Bacon, *Moral and Historical Works*, 244; the essay is titled "Dionysus, or Desire."
68. Catullus, *Poems* 64.254–55 (trans. Lee, *Catullus: The Complete Poems*, 95).
69. Thomas Tryon, *The Way to Health, Long Life and Happiness*, cited in Filipczak, *Hot Dry Men*, 73.
70. Porras, *Pieter Bruegel's Historical Imagination*, chapter 3.
71. Bakhtin, *Rabelais and His World*; on the Netherlands specifically, see Bruaene and Bouchaute, "Rederijkers, Kannekijkers," 18.
72. Vanhaelen, *Comic Print and Theatre*. On Bruegel and Bacchus, see Porras, *Pieter Bruegel's Historical Imagination*, chapter 3.
73. Tlusty, *Bacchus and Civic Order*, 5; Stewart, "Taverns in Nuremburg Prints," 97.
74. Ronsard, *Hymne de Bacchus*, cited and translated in Cave, "Ronsard's Bacchic Fresco," 66.
75. On Heinsius's critique of Ronsard's *Hymn to Bacchus*, see Becker-Cantorino, *Daniel Heinsius*, 43.
76. Coornhert, *Ethics*, 53.
77. Van Mander, *Lives*, 2:13.
78. Weststeijn, "'Painting's Enchanting Poison,'" 141–77.
79. Van Mander, *Lives*, 1:42, 2:40, 3:51, and 3:14.
80. This notion is derived from Aristotle. See, among others, Klein, "Spirito Peregrino," 65. For an overview from antiquity to the present, see Smith et al., *Animal Spirit Doctrine*, esp. 20–22.
81. Tybjerg, "Wonder-Making," 443–66, and MacDougall, "Ingegnoso artifizio," 62–64.
82. See the classic studies by Wind, *Pagan Mysteries*, and Bull, *Mirror of the Gods*.
83. Van Mander, *Lives*, 2:323, 3:33–34.
84. Magny, *Hymne de Bacchus*, cited in Cave, "Ronsard's Bacchic Fresco," 62.
85. Witkamp, *Amsterdam in Schetsen*, 83, 90–91.
86. Jolle, *Description de la ville*, 241.
87. The definition is from Christopher Plantin's *Dictionarium Tetraglotton* of 1562; see discussion in Porras, *Pieter Bruegel's Historical Imagination*, 1.

Chapter 3

1. This insight is derived from Carol Shields's novel *Larry's Party*, about a maze maker.
2. Cats, *Houwelyck*, n.p.: "Een hof vol bange vreucht, en duysent vreemde slagen." See also De Jong, *Art and Nature*, 126.
3. Schaefers and Backer, *Doolhoven en Labyrinten*, 9, and Reneman, "'Oude Doolhof,'" 5.
4. See Doob, *Idea of the Labyrinth*, esp. chapter 2.
5. Diehl, "Into the Maze," 284.
6. Kern, *Through the Labyrinth*, 247–65, and Matthews, *Mazes and Labyrinths*, 125.
7. It is uncertain when the first multicursal mazes appeared in the North. There is an archival reference to a labyrinth in the gardens at Cannenburgh Castle near Vaassen: in 1553, a gardener was paid for a "dolhoff," but there is no other record, so it is not known whether the maze was unicursal or multicursal. See Schaeffers and Backer, *Doolhoven en Labyrinten*, 26.
8. Serlio included labyrinth plans in the fourth book of *De Architectura*, published in Venice in 1537; a Flemish version was published in Anvers by Pieter Coecke van Aelst in 1539. This work was reissued in Amsterdam in 1606. Coecke van Aelst's pupil, the Dutch architect and engineer Vredeman de Vries, drew on Serlio's designs for his *Hortorum Viridariorumque* of 1583. See Brunon, "Paradigme labyrinthique," 73–74, and De Jong, "Aristocratiques ornements," 133–38.
9. Doob, *Idea of the Labyrinth*, 40–41. Curiously, visual imagery of multicursal mazes only appears in the mid-sixteenth century. All pictorial representations of classical and medieval mazes depict the unicursal type, even though contemporaneous texts describe forked paths, dead ends, and internal choices. While scholars have not offered a definitive reason for this shift in iconography, I find it suggestive that the dynamics of the multicursal labyrinth were perceived as analogous to the relationship between God's grace and human works, a contentious issue in the post-Reformation period.
10. Cats, *Houwelyck*, n.p.; Wright, *The Maze and the Warrior*, 7–15.
11. Von Zesen, *Beschreibung der Stadt Amsterdam*, 190: "Weil sich darinnen die sinne verirren und die augen vergassen zu nennen."
12. Tschumi, *Architecture and Disjunction*, 39; Bal, *Quoting Caravaggio*, 18, 24–25, 34, 130; Buci-Glucksman, *Baroque Reason*, 84–85, 93–94, 114.
13. The phrase, evoking how pedestrian movements create urban space, is from de Certeau, *Practice of Everyday Life*, 97.
14. Erasmus, *Sileni of Alcibiades*, 174–75; see also Somos, *Secularization*, 183.

15. Ovid, "Ariadne to Theseus," *Heroides and Amores* 10.71–72 (trans. Showerman, 127).
16. Comenius, *Labyrinth of the World*, 2. Comenius lived as a refugee in Amsterdam for several years.
17. Paradin, *Heroicall Devises*, 118–19.
18. Quarles, *Emblems, Divine and Moral*, 190.
19. Diehl, "Into the Maze," 286; Kern, *Through the Labyrinth*, 209; Stronks, *Negotiating Differences*, 129. The image in Quarles is derived from the earlier engraving by Boetuis van Bolsward for Hugo's *Pia Desideria*.
20. Calvin, *Institutes of the Christian Religion*, 1:83.
21. Quarles, *Emblems, Divine and Moral*, 191.
22. Ibid., 189.
23. Calvin, *Institutes of the Christian Religion*, 1:74.
24. Quarles, *Emblems, Divine and Moral*, 189.
25. Kern, *Through the Labyrinth*, 234.
26. Perrière, *Theatre des Bons Engins*. Number 35 is an emblem of Volupté, with an accompanying poem that begins: "En volupté facilement on entre: Mais on en sort à grand difficulté." Kern provides this English translation: "So to vaine pleasures it is ease to go / But to returne again it is not so" (*Through the Labyrinth*, 221).
27. Kern, *Through the Labyrinth*, 31, 226, and Matthews, *Mazes and Labyrinths*, 1.
28. Cats, *Houwelyck*, n.p.: "Het los en dom gevoel dat jonge lieden plegen / is juyst gelijck een hof met duysent omme-wegen."
29. Ibid.: "Een hof, een selsaem perck, daer yeder met verlangen / Sijn lusten onderhoudt met duysent omme-gangen."
30. Wyatt, "It May Be Good," cited in Doob, *Idea of the Labyrinth*, 1.
31. Domselaer, *Beschryvinge van Amsterdam*, 309.
32. An anonymous handwritten poem about the Oude Doolhof, dated to around 1700, is in a folder of ephemera concerning the Doolhoven in the Amsterdam City Archive, inv. no. 15009:2282–83. It directly cites lines from Cats's poem about the labyrinth. The poem is transcribed in Eisma, "Bewegende en Geklede Beelden," 80.
33. Plutarch, *Parallel Lives* 15.2 (ed. Perrin, 1:29).
34. Domselaer, *Beschryvinge van Amsterdam*, 308: "zommige gedoode lichame om hem leggende." Artists sometimes drew on visual representations of centaurs when picturing the Minotaur, which appears to be the case on the title page of the Oude Doolhof booklet thought to be authored by Jan Vos, *D'Oprechte Aenwijser*. According to Plutarch and others, Theseus joined the Lapiths in their battle against the Centaurs.
35. Passe, *Dool-hof*, n.p: "En dolt zo lang hij kan, tot dat hy dealt in 't Graf."
36. Passe, *Nieuwen en Vermaeckelijcken Dool-hof*, n.p.: "Een voorbeelt van des Menschen Leven, die oock naar veel moeyte moet komen in sijn rustplaets ende wederkeeren van waer hij gekomen is."
37. On the use of this paradox in the poems of John Donne and Michael Drayton, see Colie, *Paradoxia Epidemica*, 126–27.
38. For the association of the labyrinth with the uterus, see Kern, *Through the Labyrinth*, 30.
39. The *geranos* (crane dance) is discussed by Homer, Plutarch, and others as preserving a pattern of choreographic movement that unraveled the ancient labyrinth. Van Mander references it in his commentary on Ovid in the 1604 edition of *Het Schilderboeck* (p. 1066). On the relationship between labyrinths and ephemeral performances, see Kern, *Through the Labyrinth*, 25–27. On their relation to Renaissance dance, see Greene, "Labyrinth Dances," 1403–66.
40. Spies, "Kunst en Wetenschappen," 315, 326–27. According to Spies, the play was still being performed in 1628.
41. The chamber was named after a flower, the eglantine rose, and its motto was "In liefde bloeiende" (Blossoming/bleeding in love). Schenkeveld, *Dutch Literature*, 11–14.
42. Ibid., 13.
43. On the importance of tableaux vivants in Netherlandish civic politics, see Bussels, "Making the Most of Theatre and Painting."
44. Spies, "Rhetoric and Civic Harmony," 64–65.
45. Spies, "Kunst en Wetenschappen," 305.
46. Hooft, *Theseus ende Ariadne*, 40.
47. Ibid., 50–51.
48. Plutarch, *Parallel Lives* 24.2 (ed. Perrin, 1:53).
49. Plutarch, *Parallel Lives* 25.1 (ed. Perrin, 1:55).
50. Abrahamse, *Grote Uitleg*, 50–51.
51. Hooft, *Theseus ende Ariadne*, 50: "Voorsichticheit gaet meest al langs bebaende wegen / Maar boosheits ommewech, die Gode seer is tegen."

52. Lipsius, *On Constancy*, 31.
53. Ibid., 31; Greene, "Labyrinth Dances," 1459.
54. Van Mander, *Schilder-boeck*, 1067.
55. Hooft, *Theseus ende Ariadne*, 63: "Corcyne: Ghij cunt indien ghij wilt dees raserije stillen. Ariadne: Indien ghij wilt, dats waer, maer ick en can niet willen."
56. Doob, *Idea of the Labyrinth*, 35.
57. T. Webster, "Myth of Ariadne," 22–31.
58. Vos, *Oprechte Aenwijser*, n.p.
59. This phrase appears in the 1611 Visscher print (see fig. 5): "ghemeyne Schole van alle Menschen: met welcke name de Stadt Athenae eertijts is ghenoemt."
60. Van Mander, *Schilder-boeck*, 1067.
61. Calvin, *Institutes of the Christian Religion*, 1:74.
62. Sawday, "Self and Selfhood," 30–31.
63. Davies, "Nosce te ipsum," cited in Burke, "Representations of the Self," 19.
64. Colie, *Paradoxia Epidemica*, 388.
65. Wyatt, "It May Be Good," cited in Doob, *Idea of the Labyrinth*, 1.
66. On the concept of inoperativity, see Agamben, *Nudities*, 43–45.
67. Abrahamse, "Ruimtelijke Ontwikkeling," 29. These side streets were allocated for middle-class shops and businesses, with homes on the upper floors. Back-facing windows on the top floors were not allowed.
68. A. Vries, "Plaatsnijder," 37–43. Their father, Floris Balthasar, was a surveyor and cartographer.
69. The map was published by Philips Molevliet, and new editions were issued in 1649 and 1657. The civic government bought sixteen exemplars; it was well suited to be given as a diplomatic gift and to decorate administrative buildings, business offices, company headquarters, and private homes. Bakker, "Imago van de Stad," 60–61.
70. Kern, "Bild der Welt," 10–24, and Doob, *Idea of the Labyrinth*, 17–20, 66–70, 117. On the contrast between walking in the city and seeing it from above, see de Certeau, *Practice of Everyday Life*, 91–110.
71. The map was first published by Willem Blaeu and later appeared in volume 3 of Joan Blaeu's *Theatrum Orbis Terrarum*.
72. Doob, *Idea of the Labyrinth*, 97–100.
73. Pliny, *Natural History* 36.19 (trans. Bostock, 6:339).
74. Amsterdam cartographers would have been familiar with Gerhard Mercator's 1578 edition of Ptolemy's *Geographica*; the labyrinth appears on numerous earlier maps of Crete.
75. "Candia," in Mercator and Hondius, *Historia Mundi*, 811–12.
76. On the ubiquity of The Liar paradox in the Renaissance, see Colie, *Paradoxia Epidemica*, 5–7, 360–61, 517.
77. Abrahamse, "Ruimtelijke Ontwikkeling," 26–29; Abrahamse, *Grote Uitleg*, 35–39, 51–53.
78. This solution was practical. Rather than demolishing buildings and evicting thousands of inhabitants, *de paden* area was reordered with many buildings moved to align to the new grid plan. See Abrahamse, *Grote Uitleg*, 26.
79. Ibid., 263, 336, 353.
80. Von Zesen, *Beschreibung der Stadt Amsterdam*, cited in Abrahamse, *Grote Uitleg*, 21. Planners, surveyors, and builders made great efforts to make everything straight, especially the line that separated public areas from private properties.
81. Abrahamse, *Grote Uitleg*, 14.
82. Ibid., 26. The geometric half-moon shape was valued for its beauty but also for its functionality and economic potential to attract new inhabitants and businesses.
83. The name *Jordaan* is thought to be derived from the French *jardin*, perhaps in reference to the amount of green space in the New Town.
84. Vidler, "History of the Folly," 45–51.
85. There is a chimney-piece painting by Willem Strijcker (1657) in the Insurance Chamber of the New Town Hall that depicts Theseus, Ariadne, and the dead Minotaur. The idea in this context was that Ariadne's thread is like insurance.
86. Spierenburg, *Prison Experience*, 88.
87. Hell, "Trade, Transport and Storage," 744, and Kilian, *Paintings of Karel du Jardin*, 48–49, 208–9.
88. Coornhert, "Boeven-Tucht," 89–104, and Spierenburg, *Prison Experience*, 44–48.
89. "Schrik niet ik wreek geen quaat maar dwing tot goet / Straf is myn hant maar lieflyk myn gemoet." The portal was designed by De Keyser in 1607, and he requested Hooft to write verses for it. See De Bray, *Architectura Moderna*, 74, and Neurdenburg, *Hendrick de Keyser*, 98.

90. Dapper, *Historische Beschryving*, 417–18.
91. Van Mander, *Lives*, 2:68.
92. Coornhert, "Boeven-Tucht," 101.
93. De Bray, *Architectura Moderna*, 22.
94. Spierenburg, *Prison Experience*, 88. The citation is from Seneca, *Hercules Furens*, line 435: "Virtutis est domare quae cuncti pavent."
95. Heinsius, *On Plot*, cited in Heinen, "Huygens, Rubens, and Medusa," 155.
96. Colie, *Paradoxia Epidemica*, 22.

Chapter 4

1. Fokkens, *Beschrijvinge der Wijdt-vermaarde Koop-Stadt*, 300–302. Eisma speculates that the gallery was built in the early 1660s and that before this date, the mechanical devices had been displayed in smaller outbuildings ("Bewegende en Geklede Beelden," 78).
2. Duncan, *Civilizing Rituals*, 8–10.
3. Passe, *Nieuwen en Vermaeckelijken Dool-hof*, n.p.: "verroering doende elck Beelt, zijn actie of het leefde, tot verwonderingh en veramaeck der aenschouwers."
4. In private art collections, picture curtains served as a means of highlighting and protecting the rarest pieces and of concealing and revealing salacious works to heighten the sensual effect on selected viewers. See Vanhaelen, *Wake of Iconoclasm*, 62, 73, 171.
5. Fokkens, *Beschrijvinge der Wijdt-vermaarde Koop-Stadt*, 301–2: "self oock geluydt gheven van singhen, bij near spreeken, spleen, krayen, loyen, bleên, en queecken der dieren."
6. The David and Goliath statues were possibly made by Vinckenbrinck, who also made the Bacchus fountain. David once had a wig of human hair, enhancing the lifelike appearance. See Eisma, *David en Goliath*, 47. The Amsterdam Museum has recently acquired a wooden statue of Hercules by Vinckenbrinck. Technical examination reveals that it, too, may have been a moving statue, which raises the likelihood that it was part of the Doolhof displays. Thank you to Anne Diestelkamp and Prof. Dr. Erma Hermens for bringing this work to my attention.
7. Eisma, "Bewegende en Geklede Beelden," 84.
8. McHam, "Public Sculpture in Renaissance Florence," 149–88.
9. Passe, *Dool-hof*, n.p.
10. Knoespel, "Gazing on Technology," 117–19.
11. Aristotle, *Metaphysics* 1.982b (trans. Tredennick, 1:13). See also Summer, "Pandora's Crown," 70, and Marr, "'Gentille Curiosité,'" 149.
12. Vitruvius, *Ten Books on Architecture* 10.7.5 (trans. Morgan, 298).
13. Plato, *Euthyphro* 11c–d (trans. Jowett, 312).
14. Homer, *Iliad* 18.592 (trans. Green, 465–66); see S. Morris, *Daidalos*, 9–13, and de Solla Price, "Automata and the Origins," 10.
15. Passe, *Dool-hof*, n.p.
16. Passe, *Nieuwen en Vermaeckelijcken Dool-hof*, n.p.
17. Grafton, "Devil as Automaton," 48; Keating, *Animating Empire*; Mayr, *Authority, Liberty and Automatic Machinery*; Koeppe, *Making Marvels*.
18. Riskin, *Restless Clock*, 20.
19. Evelyn, *Diary*, cited in ibid., 39.
20. Morgan, *Nature as Model*, chapter 6. The elector was Calvinist; he and his wife moved to The Hague in 1619, where they lived in exile.
21. Montaigne, *Complete Works*, 1175.
22. Ibid. The bird device is derived from the treatise of Hero of Alexandria. See Riskin, *Restless Clock*, 29–30.
23. Maurice and Mayr, *Clockwork Universe*, vi, 27.
24. For an overview, see Riskin, *Restless Clock*, chapter 1, and Kang, "Wonders of Mathematical Magic," 113–39.
25. Aristotle, *Metaphysics* 1.982b (trans. Tredennick, 1:13).
26. Evelyn, *Diary*, 1:23.
27. On this type of knowledge production, see Long, *Openness, Secrecy, Authorship*, 249.
28. Domselaer, *Beschryvinge van Amsterdam*, 309. This device can be seen in a guidebook print of the Nieuwe Doolhof courtyard included in chapter 7 as fig. 41.
29. Principe, *Secrets of Alchemy*, 12.
30. On Hero, see Tybjerg, "Wonder-Making," 451–52. On de Caus, see Werrett, "Wonders Never Cease," 136. Secret knowledge is discussed in Long, *Openness, Secrecy, Authorship*, 1–2, and Kang, *Sublime Dreams*, 120.

31. Wilkins, *Mathematicall Magick*, 2–3.
32. Ibid., 103, 115.
33. For an excellent overview, see Riskin, *Restless Clock*, esp. chapter 1.
34. King, "Perpetual Devotion," 263–92.
35. Mayr, *Authority, Liberty, and Automatic Machinery*, 24. On earlier religious understandings of matter, see Bynum, *Christian Materiality*. On the shift away from religious wonder, see Jorink, *Reading the Book of Nature*, 6–7, and Riskin, *Restless Clock*, 2–10.
36. Riskin, *Restless Clock*, 22–25.
37. Eire, *War Against the Idols*, 3.
38. Calvin, *Institutes of the Christian Religion*, quoted and translated in Michalski, *Reformation and the Visual Arts*, 63.
39. Calvin, *Institutes of the Christian Religion*, 1:108.
40. De Jongh, "Schaduwen van Daedalus," 10–15.
41. Junius, *Painting of the Ancients*, 125.
42. Ibid., 351.
43. Van Mander, *Lives*, 1:62.
44. Ibid., 1:69.
45. Ibid., 1:365.
46. Van Mander, *Schilderboek*, 1067, 1069.
47. See Van Hoogstraten, *Inleyding*, 358–60.
48. Van Mander, *Schilderboek*, 1067.
49. Plato, *Meno* 97d–e (ed. Hamilton and Cairns, 381); see also Strauss, "Automata," 45.
50. Vitruvius, *Ten Books on Architecture* 10.7.5 (trans. Morgan, 298).
51. Wilkins, *Mathematicall Magick*, 102.
52. Bacon, "Daedalus, or Mechanical Skill," 236 (originally published in *Wisdom of the Ancients*).
53. Ovid, *Metamorphoses* 8.188–89 (trans. Johnson, 211); see also Sawday, *Engines of the Imagination*, 7, and Sawday, "'Forms Such as Never Were,'" 181–84.
54. Bacon, "Daedalus, or Mechanical Skill," 237.
55. Ibid., 236.
56. Ibid., 237.
57. Ibid.
58. S. Morris, *Daidalos*, 215.
59. Gaukroger, *Francis Bacon*, 2–3, and Jorink, *Reading the Book of Nature*, 4.
60. Bacon, "Daedalus, or Mechanical Skill," 237.
61. Bacon, *Works of Francis Bacon*, 4:18.
62. Calvin, *Institutes of the Christian Religion*, 1:83.
63. Bacon, *Advancement of Learning*, 221; see also Lewis, "Francis Bacon," 369.
64. Daston and Park, *Wonders and the Order of Nature*, 228, and Scholar, *Je-Ne-Sais-Quoi*, 109–10.
65. Mayr, *Authority, Liberty, and Automatic Machinery*, 59, and Summer, "Pandora's Crown," 51.
66. Bacon, "Clue to the Maze," 233.
67. Bacon, *Moral and Historical Works*, 87.
68. Jardine, *Francis Bacon*, 173–75.
69. Gaukroger, *Francis Bacon*, 78, and Sawday, *Engines of the Imagination*, 25–26.
70. Gaukroger, *Francis Bacon*, 6–17.
71. Descartes, *Discourse on Method and Meditations*, 68; see also Gaukroger, *Descartes*, viii, xvi–xvii. Descartes was living in Amsterdam from 1630 to 1632 and worked on *Traité de l'homme* from 1632 to 1633, a key work in the philosopher's development of mechanistic physiology, the theory of the body as an automaton.
72. Riskin, *Restless Clock*, 45.
73. Gaukroger, *Descartes*, 62–64.
74. Descartes, "La recherche de la verité," cited and translated in Werrett, "Wonders Never Cease," 141.
75. K. Morris, "Bêtes-machines," 401–14; Mayr, *Authority, Liberty, and Automatic Machinery*, 66–67.
76. Descartes, *Discourse on Method and Meditations*, 33; see also Gaukroger, *Descartes*, 288–89.
77. Descartes, *World and Other Writings*, 99. The passage is from part 1 of *Treatise on Man*.
78. Judovitz, *Culture of the Body*, 74; des Chene, *Spirits and Clocks*, 3; Gaukroger, *Descartes*, 279.
79. Domselaer, *Beschryvinge van Amsterdam*, 309.
80. Dante, *Divine Comedy*, 39.
81. Boccaccio, *Famous Women*, 23.
82. See Gibson, *Figures of Speech*, 118.
83. Ibid., 126–41. On representations of the battle for the pants, see Davis's classic essay "Women on Top." On the theme in Dutch comic culture, see Vanhaelen, *Comic Print*, chapter 3.
84. Riskin, "Introduction," 5.
85. Reneman, "'Oude Doolhof,'" 20.
86. Brereton, *Travels in Holland*, 57. The references to pissing were tellingly omitted from later editions of Brereton's text. See Wijnman, "Jan Theunisz.," 29.
87. De Roever, *Uit Onze Oude Amstelstad*, 2:25–27.
88. Reneman, "'Oude Doolhof,'" 20, 32.
89. Heinsius, for instance, states that "a man lacking mind resembles brutes" (*On Plot*, 11).
90. Brereton, *Travels in Holland*, 57.

91. The 1779 travel writings of French journalist Samuel Francois l'Honoré are cited in Jensen, *Reizigers te Amsterdam*, 140: "on voit le spectacle de quelques automates gigantesques et autres poupées qu'un homme fait mouvoir, en récitant des vers emphatiques. On y voit entre autres la Nourrice du Roi David, qui étale son sein d'une façon fort indécente. Ce spectacle, autrefois le seul que l'on connaissoit ici, est fort couru de la populace et des enfants."
92. Brereton, *Travels in Holland*, 56.
93. Wijnman, "Jan Theunisz.," 31.
94. Brereton, *Travels in Holland*, 56.
95. Evelyn, *Diary*, 1:23.
96. See the insights of Voskuhl, *Androids in the Enlightenment*, 15–21.
97. Heinsius, *On Plot*, 11–13.
98. Daston and Park, *Wonders and the Order of Nature*, 19.
99. N. Elias, *Civilizing Process*, 1:150.
100. Heinsius, *On Plot*, 14.

Chapter 5

1. Howell, *Epistolae Ho-Elianae*, 2:11, 26.
2. Van Mander, *Lives*, 1:46–49.
3. Commelin, *Beschrijvinghe van Amsterdam*, 742.
4. Fokkens, *Beschrijvinge der Wijdt-vermaarde Koop-Stadt*, 302, 305.
5. Passe, *Dool-hof*, n.p.
6. Jochum is first pictured and described in a guidebook that predates 1648. See Reneman, "'Oude Doolhof,'" 76.
7. The poem is in the pamphlet *Verklaringe van Treffelijcke Konstighe Wercken*, n.p.
8. Stoichita, *Pygmalion Effect*, 1–2. An important recent art-historical approach to automata is Keating, *Animating Empire*.
9. Cassiodorus, *Variae* 1.45, cited and translated in Schiefsky, "Art and Nature," 87.
10. Judovitz, *Culture of the Body*, 74; Wood, *Edison's Eve*, xvii–xviii; Maisano, "Descartes avec Milton," 27.
11. Wilkins, *Mathematicall Magick*, 2; see also Grafton, *Magic and Technology*, 6–18.
12. Bacon, *Works of Francis Bacon*, 4:95; see also Merchant, "Francis Bacon," 551–99, and Alpers, *Vexations of Art*.
13. Summer, "Pandora's Crown," 46.
14. Fokkens, *Beschrijvinge der Wijdt-vermaarde Koop-Stadt*, 302.
15. Simmel, "Stranger," 402–6.
16. On these paintings and ethnographic eclecticism, see Swan, "Lost in Translation," esp. 111–14.
17. Huygens mentions the painting in his autobiography of ca. 1630, cited in Swan, "Lost in Translation," 107. Lievens's painting was in the collection of Stadholder Frederick Hendrik; Huygens lauded its "inestimable value and unrivalled artistry" (ibid.).
18. On the Dutch domestication of exotic goods like porcelain, see Odell, "Delftware and the Domestication."
19. For discussion of these paintings, see Hochstrasser, *Still Life and Trade*, 204–12, and Ford, "People as Property."
20. Postma, *Dutch in the Atlantic Slave Trade*, and Boxer, *Dutch Seaborne Empire*.
21. On the repeated visual motif of African figures conveying trade goods, see Vanhaelen, "Mapping Angels."
22. On the automaton as unthinking slave, see Mayr, *Authority, Liberty, and Automatic Machinery*, 66–67, and Beaune, "Classical Age of Automata," 431.
23. Plato, *Meno* 97e (ed. Hamilton and Cairns, 381).
24. Aristotle, *Politics* 1.1253b, cited in S. Morris, *Daidalos*, 225; see also Kang, "Wonders of Mathematical Magic," 120–21.
25. Heinsius, *On Plot*, 13.
26. Spinoza, *Treatise on the Emendation of the Intellect* 48.24–25, cited in Mayr, *Authority, Liberty, and Automatic Machinery*, 68.
27. Aytes, "The 'Other' in the Machine."
28. Passe, *Dool-hof*, n.p.
29. Descartes dedicated this work to Princess Elisabeth of the Palatinate, the daughter of Frederick and Elisabeth of Bohemia, patrons of de Caus's famed garden automata at Heidelberg.
30. Descartes, *Passions of the Soul*, 20–22.
31. Ibid.
32. Von Uffenbach, *Merkwürdige Reisen*, 1:218–20.
33. Descartes, "La recherche de la verité," cited and translated in Werrett, "Wonders Never Cease," 141.
34. On early Chinese accounts of automata, see Needham, *Science and Civilisation*, 2:53–54. On

Su Sung, see Needham, Ling, and de Solla Price, *Heavenly Clockwork*, 1–3.

35. Needham, Ling, and de Solla Price, *Heavenly Clockwork*, 191–92, and Zielinski and Weibel, *Allah's Automata*, 6–11.
36. Kurz, *European Clocks and Watches*, 1–8.
37. Truitt, *Medieval Robots*, 12–39, and Daston and Park, *Wonders and the Order of Nature*, 103.
38. On clocks as diplomatic gifts, see Kurz, *European Clocks and Watches*, 20–35; Cipolla, *Clocks and Culture*, chapter 2; Keating, *Animating Empire*, 95.
39. Pagani, *Eastern Magnificence*, 29.
40. Ricci, *Storia dell'introduzione del cristianesimo in Cina*, cited and translated in Needham, Ling, and de Solla Price, *Heavenly Clockwork*, 154; see also Cipolla, *Clocks and Culture*, 81–88.
41. Needham, Ling, and de Solla Price, *Heavenly Clockwork*, 141–45.
42. Schaffer, "Enlightened Automata," 135–36, 151.
43. Corrigan et al., *Asia in Amsterdam*, 24, 36–37, 53, and Hochstrasser, *Still Life and Trade*, 153.
44. Passe, *Dool-hof*, n.p.
45. Barlaeus, *Mercator sapiens*, 57–59, 83–84.
46. Blaeu, *Toonneel der Steden*, 145, 149.
47. Barlaeus, *Mercator sapiens*, 59.
48. Jolle, *Description de la ville*, 240–46.
49. Wijnman, "Jan Theunisz.," 68.
50. Ibid., 29–31, 68. See also Jensen, *Riezigers te Amsterdam*, cat. nos. 53, 38, and 41.
51. Hegenitius, *Itinerarium Frisio-Hollandicum*, 68–69; see also Wijnman, "Jan Theunisz.," 31.
52. Vrolijk and Leeuwen, *Arabic Studies*, 17–18, 30.
53. Grafton, *Joseph Scaliger*, 2:387.
54. On the lexicon and Raphelengius's Arabic manuscript collection, see Hamilton, "'Nam Tirones Sumus,'" 557–89, and Wijnman, "Hebraïcus Jan Theunisz.," 9–10.
55. Theunisz wrote an account of his interactions with ʿAbd al-ʿAzīz. This untitled manuscript, written in Arabic and Latin, is currently in the Regionaal Archief Leiden en Omstreken (shelfmark lb 69501). Dorrit van Dalen refers to the manuscript as *Conversations*, and I follow suit. The narrative of their first meeting is on p. 13 of *Conversations*, cited and translated in Dalen, "Johannes Theunisz," 165.
56. On paintings of Muslims in Calvinist churches, see Vanhaelen, *Wake of Iconoclasm*, 38–43.
57. Wijnman speculates that he stayed at Theunisz's inn ("Jan Theunisz.," 51), but Dalen claims that he stayed in the home of Theunisz ("Johannes Theunisz," 168).
58. On connections between Theunisz and English Puritan intellectuals, see Sprunger, *Trumpets from the Tower*, 46–58.
59. Theunisz, *Conversations*, 1, cited and translated in Dalen, "Johannes Theunisz," 163.
60. Theunisz, *Conversations*, 74, cited and translated in ibid., 181.
61. Dalen, "Johannes Theunisz," 168, 188. The prayer book is preserved in the Amsterdam University Library, shelfmark OTM 111 A16.
62. Theunisz, *Conversations*, 13, cited and translated in ibid., 165.
63. The inscription is reproduced in Vrolijk and Ommen, *"All My Books in Foreign Tongues,"* cat. no. 45.
64. Wijnman, "Jan Theunisz.," 40–41, 73–76.
65. Ibid., 50, 60.
66. Sprunger, *Trumpets from the Tower*, 58, and Wijnman, "Jan Theunisz.," 56–65.
67. Sprunger, *Trumpets from the Tower*, 56–58; Dalen, "Johannes Theunisz," 166.
68. Wijnman, "Jan Theunisz.," 74.
69. Hamilton, Voolstra, and Visser, *From Martyr to Muppy*; on house churches, see Kaplan, "Fictions of Privacy," 1031–64, and Vanhaelen, "Spiritual Culture," 233–34.
70. Van der Veen, *Ian vander Veens Zinne-beelden*, cited in Kaplan, *Muslims in the Dutch Golden Age*, 5. See also Wijnman, "Jan Theunisz.," 105.
71. Wijnman speculates that the "Ian Taurens" in the poem is a misspelling of Jan Theunisz ("Jan Theunisz.," 104–6). On Torrentius, see C. Brown, "Strange Case," 227.
72. See Dickson, *Tessera of Antilla*, and Principe, *Secrets of Alchemy*.
73. Wijnman, "Jan Theunisz.," 77–101. Blok and Holhuysen refer to the controversy about "godslasterlijke opschriften" (*Nieuw Nederlandsch Biografisch Woordenboek*, 9:1121–22).
74. Possibly the inn was nicknamed *Mennisten Bruyloft* not just because of the religion of the publican but also because it sounds like *Chymische Bruyloft*, the infamous Rosicrucian alchemical tract by "Christian Rosencreutz," written by Johann Valentin

Andreae, published anonymously in 1616. See Dickson, *Tessera of Antilla*, chapter 1.

75. Grafton, *Joseph Scaliger*, 2:1–2.
76. Vrolijk and Leeuwen, *Arabic Studies*, 27.
77. While in Morocco, Golius had copies made of some of the manuscripts in the library of Sultan Mulay Zaydan. See Wikam, "Jacobus Golius," 53, and Vrolijk and Leeuwen, *Arabic Studies*, 40–44.
78. Golius's private collection was bought by Narcissus Marsh, who bequeathed it to the Bodleian in 1713. See Wakefield, "Arabic Manuscripts," 128–46.
79. Hattab, *Descartes on Forms and Mechanisms*, 156. Golius's *Conica* manuscript was a copy that he commissioned; it was later sold by Amsterdam bookseller Ratelband to Thomas Marshall, who bequeathed it to the Bodleian. See Wikam, "Jacobus Golius," 57–61, and Wakefield, "Arabic Manuscripts," 132–34.
80. Huygens, *Gedichten*, 209.
81. Golius's acquisitions for the university library are catalogued as Or. 1 to Or. 211. The al-Jazarī manuscript is Or. 117; the Banū Mūsà is Or. 168; and Hero's *Barulcus* (Mechanics) is Or. 51:1. Levinas Warner later acquired a second al-Jazarī manuscript, which he left in his will to Leiden University in 1665. It is catalogued as Or. 656. See Vrolijk and Leeuwen, *Arabic Studies*, 42–47, 58.
82. Shākir, Shākir, and Shākir, *Book of Ingenious Devices*.
83. Al-Jazarī, *Book of Knowledge*, 11–12.
84. This manuscript is catalogued as Leiden University Library, SCA 45.
85. The Vossius manuscript of *Pneumatica* is catalogued as VGQ 19. Dionysius Vossius studied with Heinsius as well as Erpenius and Golius. See Van der Aa, *Biographisch Woordenboek*, 406–7. On his association with Theunisz, see Dalen, "Johannes Theunisz," 166.
86. On Theunisz's language skills, which included Greek, see Wijnman, "Jan Theunisz.," 40, 104. Of the eleven extant al-Jazarī manuscripts, two are in Leiden University Library; of the three extant Banū Mūsà manuscripts, one is in Leiden University Library. See Hill, introduction to *Book of Knowledge*, 3–6, 9.
87. Hill, *Islamic Science and Engineering*, 122.
88. Truitt, *Medieval Robots*, 20. See also Hill, *Studies in Medieval Islamic Technology*, 222–37.
89. Wijnman, "Jan Theunisz.," 71.
90. Brereton, *Travels in Holland*, 56.
91. Ibid., 57.
92. Al-Jazarī, *Book of Knowledge*, 157–69.
93. Brereton, *Travels in Holland*, 57.
94. Evelyn, *Diary*, 1:23.
95. Al-Jazarī, *Book of Knowledge*, 3.
96. Ibid., 95–97, fig. 80.
97. Ibid., 98–106, fig. 82.
98. Dupré, *Laboratories of Art*, ix.
99. Long, "Trading Zones," 840–47.
100. Brandt and Hogendoorn, *German and Dutch Theatre*, 349.
101. Brereton, *Travels in Holland*, 56.
102. Hegenitius, *Itinerarium Frisio-Hollandicum*, 68–69.

Chapter 6

1. *Verklaringe van Treffelijcke Konstighe Wercken*, n.p.
2. Ibid.
3. Vos, *Oprechte Aenwijser*, n.p.
4. Domselaer, *Beschryvinge van Amsterdam*, 309–10.
5. Brereton, *Travels in Holland*, 57–58.
6. Daninos, "Wax Figures in Italy," 12–13.
7. Vasari, *Vasari on Technique*, 148–49.
8. Ibid., 149.
9. The differences between modeling and casting are explained by Schlosser, "History of Portraiture," 250–51. On the uses, properties, and supply of beeswax, see Guerzoni, "Use and Abuse," 43–59.
10. Brereton, *Travels in Holland*, 58, and Wilson, "Renaissance Portrait," 462.
11. Vasari, *Lives*, 240 (from the biography of Andrea del Verrocchio).
12. Schlosser, "History of Portraiture," 296.
13. Ibid., 299.
14. Panzanelli, "Compelling Presence," 15–17.
15. Pliny, *Natural History* 35.2 (trans. Bostock, 6:225).
16. Panzanelli, "Compelling Presence," 18–21.
17. Polybius, *Histories* 6.53 (trans. Scott-Kilvert, *Rise of the Roman*, 347); see also Schlosser, "History of Portraiture," 183.
18. Didi-Huberman, "Molding Image," 77–80, and Haldar, "From Dignity to Shame," 245.
19. Vasari, *Lives*, 240 (from the biography of Andrea del Verrocchio); see also Ricci, "Masks of Power," 67.

20. The votive image marked a transaction between the votary and the saint or deity. See Van der Velden, *Donor's Image*, 193.
21. Buchelius, *Iter Italicum*, cited in Freedberg, *Power of Images*, 229.
22. Panzanelli, "Compelling Presence," 15.
23. Warburg, "Art of Portraiture," 191, 206; Panzanelli, "Compelling Presence," 13–15.
24. Vos, *Oprechte Aenwijser*, n.p.
25. Warburg, "Art of Portraiture," 191, 206.
26. Holmes, "Ex-votos," 177; see also Panzanelli, "Compelling Presence," 15–17.
27. Van der Velden, *Donor's Image*, 3–4.
28. Sacchetti, *Novella*, no. 109, cited and translated in Warburg, "Art of Portraiture," 204.
29. Calvin, *Institutes of the Christian Religion*, 1:122.
30. Van der Velden, *Donor's Image*, 223, 259.
31. Schlosser, "History of Portraiture," 296; see also Altick, *Shows of London*, 50.
32. Syson, "Polychrome and Its Discontents," 26.
33. Didi-Huberman, "Viscosities and Survivals," 159–60.
34. Panzanelli, "Compelling Presence," 15–17; Haldar, "From Dignity to Shame," 247.
35. Panzanelli, "Compelling Presence," 21.
36. Polybius, *Histories* 6.53 (trans. Scott-Kilvert, *Rise of the Roman*, 347). A nineteenth-century sketch by J. G. L. Rieke shows costumed waxworks seated on a row of chairs upon a platform above the automata.
37. Ibid.
38. Polybius, *Histories* 6.54 (trans. Scott-Kilvert, *Rise of the Roman*, 347).
39. Scholten, *Sumptuous Memories*, 15–41.
40. De Rouck, *Nederlandtsche Herault*, 11–12.
41. Vossius, "De graphice, sive arte pingendi," cited and translated in Weststeijn, *Visible World*, 198.
42. Mountague, *Delights of Holland*, 182–83.
43. The Van de Passes produced an important series of printed portraits of European monarchs, including the Swedish king and the Dutch stadtholders. See Verbeek, "Zeldzame Vorstenportretten," 76–84, and Verbeek, "Naschrift bij de Vorstenportretten," 28–30.
44. *Verklaringe van Treffelijcke Konstighe Wercken*; Passe, *Dool-hof*; Passe, *Den Nieuwen en Vermaeckelijcken*; see also Veldman, *Crispijn de Passe*, 291, 375.
45. Orenstein, "Who Took the King?," 44–47.
46. On secular saints, see Vroom, *Wonderlid*.
47. Duits, "Hoofts 'Henrik de Gróte,'" 280.
48. Lipsius, *Politica*, 3.
49. Ibid., 86; see also Houdt, "*Spectacle of Power*," 16–19.
50. Duits, "Hoofts 'Henrik de Gróte,'" 279.
51. Ibid., 275.
52. Lipsius, *Politica*, 102–3.
53. Ibid., 103.
54. Schlosser, "History of Portraiture," 173.
55. Kunzle, *From Criminal to Courtier*, 415–16.
56. Guerzoni, "Use and Abuse," 59.
57. Appian, *The Civil Wars* 2.20.147; cited in Schlosser, "History of Portraiture," 184.
58. Junius, *Painting of the Ancients*, 139.
59. Panzanelli, "Compelling Presence," 23.
60. Some of these were modeled from a death mask cast directly from his face. See Schlosser, "History of Portraiture," 212–16.
61. Vitry, "Sculpteur Michel Bourdin," 290–92; Schlosser, "History of Portraiture," 265; Wilson, "Renaissance Portrait," 460.
62. Gatacre and Dru, "Portraiture in Le Cabinet," 619.
63. See the classic study by Kantorowicz, *King's Two Bodies*. See also Giesey, *Royal Funeral Ceremony*; Ricci, "Masks of Power," 61–62.
64. Brückner, *Bildnis und Brauch*, 105.
65. Giesey, *Royal Funeral Ceremony*, 80–85; Brückner, *Bildnis und Brauch*, 154.
66. Woodward, *Theatre of Death*, 88.
67. Ibid., 91.
68. *Verklaringe van Treffelijcke Konstighe Wercken*, n.p.
69. Mulier, "Willem van Oranje," and Nierop, "Oranje Boven."
70. Vanhaelen, *Wake of Iconoclasm*, chapter 4.
71. Scholten, "Sculpted Portrait," 42.
72. Scholten, "Quellinus's Burgomasters," 87.
73. Scholten, "Beheaded Bust."
74. Ray, *Travels Through the Low-Countries*, 47.
75. Arnade, *Beggars, Iconoclasts, and Civic Patriots*, esp. chapter 5, and Horst, "Duke of Alba."
76. The poster is discussed in more detail in the following chapter.
77. Pieter Janszoon Twisck, *Chronijck van den ondergane der tijrannen*, cited and translated by Schmidt, "Past *in* a Foreign Country," 22.
78. Ibid.

79. See Hochstrasser, *Still Life and Trade*, esp. chapter 5.
80. Van Mander, *Lives*, 1:189.
81. Ibid., 3:247, 3:251.
82. Ibid., 1:190. See also Jonckheere, *Willem Key*, 107–10. The original portrait by Key is in the Palacio de Liria, Madrid; an anonymous copy is in the Rijksmuseum, Amsterdam.
83. Van Mander, *Lives*, 1:190.
84. Schlosser gestures toward universalizing notions of bewitchment and demonism ("History of Portraiture," 175–76). Freedberg likewise writes of *envoûtment*, a form of sympathetic magic in which the maltreatment of a lifelike effigy can harm the sitter (*Power of Images*, 263–70). See important critiques of this approach by Van der Velden, "Medici Votive Images," 127; Kretzschmar, "Ähnlichkeit und Material," 365–75; Brückner, *Bildnis und Brauch*, 185.
85. Weststeijn, "'Painting's Enchanting Poison,'" 141–77.
86. Arnade, *Beggars, Iconoclasts, and Civic Patriots*, 166.
87. Horst, "Duke of Alba," 133; Smolderen, *Statue du Duc d'Albe*, 41–49; Arnade, *Beggars, Iconoclasts, and Civic Patriots*, 169.
88. Calvin, *On the Shunning of the Unlawful Rites of the Ungodly and Preserving the Purity of the Christian Religion*, cited in Arnade, *Beggars, Iconoclasts, and Civic Patriots*, 122.
89. The anonymous 1576 treatise is titled *Vertoog ende openinghe om een goede, salighe ende generale vrede te maken in dese Nederlanden*, cited in ibid., 208.
90. Ibid., 198.
91. On the iconography of the statue, see ibid., 198–203.
92. Horst, "Duke of Alba," 134.
93. Vondel, "Op den koperen duim van 't beelt des Hertogen van Alba," cited and translated in Horst, "Duke of Alba," 134.
94. Schlosser, "History of Portraiture," 187.
95. Edgerton, *Pictures and Punishment*, 60.
96. Ibid., 92.
97. Freedberg, "From Defamation to Mutilation," 194.
98. Fokkens, *Beschrijvinge der Wijdt-vermaarde Koop-Stadt*, 302, and Vos, *Oprechte Aenwijser*.
99. Brückner, *Bildnis und Brauch*, 147.
100. Jacob Döpler, *Theatrum poenarum, supplicorum, et executionum criminalium* (1697), cited in Freedberg, *Power of Images*, 261; see also Brückner, *Bildnis und Brauch*, 294–95.
101. Foucault, *Discipline and Punish*, 29.
102. Bont, "Bestje van Meurs," 84.
103. Bynum, *Holy Feast*, 115.
104. See especially ibid. See also Van Deth and Vandereycken, *Van Vastenwonder Tot Magerzucht*, 78–81.
105. Bont, "Bestje van Meurs," 86.
106. This trial occurred in 1607. See ibid., 86–87, and Reneman, "'Oude Doolhof,'" 52–54.
107. Wassenaer, *Vyftiende Deel of 't Vervolgh*, 64–65.
108. "Tot Meurs heeftmen ghevanghen ghenomen een Dochter ghenaemt Eva Vlieghe, dewelcke in de 30 Jaren voor den Volcke uytghegheven heft niet tee ten / ende alhoewel daer scherpelijck so van de Gravinne als Predicanten ende ander Luyden opghelet is gheweest soo en heeftmen 'tselve bedroch niet connen mercken." *Courante uyt Italien ende Duytschland*, June 24, 1628.
109. See Wassenaer, *Vyftiende Deel of 't Vervolgh*, 64–65, and Reneman, "'Oude Doolhof,'" 52–54.
110. Wassenaer, *Vyftiende Deel of 't Vervolgh*, 64–65.
111. Calvin, *Institutes of the Christian Religion*, 1:27–28.
112. Huet, *Monstrous Imagination*, 188.
113. Pilbeam, *Madame Tussaud*, 10–11.
114. Coornhert, *Ethics*, 62.
115. Haldar, "From Dignity to Shame," 240.
116. Schlosser, "History of Portraiture," 173, 247.

Chapter 7

1. Borges, "Garden of Forking Paths," 26–28; see also Weiss, *Mirrors of Infinity*, 9.
2. Sipiora, "Introduction," 3–7; Smith, "Time and Qualitative Time," 56.
3. Passe, *Dool-hof*, n.p.
4. Ibid.
5. On precinematic aspects of automata shows, see Bruno, *Atlas of Emotion*, 147–51, and Grootenboer, "Clock Picture." On temporality and Dutch art, see Weststeijn, *Visible World*, 185; Brusati, "Perspectives in Flux," 908–33; Massey, "Reflections on Temporality," 1050–57.
6. Passe, *Dool-hof*, n.p.: "Die klimmen uyt des Zee en u ten toon gestelt, Gelijck het Daniel uyt sijn gesicht vertelt."
7. Ibid., n.p.

8. Grafton, *Joseph Scaliger*, 1:2–4; Grafton, "Renaissance Histories," 196–98, 202; Mayr, *Authority, Liberty, and Automatic Machinery*; Sawday, *Engines of the Imagination*, 233; Helms, *Craft and the Kingly Ideal*, 69.
9. Maurice and Mayr, *Clockwork Universe*, vii.
10. Zielinski and Weibel, *Allah's Automata*.
11. E. Brown, *Naauwkeurige en Gedenkwaardige*, 24: "een goede harmony van klokke-spel."
12. Ibid., 24: "Men ziet 'er veele treffelijke Water-werken en verscheyde bewegingen by klok-werk / schilderyen / ende andere divertissementen en vermakelijkheden."
13. Fokkens, *Beschrijvinge der Wijdt-vermaarde Koop-Stadt*, 302–3, and Domselaer, *Beschryvinge van Amsterdam*, 309.
14. Passe, *Dool-hof*, n.p.
15. Mayr, *Authority, Liberty, and Automatic Machinery*, 26, 121.
16. Landes, *Revolution in Time*, 93.
17. Mayr, *Authority, Liberty, and Automatic Machinery*, 21.
18. Passe, *Dool-hof*, n.p.
19. Schama, *Embarrassment of Riches*, 101–3, and Nadler, *Rembrandt's Jews*, 98–100.
20. Sipiora, "Introduction," 6–15.
21. Passe, *Dool-hof*, n.p.
22. Luther, *Military Sermon* (1529), cited and translated in Proksch, "Luther's Eschatology," 2.
23. Luther, *Widmungsbrief zu seiner Danielübersetzung*, cited and translated in Francisco, *Martin Luther*, 82–83.
24. Perlove and Silver, *Rembrandt's Faith*.
25. Frank Kermode, *Sense of an Ending*, cited in Sipiora, "Introduction," 2.
26. C. Webster, *Great Instauration*, 24.
27. See Dickson, *Tessera of Antilla*, 3–4.
28. C. Webster, *Great Instauration*, 3–5.
29. Mayr, *Authority, Liberty, and Automatic Machinery*, 105.
30. Bacon, *Works of Francis Bacon*, 4:12.
31. Gaukroger, *Francis Bacon*, 7–9, 17, and Eamon, *Science and the Secrets*, 327.
32. Bacon, *Works of Francis Bacon*, 4:114.
33. Eamon, *Science and the Secrets*, 310.
34. Bacon, *New Atlantis*, 26.
35. Bacon, *Works of Francis Bacon*, 4:12.
36. This part of the show is only discussed in the 1657 Mansvelt edition of the pamphlet *Verklaringe van Verscheyden Kunst-Rijcke*, n.p.
37. Ibid., n.p.
38. On the characterization of the Princes of Orange as Solomon, see Schama, *Embarrassment of Riches*, 122, and the painting by Jan van Orley, *William III as Solomon* (ca. 1685) in the Government Art Collection, UK.
39. C. Webster, *Great Instauration*, 2, 8–9.
40. Bacon, *Advancement of Learning*, 71.
41. C. Webster, *Great Instauration*, 24.
42. There were numerous Reformed societies and private academies across Europe that were inspired by Bacon's teachings and by common ideals about the millenarian advancement of experimental philosophy; they frequently met in taverns. See Young, *Faith, Medical Alchemy and Natural Philosophy*, 102–7, and Dickson, *Tessera of Antilla*, 17, 182.
43. Passe, *Dool-hof*, n.p.: "nochtans en konnen wy niet bemercken, dat soo wel by den Griecksche of Romeynsche Meesters, ennige dier-gelijck als dese Konstige Vindingh te sien is geweest."
44. Passe, *Nieuwen en Vermaeckelijcken Dool-hof*, n.p.
45. Vos, *Oprechte Aenwijser*, n.p.
46. Comenius's ideals about universal wisdom and education are articulated in *General Consultation on an Improvement of All Things Human*, cited in Young, *Faith, Medical Alchemy and Natural Philosophy*, 250. Comenius joined forces with Samuel Hartlib and John Dury to work toward universal knowledge and Christian unity. See Dickson, *Tessera of Antilla*, 147–82; Trevor-Roper, *Crisis of the Seventeenth Century*, 250–71; Clucas, "In Search of 'The True Logick,'" 52–58.
47. Many thanks to Meredith Martin for her insights at the 2019 CAA conference. Thank you also to Shelley Perlove, Walter Melion, and Joseph Koerner for their comments.
48. Probably published around 1630, this poster advertised key attractions at the first Nieuwe Doolhof on Looiersgracht adjacent to the Oude Doolhof. Before it came under the proprietorship of the Lingelbachs, this establishment was run by Jonas Bargois and then by Bargois's widow, Maria de Gomme, and her second husband, Lubbert Jansz Root, who is portrayed in the poster's bottom left corner.

49. C. Webster, *Great Instauration*, 20.
50. Ovid, *Metamorphoses* 1.96 (trans. Johnson, 7).
51. Scholten, "Beheaded Bust."
52. Bacon, *Advancement of Learning*, 39.
53. Passe, *Nieuwen en Vermaecklijcken Dool-hof*, n.p.
54. Jensen, *Reizigers te Amsterdam*, 122.
55. Hooft, *Granida*; on paintings of Granida, see Witt, *Jan van Noordt*, 53–56.
56. *Verklaringe van Treffelijcke Konstige Wercken*, n.p.
57. Kettering, *Dutch Arcadia*, 8–11; see also Van den Brink, *Gedroomde Land*.
58. Marx, *Machine in the Garden*, 3–10.
59. Ibid., 15–25, and Sawday, *Engines of the Imagination*, 7–15.
60. Knoespel, "Gazing on Technology," 99–121.
61. Ibid., 117–19.
62. Price, "Automata and the Origins," 13.
63. Mayr, *Authority, Liberty, and Automatic Machinery*, 26.
64. Schaffer, "Enlightened Automata," 126–27, 135–36.
65. Dear, "Mechanical Microcosm," 61, 75.
66. Jonston, *History of the Constancy*, 3.
67. Ibid., dedication.
68. See C. Webster, *Great Instauration*, 15–18.
69. Jonston, *History of the Constancy*, 3.
70. A number of Dutch museums no longer use the term "Golden Age," as announced in this Amsterdam Museum press release: https://www.amsterdammuseum.nl/sites/default/files/press_release_the_amsterdam_museum_to_stop_using_term_golden_age.pdf. See also the historical research project led by Jan Blanc at the University of Geneva: https://www.dutch-golden-ages.com.
71. Boxer, *Dutch Seaborne Empire*, first published in 1965, provides ample historical evidence of these practices.

Epilogue

1. Bacon, "Daedalus, or Mechanical Skill," 238.
2. Ibid., 237.
3. Sawday, *Engines of the Imagination*, 2–4, 19–20.
4. Bacon, "Daedalus, or Mechanical Skill," 238.
5. Breeman, *Waanhoopstoonen van het Oude Doolhof*; the auction took place on June 30, 1862, and many of the works disappeared into private collections. See Witkamp, *Amsterdam in Schetsen*, 88, and Reneman, "'Oude Doolhof,'" 37–39.
6. Breeman, *Waanhoopstoonen van het Oude Doolhof*, 2–3.
7. Ibid., 5: "Treurt Herders, herderinnen / 't Vooruitzigt, dat u wacht, is treurig voor uw geest! / Is eens mijn erf verkocht, wat moet gij dan beginnen?"
8. Reneman, "'Oude Doolhof,'" 85.
9. Stieber notes that the most scathing critiques were printed in the progressive journal *De Kroniek* ("Marking Time and Space," 197).
10. Comments of Amsterdam city councilor Emanuel Boekman, who served on the municipal board that oversaw education and culture; cited in Wiggers, "Stad Amsterdam," 73.
11. Ibid.
12. See Eisma, *David en Goliath*, and Eisma, "Bewegende en Geklede Beelden." Today, the mechanisms of the works have been restored, and the statue group is displayed in the Schuttersgalerij, a covered walkway in the Amsterdam Museum.
13. The Athenian adage is quoted by Erasmus, *Sileni of Alcibiades*. See also Somos, *Secularization and the Leiden Circle*, 183, and the essays in Winkler and Zeitlin, *Nothing to Do with Dionysos?*

BIBLIOGRAPHY

Aa, Abraham Jakob van der. *Biographisch Woordenboek der Nederlanden*. Haarlem: Brederode, 1876.

Abrahamse, Jaap Evert. *De Grote Uitleg van Amsterdam: Stadsontwikkeling in de Zeventiende Eeuw*. Bussum: Thoth Uitgeverij, 2011.

———. "De Ruimtelijke Ontwikkeling van Amsterdam in de Zeventiende Eeuw ende Opkomst van de Stedenbouw als Wetenschap." In *Het Aanzien van Amsterdam: Panorama's, Plattengronden en Profielen uit de Gouden Eeuw*, edited by Bram Bakker and E. Schmitz, 24–41. Bussum: Thoth Uitgeverij, 2007.

Agamben, Giorgio. *Nudities*. Translated by David Kishik and Stefan Pedatella. Stanford: Stanford University Press, 2010.

Ailly, A. E. d'. *Catalogus van Amsterdamsche Plattegronden*. Amsterdam: Archief der Gemeente Amsterdam, 1934.

Alpers, Svetlana. *The Art of Describing: Dutch Art in the Seventeenth Century*. Chicago: University of Chicago Press, 1984.

———. *The Vexations of Art: Velázquez and Others*. New Haven: Yale University Press, 2005.

Altick, Richard. *The Shows of London*. Cambridge: Belknap, 1978.

Aristotle. *Metaphysics. Books I–IX*. Translated by Hugh Tredennick. Vol. 1. Cambridge: Harvard University Press, 2003.

Arnade, Peter. *Beggars, Iconoclasts, and Civic Patriots: The Political Culture of the Dutch Revolt*. Ithaca: Cornell University Press, 2008.

Aytes, Ayhan. "The 'Other' in the Machine: Oriental Automata and the Mechanization of the Mind." PhD diss., University of California, San Diego, 2012.

Bacon, Francis. *The Advancement of Learning*. Edited by Michael Kiernan. Oxford: Clarendon, 2000.

———. "The Clue to the Maze (Filum Labyrinth)." In *The Works of Lord Bacon: With an Introductory Essay, and a Portrait*, 1:232–35. London: Henry G. Bohn, 1850.

———. "Daedalus, or Mechanical Skill." In *Moral and Historical Works*, 236–38.

———. *The Moral and Historical Works of Lord Bacon Including His Essays, Apophthegms, Wisdom of the Ancients, New Atlantis and the Life of Henry the Seventh*. London: Henry G. Bohn, 1860.

———. *New Atlantis*. London: Newcomb, 1659.

———. *The Works of Francis Bacon*. Edited by James Spedding, Robert Leslie Ellis, and Douglas Denon Heath. 14 vols. London: Longman, 1861–79.

Bakhtin, Mikhail. *Rabelais and His World*. Translated by Hélène Iswolsky. Bloomington: Indiana University Press, 1984.

Bakker, Boudewijn. "Het Imago van de Stad: Zelf Portret als Propaganda." In *Het Aanzien van Amsterdam: Panorama's, Plattengronden en Profielen uit de Gouden Eeuw*, edited by Bram Bakker and E. Schmitz, 56–78. Bussum: Thoth Uitgeverij, 2007.

Bal, Mieke. *Quoting Caravaggio: Contemporary Art, Preposterous History*. Chicago: University of Chicago Press, 1999.

Balfour, Ian. "'The Whole Is the Untrue': On the Necessity of the Fragment (After Adorno)." In *The Fragment: An Incomplete History*, edited by William Tronzo, 83–92. Los Angeles: Getty Research Institute, 2009.

Baltrušaitis, Jurgis. *Aberrations: An Essay on the Legend of Forms*. Translated by Richard Miller. Cambridge: MIT Press, 1989.

Bann, Stephen. "Un retour de l'art au jardin: Autour de quelques labyrinthes modernes et postmodernes." In Brunon, *Jardin comme labyrinthe*, 119–215.

Barlaeus, Caspar. *Mercator sapiens*. Translated and edited by S. Van der Woude. Amsterdam: Universiteitsbibliotheek, 1967.

Bass, Marisa, Anne Goldgar, Hanneke Grootenboer, and Claudia Swan, eds. *Conchophilia: Shells, Art, and Curiosity in Early Modern Europe*. Princeton: Princeton University Press, 2021.

Beaune, Jean-Claude. "The Classical Age of Automata: An Impressionistic Survey from the Sixteenth to the Nineteenth Century." In *Fragments for a History of the Human Body*, edited by Michel

Feher, Ramona Naddaff, and Nadia Tazi, 1:431–80. New York: Zone, 1989.

Becker-Cantarino, Barbara. *Daniel Heinsius*. Boston: Twayne, 1978.

Bedini, Silvio A. "The Role of Automata in the History of Technology." *Technology and Culture* 5, no. 1 (1964): 24–42.

Behrmann, Carolin, ed. *Images of Shame, Infamy, Defamation, and the Ethics of Oeconomia*. Berlin: De Gruyter, 2016.

Bennet, Jane. *Vibrant Matter: A Political Ecology of Things*. Durham: Duke University Press, 2010.

Bensaude-Vincent, Bernadette, and William R. Newman, eds. *The Artificial and the Natural: An Evolving Polarity*. Cambridge: MIT Press, 2007.

Berkel, Klaas van. *Isaac Beeckman en de Mechanisering van het Wereldbeeld*. Amsterdam: Rodopi, 1983.

Blaeu, Joan. *Theatrum Orbis Terrarum, sive Atlas Novus*. Amsterdam: Blaeu, 1640.

———. *Toonneel der Steden van de Vereenighde Nederlanden, met hare Beschrijvingen*. Amsterdam: Blaeu, 1652.

Bloemendal, Jan. "Mythology on the Early Modern Humanists' and Rhetoricians' Stage in the Netherlands." In Van de Velde, *Classical Mythology in the Netherlands*, 141–62.

Blok, P. J., and P. C. Molhuysen. *Nieuw Nederlandsch Biografisch Woordenboek*. 10 vols. Leiden: Brill, 1911–33.

Boas, Marie. "Hero's *Pneumatica*: A Study of Its Transmission and Influence." *Isis* 40, no. 1 (1949): 38–48.

Boccaccio, Giovanni. *Famous Women*. Translated by Virginia Brown. Cambridge: Harvard University Press, 2001.

Bont, B. J. M. de. "Bestje van Meurs." *Amsterdamsch Jaarboekje* 1 (1888): 81–89.

Borges, Jorge Luis. "The Garden of Forking Paths." In *Labyrinths: Selected Stories and Other Writings*, edited and translated by Donald A. Yates and James E. Irby, 19–29. New York: New Directions, 1964.

Boxer, C. R. *The Dutch Seaborne Empire, 1600–1800*. New York: Knopf, 1965.

Brandt, George W., and Wiebe Hogendoorn. *German and Dutch Theatre, 1600–1848*. Cambridge: Cambridge University Press, 1993.

Bray, Salomon de. *Architectura Moderna ofte Bouwinge van Onsen Tyt*. Amsterdam: Cornelis Danckerts, 1631.

Breeman, J. *Wanhoopstoonen van het Oude Doolhof toen het de Treurmare Vernam van in Publieke Veiling Verkocht te Worden*. Amsterdam: J. Vetten, 1852.

Brereton, William. *Travels in Holland, the United Provinces, England, Scotland, and Ireland, 1634–1635*. Edited by Edward Hawkins. London: Chetham Society, 1844.

Brink, Peter van den, ed. *Het Gedroomde Land: Pastorale Schilderkunst in de Gouden Eeuw*. Zwolle: Centraal Museum, 1993.

Brown, Bill. "Thing Theory." *Critical Inquiry* 28, no. 1 (2001): 1–22.

Brown, Christopher. "The Strange Case of Jan Torrentius: Art, Sex, and Heresy in Seventeenth-Century Haarlem." In *Rembrandt, Rubens, and the Art of Their Time: Recent Perspectives*, edited by Roland Fleischer, 225–33. University Park: Penn State University Press, 1997.

Brown, Edward. *Naauwkeurige en Gedenkwaardige reysen van Edward Brown*. Amsterdam: J. ten Hoorn, 1696.

Bruaene, Anne-Laure van, and Sarah van Bouchaute. "Rederijkers, Kannenkijkers: Drinking and Drunkenness in the Sixteenth and Seventeenth-Century Low Countries." *Early Modern Low Countries* 1, no. 1 (2017): 1–29.

Brückner, Wolfgang. *Bildnis und Brauch: Studien zur Bildfunktion der Effigies*. Berlin: Erich Schmidt Verlag, 1966.

Bruno, Giuliana. *Atlas of Emotion: Journeys in Art, Architecture, and Film*. New York: Verso, 2002.

Brunon, Hervé, ed. *Le jardin comme labyrinthe du monde: Métamorphoses d'un imaginaire de la Renaissance à nos jours*. Paris: Musée du Louvre, 2008.

———. "Le paradigme labyrinthique." In Brunon, *Jardin comme labyrinthe*, 69–131.

Brusati, Celeste. *Artifice and Illusion: The Art and Writing of Samuel van Hoogstraten*. Chicago: University of Chicago Press, 1995.

———. "Perspectives in Flux: Viewing Dutch Pictures in Real Time." *Art History* 35, no. 5 (2012): 908–33.

Bryson, Norman. *Looking at the Overlooked: Four Essays on Still Life Painting*. Cambridge: Harvard University Press, 1990.

Buci-Glucksman, Christine. *Baroque Reason: The Aesthetics of Modernity*. Translated by Patrick Camiller. London: Sage, 1994.

Bull, Malcolm, *The Mirror of the Gods*. Oxford: Oxford University Press, 2005.

Burke, Peter. "Representations of the Self from Petrarch to Descartes." In Porter, *Rewriting the Self*, 17–28.

Bussels, Stijn. "Making the Most of Theatre and Painting: The Power of *Tableaux Vivants* in Joyous Entries from the Southern Netherlands (1485–1635)." *Art History* 33, no. 2 (2010): 236–47.

Bynum, Caroline Walker. *Christian Materiality: An Essay on Religion in Late Medieval Europe*. New York: Zone, 2011.

———. *Holy Feast and Holy Fast: The Religious Significance of Food to Medieval Women*. Berkeley: University of California Press, 1987.

Calvin, John. *Institutes of the Christian Religion by John Calvin*. Translated by John Allen. 7th American ed. 2 vols. Philadelphia: Presbyterian Board of Christian Education, 1900.

Cats, Jacob. *Houwelyck dat Is de Gansche Gelegtheydt des Echten Staets*. Middelburg: Jan Pietersz van de Venne, 1625.

———. *Silenus Alcibiadis, sive Proteus*. Middelburg: Iohannis Hellenij, 1618.

Catullus, Gaius Valerius. *Catullus: The Complete Poems*. Edited and translated by Guy Lee. Oxford: Oxford University Press, 1991.

Caus, Isaac de, and Salomon de Caus. *New and Rare Inventions of Water-Works*. Translated by John Leak. London: Joseph Moxon, 1659.

Caus, Salomon de. *Les raisons des forces mouvantes avec diverses machines tant utiles que plaisantes: Ausquelles ont adjoints plusieurs desseins de grotes et fontaines*. Paris: H. Droüart, 1624.

Causé, Hendrik. *De Koninglycke Hovenier*. Amsterdam: Marcus Doornick, 1676.

Cave, Terence. "Ronsard's Bacchic Fresco." In *Retrospectives: Essays in Literature, Poetics, and Cultural History*, edited by Neil Kenny and Wes Williams, 50–75. London: Routledge, 2009.

Certeau, Michel de. *The Practice of Everyday Life*. Berkeley: University of California Press, 1984.

Chapuis, Alfred, and Edmond Droz. *Automata: A Historical and Technological Study*. Translated by Alec Reid. London: Batsford, 1958.

Chartier, Roger. *The Cultural Uses of Print in Early Modern France*. Princeton: Princeton University Press, 1987.

Chene, Dennis des. *Spirits and Clocks: Machine and Organism in Descartes*. Ithaca: Cornell University Press, 2001.

Cicero. *Treatise on the Best Style of Orators*. In *The Orations of Marcus Tullius Cicero*, translated by Charles Duke Yonge, 4:527–33. London: Bell and Daldy, 1871.

Cipolla, Carlo M. *Clocks and Culture, 1300–1700*. London: Collins, 1967.

Clark, Stuart. *Vanities of the Eye: Vision in Early Modern European Culture*. Oxford: Oxford University Press, 2007.

Cleef, Pierre van. *Les amusemens de la Hollande*. The Hague, 1739.

Clippel, Karolien de. "Rough Bacchus, Neat Andromeda: Rubens's Mythologies and the Meaning of Manner." In Van de Velde, *Classical Mythology in the Netherlands*, 95–112.

Clucas, Stephen. "In Search of 'The True Logick:' Methodological Eclecticism Among the Baconian Reformers." In *Samuel Hartlib and the Universal Reformation: Studies in International Communication*, edited by Mark Greengrass, Michael Leslie, and Timothy Raylor, 51–74. Cambridge: Cambridge University Press, 1994.

Colie, Rosalie. *Paradoxia Epidemica: The Renaissance Tradition of Paradox*. Princeton: Princeton University Press, 1966.

Comenius, John Amos. *The Labyrinth of the World and the Paradise of the Heart*. Translated by Matthew Spinka. 1623. Reprint, Ann Arbor: University of Michigan, 1972.

Commelin, Casparus. *Beschryvinge van Amsterdam*. Amsterdam: Andries van Damme, 1726.

Conan, Michel. *Baroque Garden Cultures: Emulation, Sublimation, Subversion*. Washington, DC: Dumbarton Oaks Research Library and Collection, 2005.

Cook, Harold John. *Matters of Exchange: Commerce, Medicine, and Science in the Dutch Golden Age*. New Haven: Yale University Press, 2007.

Coornhert, Dirck Volckertzoon. "Boeven-Tucht: The Discipline of Misbehaviour or the Means to Reduce the Number of Harmful Idlers."

Translated by William Freund, Roger Deacon, and Eddy Verbaan. *Theoria: A Journal of Social and Political Theory* 56, no. 118 (2009): 89–104.

———. *Ethics or the Art of Living Well, by Means of Knowledge of the Truth About Man, Sin, and Virtue*. Translated and edited by Gerrit Voogt. 1586. Reprint, Hilversum: Veloren, 2015.

Corrigan, Karina H., Jan van Campen, Femke Diercks, and Janet C. Blyberg, eds. *Asia in Amsterdam: The Culture of Luxury in the Golden Age*. Amsterdam: Rijksmuseum, 2015.

Crary, Jonathan. *Techniques of the Observer: On Vision and Modernity in the Nineteenth Century*. Cambridge: MIT Press, 1990.

Dalen, Dorrit van. "Johannes Theunisz and ʿAbd al-ʿAzīz: A Friendship in Arabic Studies in Amsterdam, 1609–1610." *Lias: Journal of Early Modern Intellectual Culture and Its Sources* 43, no. 1 (2016): 161–89.

Daninos, Andrea. "Wax Figures in Italy: A Brief History." In Daninos, *Waxing Eloquent*, 13–42.

———, ed. *Waxing Eloquent: Italian Portraits in Wax*. Translated by Catherine Bolton. Milan: Officina Libraria, 2012.

Dante Alighieri. *The Divine Comedy*. Translated by Henry Wadsworth Longfellow. Edited by Charles Welsh. New York: Davoe, 1909.

Dapper, Olfert. *Historische Beschryving der Stadt Amsterdam*. Amsterdam: Jacob van Meurs, 1663.

Daston, Lorraine, and Katharine Park. *Wonders and the Order of Nature, 1150–750*. New York: Zone, 2001.

Davis, Natalie Zemon. "Women on Top." In *Society and Culture in Early Modern France: Eight Essays*, 124–51. Stanford: Stanford University Press, 1975.

Dear, Peter. "A Mechanical Microcosm: Bodily Passions, Good Manners, and Cartesian Mechanism." In *Science Incarnate*, edited by C. Lawrence and S. Shapin, 51–82. Chicago: University of Chicago Press, 1998.

Descartes, René. *Discourse on Method and Meditations on First Philosophy*. Translated by Donald A. Cress. Indianapolis: Hackett, 1998.

———. *The Passions of the Soul and Other Late Philosophical Writings*. Translated by Michael Moriarty. Oxford: Oxford University Press, 2015.

———. *The Philosophical Writings of Descartes*. Vol. 2. Translated by John Cottingham, Robert Stoothoff, and Dugald Murdoch. Cambridge: Cambridge University Press, 1984.

———. *The World and Other Writings*. Translated and edited by Stephen Gaukroger. Cambridge: Cambridge University Press, 1998.

Deth, Ron van, and Walter Vandereycken. *Van Vastenwonder Tot Magerzucht: Anorexia Nervosa in Historisch Perspectief*. Meppel: Boom, 1988.

Dickey, Stephanie S., and Herman Roodenburg, eds. "The Passions in the Arts of the Early Modern Netherlands." Special issue, *Nederlands Kunsthistorisch Jaarboek* 60, no. 1 (2010).

Dickey, Stephanie S., and Herman Roodenburg. "Introduction: The Motions of the Mind." *Nederlands Kunsthistorisch Jaarboek* 60, no. 1 (2010): 7–18.

Dickson, Donald R. *The Tessera of Antilla: Utopian Brotherhoods and Secret Societies in the Early Seventeenth Century*. Leiden: Brill, 1998.

Didi-Huberman, Georges. "The Molding Image: Genealogy and the Truth of Resemblance in Pliny's *Natural History*, Book 35, 1–7." In *Law and the Image: The Authority of Art and the Aesthetics of Law*, edited by Costas Douzinas and Lynda Nead, 71–88. Chicago: University of Chicago Press, 1999.

———. "Viscosities and Survivals: Art History Put to the Test by the Material." In Panzanelli, *Ephemeral Bodies*, 154–70.

Diehl, Huston. "Into the Maze of the Self: The Protestant Transformation of the Image of the Labyrinth." *Journal of Medieval and Renaissance Studies* 16, no. 1 (1986): 281–301.

Dijkshoorn, Willemien, and Erik de Jong. *Herengracht*. Zwolle: Waanders, 1998.

———. *Prinsengracht*. Zwolle: Waanders, 2000.

Domselaer, Tobias van. *Beschryvinge van Amsterdam*. Amsterdam: Marcus Willemsz. Doornick, 1665.

Doob, Penelope Reed. *The Idea of the Labyrinth: From Classical Antiquity Through the Middle Ages*. Ithaca: Cornell University Press, 1994.

Duits, H. "Hoofts 'Henrik de Gróte': Een neostoïches vorstenspiegel." In *Mensen van de Nieuwe tijd: Een liber amicorum voor A. Th. Van Deursen*, edited by M. Bruggeman and Arie van Deursen, 266–83. Amsterdam: Bert Bakker, 1996.

Dulk, Leo den, and Erik de Jong. *Keizersgracht*. Zwolle: Waanders, 1997.

Duncan, Carol. "Art Museums and the Ritual of Citizenship." In *Exhibiting Cultures: The Poetics and Politics of Museum Display*, edited by I. Karp and S. Lavine, 88–103. Washington, DC: Smithsonian Institution, 1991.

———. *Civilizing Rituals: Inside Public Art Museums*. London: Routledge, 2007.

Duncan, Carol, and Alan Wallach. "The Museum of Modern Art as Late Capitalist Ritual: An Iconographic Analysis." *Marxist Perspectives* 4 (1978): 28–51.

Dupré, Sven. *Laboratories of Art: Alchemy and Art Technology from Antiquity to the 18th Century*. Cham: Springer, 2014.

Eamon, William. *Science and the Secrets of Nature: Books of Secrets in Medieval and Early Modern Culture*. Princeton: Princeton University Press, 1994.

Eck, Caroline van. *Art, Agency and Living Presence: From the Animated Image to the Excessive Object*. Leiden: De Gruyter, 2015.

Eck, Caroline van, and Stijn Bussels, eds. "Theatricality in Early Modern Visual Art and Architecture." Special issue, *Art History* 33, no. 2 (2010).

Edgerton, Samuel. *Pictures and Punishment: Art and Criminal Prosecution During the Florentine Renaissance*. Ithaca: Cornell University Press, 1985.

Eire, Carlos M. N. *War Against the Idols: The Reformation of Worship from Erasmus to Calvin*. Cambridge: Cambridge University Press, 2003.

Eisma, Marianne. "Albert Jansz Vinckenbrink, Ontwerper en Beeldsnijder." *Maandblad Amstelodamum* 83 (1996): 33–43.

———. "Bewegende en Geklede Beelden in het Amsterdams Historisch Museum." In Jonker, *In Beeld Gebracht*, 76–88.

———. *David en Goliath met Zijn Schilddrager: Een Beeldengroep uit het Oude Doolhof*. Wormer: Inmerc, 1996.

———. "De Fonteinmaker Jonas Bargois en het Nieuwe Doolhof 'In de Orange Pot.'" *Amstelodamum* 84, no. 2 (1997): 55–62.

Elias, Johan E. *De Vroedschap van Amsterdam: 1578–1795*. Amsterdam: N. Israel, 1963.

Elias, Norbert. *The Civilizing Process*. Vol. 1, *The History of Manners*. Translated by Edmund Jephcott. Oxford: Blackwell, 1982.

Erasmus, Desiderius. *All the Familiar Colloquies of Desiderius Erasmus, of Roterdam: Concerning Men, Manners, and Things*. Translated by N. Bailey. London: J. J. and P. Knapton et al., 1733.

———. *The Praise of Folly*. Translated by Hoyt Hopewell Hudson. Edited with foreword by Anthony Grafton. Princeton: Princeton University Press, 2015.

———. *The Sileni of Alcibiades*. Edited by William Barker. Toronto: University of Toronto Press, 2001.

Evelyn, John. *The Diary of John Evelyn*. Edited by William Bray. 2 vols. London: M. W. Dunne, 1901.

Filipczak, Zirka Zaremba. *Hot Dry Men, Cold Wet Women: The Theory of Humors in Western European Art, 1575–1700*. New York: American Federation of Arts, 1997.

Fisher, Adrian. *The Amazing Book of Mazes*. New York: Abrams, 2006.

Fokkens, Melchior. *Beschrijvinge der Wijdt-vermaarde Koop-Stadt Amstelredam*. Amsterdam: Abraham en Jan de Wees, 1662.

Ford, Charles. "People as Property." *Oxford Art Journal* 25, no. 1 (2002): 3–16.

Foucault, Michel. *Death and the Labyrinth: The World of Raymond Roussel*. Translated by Charles Ruas. London: Athlone, 1986.

———. *Discipline and Punish: The Birth of the Prison*. Translated by Alan Sheridan. New York: Vintage, 1977.

Francisco, Adam. *Martin Luther and Islam: A Study in Sixteenth-Century Polemics and Apologetics*. Leiden: Brill, 2007.

Franken, D. "Albert Jansz Vinckenbrink." *Oud Holland* 5, no. 1 (1887): 72–92.

Franken, Daniel. *L'Œuvre gravé des van de Passe: Catalogue raisonné des estampes de Chrispijn senior et junior, Simon, Willem, Magdalena et Chrispijn III van de Passe, graveurs néerlandais des XVIe et XVIIe siècles*. Amsterdam: G. W. Hissink, 1975.

Freedberg, David. "From Defamation to Mutilation: Gender Politics and Reason of State in Africa." In Behrmann, *Images of Shame*, 193–216.

———. *The Power of Images: Studies in the History and Theory of Response*. Chicago: University of Chicago Press, 1989.

Gatacre, Edward V., and Laura Dru. "Portraiture in Le Cabinet de Cire de Curtius and Its Successor Madame Tussaud's Exhibition." In *La ceroplastica nella scienza e nell'arte: Atti del i Congresso Internationale; Firenze, 3–7 giugno 1975*, 617–35. Florence: Leo S. Olschki, 1977.

Gaukroger, Stephen. *Descartes: An Intellectual Biography*. Oxford: Clarendon, 1995.

———. *Francis Bacon and the Transformation of Early-Modern Philosophy*. Cambridge: Cambridge University Press, 2001.

Gell, Alfred. *Art and Agency: An Anthropological Theory*. Oxford: Clarendon, 1998.

Geyl, A. *Nieuw Nederlandsch Biografisch Woordenboek*. 10 vols. Leiden: Sijthoff, 1911–37.

Gibson, Walter S. *Figures of Speech: Picturing Proverbs in Renaissance Netherlands*. Berkeley: University of California Press, 2010.

Giesey, Ralph E. *The Royal Funeral Ceremony in Renaissance France*. Geneva: Droz, 1960.

Ginzburg, Carlo. "Representation, as the Work, the Idea and the Thing." In *Wooden Eyes: Nine Reflections on Distance*, 63–76. Translated by Martin Ryle and Kate Soper. New York: Columbia University Press, 2001.

Goldgar, Anne. *Tulipmania: Money, Honor, and Knowledge in the Dutch Golden Age*. Chicago: University of Chicago Press, 2007.

Gorp, P. J. M. van, E. de Jong, and D. P. Snoep. *Zijdebalen, Lusthof aan de Vecht*. Utrecht: Centraal Museum, 1981.

Gorski, Philip S. *The Disciplinary Revolution: Calvinism and the Rise of the State in Early Modern Europe*. Chicago: University of Chicago Press, 2003.

Gouw, Jan ter, and Jacob van Lennep. *De Uithangteekens in Verband met Geschiedenis en Volksleven Beschouwd*. Illustrated by G. W. Zürcher. 2 vols. Leiden: Kraay, 1867–69.

Grafton, Anthony. "The Devil as Automaton: Giovanni Fontana and the Meanings of a Fifteenth-Century Machine." In Riskin, *Genesis Redux*, 46–62.

———. *Joseph Scaliger: A Study in the History of Classical Scholarship*. 2 vols. Oxford: Oxford University Press, 1983.

———. *Magic and Technology in Early Modern Europe*. Washington, DC: Smithsonian Institution, 2005.

———. "Renaissance Histories of Art and Nature." In Bensaude-Vincent and Newman, *The Artificial and the Natural*, 185–210.

Greene, Thomas M. "Labyrinth Dances in the French and English Renaissance." *Renaissance Quarterly* 54, no. 1 (2001): 1403–66.

Groen, Jan van der. *Den Nederlandtsen Hovenier*. Edited by C. S. Oldenburger-Ebbers and D. O. Wijnands. Utrecht: Stichting Matrijs, 1988.

Grootenboer, Hanneke. "A Clock Picture as Philosophical Experiment: The Tableau Mécanique in the Physics Cabinet of Bonnier de la Mosson." *Art History* 39, no. 2 (2016): 340–55.

———. "Introduction: On the Substance of Wax." *Oxford Art Journal* 36, no. 1 (2013): 1–12.

———. "Sublime Still Life: On Adriaen Coorte, Elias van den Broeck, and the *Je ne sais quoi* of Painting." *Journal of Historians of Netherlandish Art* 8, no. 2 (2016). http://doi.org/10.5092/jhna.2016.8.2.10.

———, ed. "Theorizing Wax: On the Meaning of a Disappearing Medium." Special issue, *Oxford Art Journal* 36, no. 1 (2013).

Gross, Kenneth. *The Dream of the Moving Statue*. Ithaca: Cornell University Press, 1992.

Guerzoni, Guido. "Use and Abuse of Beeswax in the Early Modern Period." In Daninos, *Waxing Eloquent*, 43–59.

Le Guide, ou, Nouvelle Description d'Amsterdam. Amsterdam: Covens & Mortier, 1753.

Haldar, Piyal. "From Dignity to Shame: X-rays and Legal Persona." In Behrmann, *Images of Shame*, 239–64.

Hamilton, Alistair. "'Nam Tirones Sumus': Franciscus Raphelengius' *Lexicon Arabico-Latinum* (Leiden 1613)." In *Studia in Memoriam Christophori Plantini (ca. 1520–1589)*, edited by Marcus de Schepper and Francine de Nave, 557–89. Antwerp: Vereeniging der Antwerpsche Bibliophielen, 1989.

Hamilton, Alastair, Sjouke Voolstra, and Pieter Visser, eds. *From Martyr to Muppy (Mennonite Urban Professionals): A Historical Introduction to Cultural Assimilation Processes of a Religious Minority in the Netherlands*. Amsterdam: Amsterdam University Press, 1994.

Harrison, Robert Pogue. *Gardens: An Essay on the Human Condition*. Chicago: University of Chicago Press, 2008.

Hattab, Helen. *Descartes on Forms and Mechanisms*. Cambridge: Cambridge University Press, 2009.

Hegenitius, Gottfried. *Itinerarium Frisio-Hollandicum*. Leiden: Elsevier, 1630.

Heinen, Ulrich. "Huygens, Rubens, and Medusa: Reflecting the Passions in Paintings, with Some Consideration of Neuroscience in Art History." *Nederlands Kunsthistorisch Jaarboek* 60, no.1 (2010): 151–78.

Heinsius, Daniel. *Nederduytsche Poemata*. Vol. 3, *Hymnus oft Lof-sanck van Bacchus, Waerin 'tgebruyck ende Misbruyck vande Wijn Beschreven Wort*. Edited by Petrus Scriverius. Amsterdam: Willem Jansz Blaeu, 1616.

———. *On Plot in Tragedy*. Translated by Paul R. Sellin and John J. McManmon. Northridge: San Fernando Valley State College, 1971.

Hell, Maarten. "Trade, Transport and Storage in Amsterdam Inns (1450–1800)." *Journal of Urban History* 40, no. 4 (2014): 742–61.

Helms, Mary. *Craft and the Kingly Ideal: Art, Trade, and Power*. Austin: University of Texas Press, 1993.

Heuer, Christopher P. *The City Rehearsed: Object, Architecture, and Print in the Worlds of Hans Vredeman de Vries*. London: Routledge, 2009.

Hill, Donald R. Introduction to *The Book of Knowledge of Ingenious Mechanical Devices*, by Ismāʿīl Ibn al-Razzāz al-Jazarī, 3–12. Translated and annotated by Donald R. Hill. Dordrecht: Reidel, 1974.

———. *Islamic Science and Engineering*. Edinburgh: Edinburgh University Press, 1993.

———. *Studies in Medieval Islamic Technology: From Philo to al-Jazarī—from Alexandria to Diyār Bakr*. Edited by David A. King. Aldershot: Ashgate, 2010.

Hochstrasser, Julie Berger. *Still Life and Trade in the Dutch Golden Age*. New Haven: Yale University Press, 2007.

Hollander, August den, Mirjam van Veen, Anna Voolstra, and Alex Nord, eds. *Religious Minorities and Cultural Diversity in the Dutch Republic*. Leiden: Brill, 2014.

Holly, Michael Ann. "Back of the Painted Beyond." *Art History* 42, no. 5 (2019): 892–913.

Holmes, Megan. "Ex-votos: Materiality, Memory, and Cult." In *The Idol in the Age of Art: Objects, Devotions and the Early Modern World*, edited by Michael Cole and Rebecca Zorach, 165–88. Aldershot: Ashgate, 2009.

Homer. *The Iliad*. Translated by Peter Green. Oakland: University of California Press, 2015.

Hooft, P. C. *Granida*. Edited by Lia van Gemert. 1605. Reprint, Amsterdam: Amsterdam University Press, 1998.

———. *Theseus ende Ariadne*. Amsterdam: Willem Jansz. Wijngaerts, 1628.

Hoogstraten, Samuel van. *Inleyding tot de Hooge Schoole der Schilderkonst*. 1678. Reprint, Doornspijk: Davaco, 1969.

Horst, Daniel R. "Duke of Alba: The Ideal Enemy." *Arte Nuevo: Revista de Estudios Áureos* 1 (2014): 130–54.

Houdt, Toon van. "The Spectacle of Power: Lipsius' Model of Princely (and Humanist) Conduct in His *Monita et exempla politica* (1605)." In *Miraculum Eruditionis: Neo-Latin Studies in Honour of Hans Helander*, edited by M. Berggren and C. Henriksén, 13–30. Uppsala: Uppsala Universitet, 2007.

Howell, James. *Epistolae Ho-Elianae: Familiar Letters Domestic and Forren*. 3 vols. 2nd ed. London: Humphrey Moseley, 1650.

Huet, Marie-Hélène. *Monstrous Imagination*. Cambridge: Harvard University Press, 1993.

Hugo, Herman. *Pia Desideria*. Antwerp: Hendrick Aertssens, 1624.

Hunt, John Dixon. *The Dutch Garden in the Seventeenth Century*. Washington, DC: Dumbarton Oaks Research Library and Collection, 1990.

———. *Greater Perfections: The Practice of Garden Theory*. Philadelphia: University of Pennsylvania Press, 2000.

Hunt, John Dixon, and E. de Jong. *The Anglo-Dutch Garden in the Age of William and Mary*. London: Taylor & Francis, 1988.

Huygens, Constantijn. *Gedichten van Constantijn Huygens*. Edited by J. A. Worp. Groningen: Wolters, 1893.

———. *Hofwyck*. Wassenaar: Servire, 1967.

———. *A Selection of the Poems of Sir Constantijn Huygens (1596–1687)*. Translated and edited

by Peter Davidson and Adriaan van der Weel. Amsterdam: Amsterdam University Press, 1996.
Jacobs, Fredrika H. *The Living Image in Renaissance Art*. Cambridge: Cambridge University Press, 2005.
Jardine, Lisa. *Francis Bacon: Discovery and the Art of Discourse*. Cambridge: Cambridge University Press, 2009.
———. *Hostage to Fortune: The Troubled Life of Francis Bacon*. New York: Hill and Wang, 1999.
Al-Jazarī, Ismāʿīl ibn al-Razzāz. *The Book of Knowledge of Ingenious Mechanical Devices*. Translated and annotated by Donald R. Hill. Dordrecht: Reidel, 1974.
Jensen, J. N. Jacobsen. *Reizigers te Amsterdam: Beschrijvende Lijst van Reizen in Nederland door Vreemdelingen vóór 1850*. Amsterdam: Genootschap Amstelodamum, 1919.
Jolle, Pierre le. *Description de la ville d'Amsterdam, en vers burlesques*. Amsterdam: Chés Jaques le Curieux [Elsevier], 1666.
Jonckheere, Koenraad. *Willem Key (1516–1568): Portrait of a Humanist Painter*. Turnhout: Brepols, 2011.
Jones, Caroline A., Peter Galison, and Amy E. Slaton. *Picturing Science, Producing Art*. New York: Routledge, 1998.
Jong, Erik de. "Aristocratiques ornements et divertissements populaires: Le labyrinthe dans la Renaissance du Nord (1580–1660)." In Brunon, *Jardin comme labyrinthe*, 133–50.
———. *Nature and Art: Dutch Garden and Landscape Architecture, 1650–1740*. Translated by Ann Langenakens. Philadelphia: University of Pennsylvania Press, 2000.
Jong, Erik de, Marleen Dominicus-van Soest, and Willemien B. de Vries. *Aardse Paradijzen: De Tuin in de Nederlandse Kunst*. Ghent: Snoeck-Ducaju en Zoon, 1996.
Jong, Erik de, and Claudia Schellekens. *Het Beeld Buiten: Vier Eeuwen Tuinsculptuur in Nederland*. Wijhe: Hannema de Stuers Fundatie, 1994.
Jong, Erik de, Caroline van Eck, and H. J. Scheepmaker. *Tuinkunst: Nederlands Jaarboek voor de Geschiedenis van Tuin- en Landschapsarchitectuur*. Amsterdam: Architectura & Natura, 1996.
Jonge, H. J. de. *The Auction Catalogue of the Library of J. J. Scaliger*. Utrecht: HES, 1977.
Jongh, Eddy de. "De Schaduwen van Daedalus." *Kunstschrift* 91, no. 3 (1991): 10–15.
———. *Tot Lering en Vermaak: Betekenissen van Hollandse Genrevoorstellingen uit de Zeventiende Eeuw*. Amsterdam: Rijksmuseum, 1976.
Jonker, Michiel, ed. *In Beeld Gebracht: Beeldhouwkunst uit de Collectie van het Amsterdams Historisch Museum*. Zwolle: Waanders, 1995.
Jonston, John. *A History of the Constancy of Nature*. London: John Streater, 1657.
Jorink, Eric. *Reading the Book of Nature in the Dutch Golden Age, 1575–1715*. Translated by Peter Mason. Leiden: Brill, 2010.
Judovitz, Dalia. *Culture of the Body: Genealogies of Modernity*. Ann Arbor: University of Michigan Press, 2011.
Junius, Franciscus. *The Painting of the Ancients, in Three Bookes*. London: R. Hodgkinsonne, 1638.
Kang, Minsoo. *Sublime Dreams of Living Machines: The Automaton in the European Imagination*. Cambridge: Harvard University Press, 2011.
———. "Wonders of Mathematical Magic: Lists of Automata in the Transition from Magic to Science, 1533–1662." *Comitatus: A Journal of Medieval and Renaissance Studies* 33 (2002): 113–39.
Kantorowicz, Ernst Hartwig. *The King's Two Bodies: A Study in Mediaeval Political Theology*. 1957. Reprint, Princeton: Princeton University Press, 1997.
Kaplan, Benjamin J. "Fictions of Privacy: House Chapels and the Spatial Accommodation of Religious Dissent in Early Modern Europe." *American Historical Review* 107, no. 4 (2002): 1031–64.
———. *Muslims in the Dutch Golden Age: Representations and Realities of Religious Toleration*. Amsterdam: Amsterdam University Press, 2006.
Keating, Jessica. *Animating Empire: Automata, the Holy Roman Empire, and the Early Modern World*. University Park: Penn State University Press, 2018.
———. "The Machinations of German Court Culture: Early Modern Automata." PhD diss., Northwestern University, 2010.
Kern, Hermann. "Bild der Welt und heiliger Bezirk: Labyrinthstädte-Satdlabyrinthe; Image of the World and Sacred Realms: Labyrinth-Cities—City-Labyrinths." *Daidalos* 3 (1982): 10–24.

———. *Through the Labyrinth: Designs and Meanings over 5,000 Years*. Edited by Robert Ferré. Translated by Abigail H. Clay and Sandra Burns Thomson. Munich: Prestel, 2000.

Kettering, Alison McNeil. *The Dutch Arcadia: Pastoral Art and Its Audience in the Golden Age*. Montclair: Boydell, 1982.

Kijk in de Wereld, Jan. *Omzwervingen door Neêrlands Hoofdstad*. Amsterdam: Hendrik Frijlink, 1845.

Kilian, Jennifer M. *The Paintings of Karel du Jardin, 1626–1678*. Amsterdam: John Benjamins, 2005.

King, Elizabeth. "Perpetual Devotion: A Sixteenth-Century Machine That Prays." In Riskin, *Genesis Redux*, 263–92.

Klein, Robert. "Spirito Peregrino." In *Form and Meaning: Essays on the Renaissance and Modern Art*, translated by Madeline Jay and Leon Wieseltier, 62–85. New York: Viking, 1979.

Knoespel, Kenneth. "Gazing on Technology: *Theatrum Mechanorum* and the Assimilation of Renaissance Machinery." In *Literature and Technology*, edited by Mark Greenberg and Lance Schachterle, 99–121. Bethlehem: Lehigh University Press, 1992.

Koeppe, Wolfram. *Making Marvels: Science and Splendor at the Courts of Europe*. New York: Metropolitan Museum of Art, 2019.

Koyré, Alexandre. *From the Closed World to the Infinite Universe*. Baltimore: Johns Hopkins University Press, 1957.

Kretzschmar, Marthe. "Ähnlichkeit und Material: Überlegungen zum Bildzauber lebensgroßer Wachsporträts in der Frühen Neuzeit." *Zeitschrift für Kunstgeschichte* 77, no. 4 (2014): 565–75.

Kümin, Beat A. *Drinking Matters: Public Houses and Social Exchange in Early Modern Central Europe*. Basingstoke: Palgrave Macmillan, 2007.

Kunzle, David. *From Criminal to Courtier: The Soldier in Netherlandish Art, 1550–1672*. Leiden: Brill, 2002.

Kurz, Otto. *European Clocks and Watches in the Near East*. London: Warburg Institute; Leiden: Brill, 1975.

Landes, David. *Revolution in Time: Clocks and the Making of the Modern World*. Cambridge: Harvard University Press, 2000.

Lefebvre, Henri. *The Production of Space*. Translated by Donald Nicholson-Smith. Oxford: Blackwell, 1991.

Lewis, Rhodri. "Francis Bacon: Allegory and the Uses of Myth." *Review of English Studies* 61 (2010): 360–89.

Lipsius, Justus. *On Constancy*. Edited and translated by R. V. Young. 1583. Reprint, Tempe: Arizona Center for Medieval and Renaissance Studies, 2011.

———. *Politica: Six Books on Politics or Political Instruction*. Edited and translated by Jan Waszink. Assen: Royal Van Gorcum BV, 2004.

Loh, Maria. "Introduction: Early Modern Horror." *Oxford Art Journal* 34, no. 3 (2011): 321–33.

Long, Pamela O. *Openness, Secrecy, Authorship: Technical Arts and the Culture of Knowledge from Antiquity to the Renaissance*. Baltimore: Johns Hopkins University Press, 2001.

———. "Trading Zones in Early Modern Europe." *Isis* 106, no. 4 (2015): 840–47.

Luijten, Hans, Jan Bloemendal, and Grieta Felix. "*Ad Nova Semper Hians (Steeds Naar iets Nieuws Verlangend)*: Het Latijn in Jacob Cats' *Sinne- en Minnebeelden*." *De Zeventiende Eeuw* 10 (1994): 292–305.

MacDougall, Elisabeth B., ed. *Fons Sapientiae: Renaissance Garden Fountains*. Washington, DC: Dumbarton Oaks, 1978.

———. "L'ingegnoso artifizio: Sixteenth-Century Garden Fountains in Rome." In *Fountains, Statues, and Flowers: Studies in Italian Gardens of the Sixteenth and Seventeenth Centuries*, edited by Elisabeth B. MacDougall, 57–87. Washington, DC: Dumbarton Oaks, 1994.

Maisano, Scott. "Descartes avec Milton: The Automata in the Garden." In *The Automaton in English Renaissance Literature*, edited by Wendy Beth Hyman, 21–44. Farnham: Ashgate, 2011.

Mander, Karel van. *Het Schilder-Boeck*. 1604. Reprint, Utrecht: Davaco, 1969.

———. *The Lives of the Illustrious Netherlandish and German Painters: From the First Edition of the Schilderboek (1603–1604)*. Edited and translated by Hessel Miedema. 6 vols. Doornspijk: Davaco, 1994.

Marr, Alexander. "'Gentille Curiosité': Wonder-Working and the Culture of Automata in the Late Renaissance." In *Curiosity and Wonder from the Renaissance to the Enlightenment*, edited by

Robert John Weston Evans and Alexander Marr, 149–70. London: Routledge, 2006.
———. "Understanding Automata in the Late Renaissance." *Le Journal de la Renaissance* 2 (2004): 205–22.
Marx, Leo. *The Machine in the Garden: Technology and the Pastoral Ideal in America*. Oxford: Oxford University Press, 1964.
Massey, Lyle. "Reflections on Temporality in Netherlandish Art." *Art History* 35, no. 5 (2012): 1050–57.
Matthews, William Henry. *Mazes and Labyrinths*. New York: Dover, 1970.
Maurice, Klaus, and Otto Mayr, eds. *The Clockwork Universe: German Clocks and Automata, 1550–1650*. Washington, DC: Smithsonian Institution, 1980.
Maxwell, Susan. "The Pursuit of Art and Pleasure in the Secret Grotto of Wilhelm V of Bavaria." *Renaissance Quarterly* 61, no. 2 (2008): 414–62.
Mayr, Otto. *Authority, Liberty, and Automatic Machinery in Early Modern Europe*. Baltimore: Johns Hopkins University Press, 1986.
McHam, Sarah Blake. "Public Sculpture in Renaissance Florence." In *Looking at Italian Renaissance Sculpture*, edited by Sarah Blake McHam, 149–88. Cambridge: Cambridge University Press, 1998.
Meijer, D. C. "Het Oude Doolhof te Amsterdam." *Oud-Holland* 1 (1883): 31–36, 119–35.
Mercator, Gerhard, and Jodocus Hondius. *Historia Mundi: or Mercator's Atlas Containing his Cosmographicall Description of the Fabricke and Figure of the World*. Translated by Wye Saltonstall. London: T. Cotes, for Michael Sparke and Samuel Cartwright, 1635.
Mercator, Gerhard, and Ptolemy. *Tabulae Geographicae*. Cologne: Typus Godefridi Kempensis, 1578.
Merchant, Carolyn. "Francis Bacon and the 'Vexations of Art': Experimentation as Intervention." *British Journal for the History of Science* 46, no. 4 (2013): 551–99.
Meyer, Maurits de. *De Volks- en Kinderprent in de Nederlanden van de 15e tot de 20e Eeuw*. Antwerpen: Standaard-Boekhandel, 1962.
Michalski, Serge. *The Reformation and the Visual Arts: The Protestant Image Question in Western and Eastern Europe*. London: Routledge, 1993.
Mieris, Frans van. *Beschryving der Stad Leyden*. Leiden: W. A. Honkoop, C. van Hoogeveen, 1762.
Montaigne, Michel de. *The Complete Works: Essays, Travel Journal, Letters*. Translated by Donald Murdoch Frame. New York: Knopf, 2003.
Morgan, Luke. *Nature as Model: Salomon de Caus and Early Seventeenth-Century Landscape Design*. Philadelphia: University of Pennsylvania Press, 2007.
Morris, Katherine. "Bêtes-machines." In *Descartes' Natural Philosophy*, edited by Stephen Gaukroger, John Andrew Schuster, and John Sutton, 401–19. London: Routledge, 2000.
Morris, Sarah P. *Daidalos and the Origins of Greek Art*. Princeton: Princeton University Press, 1995.
Mosser, Monique, and Georges Teyssot. *The Architecture of Western Gardens: A Design History from the Renaissance to the Present Day*. Cambridge: MIT Press, 1991.
Mountague, William. *The Delights of Holland*. London: John Sturton, 1696.
Mulier, E. O. G. Haitsma. "Willem van Oranje in de Historiografie van de Zeventiende Eeuw." In *Willem van Oranje in de Historie, 1584–1984*, edited by E. O. G. Haitsma Mulier and A. E. M. Janssen, 32–62. Utrecht: HES, 1984.
Nadler, Steven. *Rembrandt's Jews*. Chicago: University of Chicago Press, 2003.
Needham, Joseph. *Science and Civilisation in China*. Vol. 2, *History of Scientific Thought*. Cambridge: Cambridge University Press, 1956.
Needham, Joseph, Wang Ling, and Derek J. de Solla Price. *Heavenly Clockwork: The Great Astronomical Clocks of Medieval China*. Cambridge: Cambridge University Press, 1960.
Neurdenburg, Elisabeth. *Hendrick de Keyser: Beeldhouwer en Bouwmeester van Amsterdam*. Amsterdam: Scheltema & Holkema, 1931.
Nierop, Henk van. "Oranje Boven: Willem van Oranje als Zinnebeeld van de Natie." In *Willem van Oranje Lezing 2001: Cultuur, Samenleving en Bestuur*, edited by Henk van Nierop et al., 7–27. Delft: NIVO, 2001.
Odell, Dawn. "Delftware and the Domestication of Chinese Porcelain." In *EurAsian Matters: China, Europe, and the Transcultural Object, 1600–1800*, edited by Monica Juneja and Anna Grasskamp,

175–202. Transcultural Research: Heidelberg Studies on Asia and Europe. Berlin: Springer, 2018.

Oestreich, Gerhard. *Neostoicism and the Early Modern State*. Translated by D. McLintock. Cambridge: Cambridge University Press, 1982.

Oostveldt, Bram van, and Stijn Bussels, eds. "The Sublime and Seventeenth-Century Netherlandish Art." Special issue, *Journal of Historians of Netherlandish Art* 8, no. 2 (2016). http://doi.org/10.5092/jhna.2016.8.2.1.

Oostveldt, Bram van, and Stijn Bussels. "Introduction: The Sublime and Seventeenth-Century Netherlandish Art." *Journal of Historians of Netherlandish Art* 8, no. 2 (2016). http://doi.org/10.5092/jhna.2016.8.2.1.

Orenstein, Nadine. "Who Took the King of Sweden to Bed?" *Print Quarterly* 8, no. 1 (1991): 44–47.

Orlers, Jan. *Beschryvinge der Stad Leyden*. Leiden: Henrick Haestens, Jan Orlers en Jan Maire, 1614.

Ovid. *Heroides and Amores*. Translated by Grant Showerman. Cambridge: Harvard University Press, 1931.

———. *Metamorphoses*. Edited by W. R. Johnson. Translated by Stanley Lombardo. Indianapolis: Hackett, 2010.

Pagani, Catherine. *Eastern Magnificence and European Ingenuity: Clocks of Late Imperial China*. Ann Arbor: University of Michigan Press, 2001.

Panzanelli, Roberta, ed. "Compelling Presence." In Panzanelli, *Ephemeral Bodies*, 13–39.

———. *Ephemeral Bodies: Wax Sculpture and the Human Figure*. Los Angeles: Getty Research Institute, 2008.

———. "Introduction: The Body in Wax, the Body of Wax." In Panzanelli, *Ephemeral Bodies*, 1–12.

Paradin, Claude. *The Heroicall Devises of M. Claudius Paradin*. London: William Kearney, 1591.

Passe, Crispijn van de. *Dool-hof, Staende op de Roose-Gracht*. Amsterdam: voor Philips Lingelbach, 1660.

———. *Hortus Floridus: The Four Books of Spring, Summer, Autumn, and Winter Flowers*. Translated by Spencer Savage, with an introduction by Eleanor Sinclair Rohde. London: Minerva, 1974.

———. *Den Nieuwen en Vermaeckelijcken Dool-hof, Staende op de Roose-Gracht*. Amsterdam: voor David Lingelbach, 1648.

Perlove, Shelley, and Larry A. Silver. *Rembrandt's Faith: Church and Temple in the Dutch Golden Age*. University Park: Penn State University Press, 2009.

Perrault, Charles. *Labyrinthe de Versailles*. Amsterdam: N. Visscher, 1682.

Perrière, Guillaume de la. *Theatre des Bons Engins*. Paris: Denis Janot, 1544.

Pers, Dirck Pietersz. *Bacchus Wonder-Wercken*. Amsterdam, 1628.

Petter, Nicolaes. *Klare Onderrichtinge der Voortreffelijke Worstel-konst*. Amsterdam: Johannes Janssonius van Waesberge, 1674.

Pilbeam, Pamela M. *Madame Tussaud and the History of Waxworks*. London: Hambledon and London, 2002.

Plato. *Euthyphro*. In *The Dialogues of Plato*, 1:189–250. Translated by B. Jowett. Oxford: Clarendon, 1871.

———. *Meno*. Translated by W. K. C. Guthrie. In *The Collected Dialogues of Plato*, edited by Edith Hamilton and Huntington Cairns, 353–84. New York: Pantheon, 1961.

———. *Symposium*. Edited by Robin Waterfield. Oxford World's Classics. Oxford: Oxford University Press, 1998.

Pliny the Elder. *The Natural History*. Edited and translated by John Bostock. Vol. 6. London: Taylor and Francis, 1855.

Plutarch. *The Parallel Lives*. Edited by Bernadotte Perrin. Vol. 1. Cambridge: Harvard University Press, 1914.

Polybius. *The Rise of the Roman Empire*. Translated by Ian Scott-Kilvert. London: Penguin, 1979.

Porras, Stephanie. *Pieter Bruegel's Historical Imagination*. University Park: Penn State University Press, 2017.

Porter, Roy, ed. *Rewriting the Self: Histories from the Middle Ages to the Present*. London: Routledge, 1997.

Postma, Johannes. *The Dutch in the Atlantic Slave Trade, 1600–1815*. Cambridge: Cambridge University Press, 1990.

Prest, John M. *The Garden of Eden: The Botanic Garden and the Re-creation of Paradise*. New Haven: Yale University Press, 1981.

Price, Derek J. de Solla. "Automata and the Origins of Mechanism and Mechanistic Philosophy." *Technology and Culture* 5, no. 1 (1964): 9–23.

Principe, Lawrence. *The Secrets of Alchemy*. Chicago: University of Chicago Press, 2013.

Proksch, Nicholas D. "Luther's Eschatology and the Turks." Paper presented at the Bethany Lutheran Theological Seminary, International Congress on Medieval Studies, Kalamazoo, MI, 2010.

Quarles, Francis. *Emblemes* London: G. M. [George Miller?], 1635.

———. *Emblems, Divine and Moral*. London: J. and E. Nutt, 1634.

Rabelais, François. *The Histories of Gargantua and Pantagruel*. Translated by J. M. Cohen. London: Penguin, 1955.

Ray, John. *Travels Through the Low-Countries, Germany, Italy and France*. London: J. Walthoe, 1738.

Reneman, Hanny. "Het 'Oude Doolhof' te Amsterdam." PhD diss., Leiden University, 1975.

Ricci, Giovanni. "Masks of Power: Funeral Effigies in Early Modern Europe." In Daninos, *Waxing Eloquent*, 61–67.

Riskin, Jessica, ed. *Genesis Redux: Essays in the History and Philosophy of Artificial Life*. Chicago: University of Chicago Press, 2007.

———. "Introduction: The Sistine Gap." In Riskin, *Genesis Redux*, 1–27.

———. *The Restless Clock: A History of the Centuries-Long Argument over What Makes Living Things Tick*. Chicago: University of Chicago Press, 2016.

Roever, N. de. "Het Nieuwe Doolhof 'in de Oranje Pot' te Amsterdam." *Oud Holland* 6 (1888): 103–12.

———. *Uit Onze Oude Amstelstad*. Vol. 2. Amsterdam: Van Looy, 1890.

Roodenburg, Herman. "*Beweeglijkheid* Embodied: On the Corporeal and Sensory Dimensions of a Famous Emotion Term." *Nederlands Kunsthistorisch Jaarboek* 60, no. 1 (2010): 307–19.

Rouck, Thomas de. *Nederlandtsche Herault, of, Adelik Toneel*. Amsterdam: Jacob Volckertsz, 1672.

San Juan, Rose Marie. *Rome: A City Out of Print*. Minneapolis: University of Minnesota Press, 2001.

Sawday, Jonathan. *Engines of the Imagination: Renaissance Culture and the Rise of the Machine*. London: Routledge, 2007.

———. "'Forms Such as Never Were in Nature:' The Renaissance Cyborg." In *At the Borders of the Human: Beasts, Bodies and Natural Philosophy in the Early Modern Period*, edited by Erica Fudge, Ruth Gilbert, and Susan Wiseman, 171–95. New York: Palgrave, 2002.

———. "Self and Selfhood in the Seventeenth Century." In Porter, *Rewriting the Self*, 29–48.

Schaefers, Fons, and Anne Mieke Backer. *Doolhoven en Labyrinten in Nederland: Gids*. Rotterdam: De Hef, 2007.

Schaffer, Simon. "Enlightened Automata." In *The Sciences in Enlightened Europe*, edited by William Clark, Jan Golinski, and Simon Schaffer, 126–65. Chicago: University of Chicago Press, 1999.

Schama, Simon. *The Embarrassment of Riches: An Interpretation of Dutch Culture in the Golden Age*. New York: Knopf, 1987.

Schenkeveld, Maria A. *Dutch Literature in the Age of Rembrandt*. Amsterdam: John Benjamins, 1991.

Schiefsky, Mark J. "Art and Nature in Ancient Mechanics." In Bensaude-Vincent and Newman, *The Artificial and the Natural*, 67–108.

Schlosser, Julius von. "History of Portraiture in Wax." In Panzanelli, *Ephemeral Bodies*, 171–314.

Schmidt, Benjamin. "The Past *in* a Foreign Country: Patriotic History and Republican Geography in the Netherlands, ca. 1600–1648." In *Narratives of Low Countries History and Culture: Reframing the Past*, edited by J. Fenoulhet and L. Gilbert, 18–25. London: UCL Press, 2016.

Scholar, Richard. *The Je-Ne-Sais-Quoi in Early Modern Europe: Encounters with a Certain Something*. Oxford: Oxford University Press, 2005.

Scholten, Frits. "A Beheaded Bust and a Fountain-Statue by Hendrick de Keyser." *Burlington Magazine* 137, no. 1113 (1995): 838–41.

———. "Quellinus's Burgomasters: A Portrait Gallery of Amsterdam Republicanism." *Simiolus* 32, no. 1 (2006): 87–125.

———. "The Sculpted Portrait in the Dutch Republic, 1600–1700." In *Heads on Shoulders: Portrait Busts in the Low Countries 1600–1800*, edited by Valerie Herremans, Jeffrey M. Muller, Frits Scholten, and Carolien van der Star, 40–51. Gent: Snoeck, 2008.

———. *Sumptuous Memories: Studies in Seventeenth-Century Dutch Tomb Sculpture*. Zwolle: Waanders, 2003.

Serlio, Sebastiano. *Generale Reglen der Architecturen*. Anvers: Pieter Coecke van Aelst, 1539.

Shakespeare, William. *The Taming of the Shrew*. Edited by Ann Thompson. 3rd ed. Cambridge: Cambridge University Press, 2017.

Shākir, Abū Jaʿfar Muḥammad ibn Mūsā ibn, Abū al-Qāsim Aḥmad ibn Mūsā ibn Shākir, and Al-Ḥasan ibn Mūsā ibn Shākir. *The Book of Ingenious Devices (Kitāb al-Ḥiy al) by the Banú (sons of) Músà bin Shákir*. Translated and annotated by Donald R. Hill. Dordrecht: Reidel, 1979.

Sherwood, Merriam. "Magic and Mechanics in Medieval Fiction." *Studies in Philology* 44, no. 4 (1947): 567–92.

Shields, Carol. *Larry's Party*. London: Harper Perennial, 2006.

Simmel, Georg. "The Stranger." In *The Sociology of Georg Simmel*, edited by Kurt H. Wolff, 402–6. Glencoe: Free Press, 1950.

Sipiora, Phillip. "Introduction: The Ancient Concept of *Kairos*." In Sipiora and Baumlin, *Rhetoric and 'Kairos,'* 1–22.

Sipiora, Phillip, and James Baumlin, eds. *Rhetoric and "Kairos": Essays in History, Theory and Praxis*. Albany: State University of New York Press, 2002.

Sluijter, Eric Jan. "Rembrandt's Portrayal of the Passions and Vondel's 'Staetveranderinge.'" *Nederlands Kunsthistorisch Jaarboek* 60, no. 1 (2010): 285–306.

Smith, C. U. M., Eugenio Frixione, Stanley Finger, and William Clower, eds. *Animal Spirit Doctrine and the Origins of Neurophysiology*. Oxford: Oxford University Press, 2012.

Smith, John E. "Time and Qualitative Time." In *Rhetoric and Kairos: Essays in History, Theory, and Praxis*, edited by Phillip Sipiora and James S. Baumlin, 46–57. Albany: State University of New York Press, 2002.

Smolderen, Luc. *La Statue du Duc d'Albe à Anvers par Jacques Jonghelinck*. Brussels: Academie Royale de Belgique, 1971.

Somos, Mark. *Secularization and the Leiden Circle*. Leiden: Brill, 2011.

Spierenburg, Pieter. *The Prison Experience: Disciplinary Institutions and Their Inmates in Early Modern Europe*. London: Rutgers University Press, 1991.

Spies, Marijke. "De Amsterdamse Doolhoven: Populair Cultureel Vermaak in de Zeventiende Eeuw." *Literatuur* 18 (2001): 70–78.

———. "Kunst en Wetenschappen op de Troon: Culturele Hoofdstad, 1578–1713." In *Geschiedenis van Amsterdam: Centrum van de Wereld, 1578–1650*, edited by Willem Frijhoff and Maarten Prak, 299–383. Amsterdam: Boom, 2004.

———. "Rhetoric and Civic Harmony in the Dutch Republic of the Late Sixteenth and Early Seventeenth Century." In *Rhetorica Movet: Studies in Historical and Modern Rhetoric in Honour of Heinrich F. Plett*, edited by P. L. Oesterreich, 57–72. Leiden: Brill, 1999.

———. *Rhetoric, Rhetoricians, and Poets: Studies in Renaissance Poetry and Poetics*. Edited by H. Duits and A. van Strien. Amsterdam: Amsterdam University Press, 1999.

Sprunger, Keith L. *Trumpets from the Tower: English Puritan Printing in the Netherlands, 1600–1640*. Leiden: Brill, 1993.

Stafford, Barbara Maria. *Artful Science: Enlightenment, Entertainment, and the Eclipse of Visual Education*. Cambridge: MIT Press, 1994.

Stafford, Barbara Maria, and Frances Terpak. *Devices of Wonder: From the World in a Box to Images on a Screen*. Los Angeles: Getty Research Institute, 2001.

Stewart, Alison. "Taverns in Nuremburg Prints at the Time of the Reformation." In *The World of the Tavern: Public Houses in Early Modern Europe*, edited by Beat Kümin and B. Ann Tlusty, 95–115. Aldershot: Ashgate, 2002.

Stieber, Nancy. "Marking Time and Space in the City: Kromhout's Decorations for the Investiture of Wilhelmina in Amsterdam." In *Festival Architecture*, edited by Sarah Bonnemaison and Christine Macy, 181–214. London: Routledge, 2008.

Stoichita, Victor. *The Pygmalion Effect: From Ovid to Hitchcock*. Chicago: University of Chicago Press, 2008.

———. *The Self-Aware Image: An Insight into Early Modern Meta-Painting*. Cambridge: Cambridge University Press, 1997.

Strauss, Linda. "Automata: A Study in the Interface of Science, Technology, and Popular Culture, 1730–1885." PhD diss., University of California, San Diego, 1987.

Strien, A. van, Kees van der Leer, and Ad Leerintveld. *Hofwijck: Het Gedicht en de Buitenplaats van Constantijn Huygens*. Zutphen: Walburg Pers, 2002.

Strien, C. D. van. *De Ontdekking van de Nederlanden: Britse en Franse Reizigers in Holland en Vlaanderen, 1750–1795*. Utrecht: Het Spectrum, 2001.

Stronks, Els. *Negotiating Differences: Word, Image and Religion in the Dutch Republic*. Leiden: Brill, 2011.

Stroom, Gerrold van der. *P. C. Hooft en de Ongeziene: Eerste "Fontein" van Amsterdam*. Amsterdam: Nodus Publikationen, 2017.

Sturm, Leonhard Christoph. *Leonhard Christoph Sturms Durch Einen Grossen Theil von Teutschland und den Niederlanden*. Augsburg: J. Wolffen, 1719.

Summer, David. "Pandora's Crown: On Mechanism in Western Art." In *Marvels, Monsters, and Miracles: Studies in the Medieval and Early Modern Imaginations*, edited by Timothy F. Jones and David A. Sprunger, 45–75. Kalamazoo: Medieval Institute, 2002.

Swan, Claudia. "Lost in Translation: Exoticism in Early Modern Holland." In *Art in Iran and Europe in the Seventeenth Century: Exchange and Reception*, edited by Axel Langer, 100–116. Zürich: Museum Rietberg, 2013.

Syson, Luke. "Polychrome and Its Discontents: A History." In *Like Life: Sculpture, Color, and the Body*, edited by Luke Syson, Sheena Wagstaff, Emerson Bowyer, and Brinda Kumar, 14–41. New York: Metropolitan Museum of Art, 2018.

Tchikine, Anatole. "'L'Anima del Giardino': Water, Gardens and Hydraulics in Sixteenth-Century Florence and Naples." In *Technology and the Garden*, edited by Michael G. Lee and Kenneth I. Helphand, 129–53. Washington, DC: Dumbarton Oaks, 2014.

Tillich, Paul. *Systematic Theology*. Vol. 3. Chicago: University of Chicago Press, 1976.

Tlusty, Ann B. *Bacchus and Civic Order*. Charlottesville: University of Virginia Press, 2001.

Trevor-Roper, H. R. *The Crisis of the Seventeenth Century: Religion, the Reformation, and Social Change*. New York: Harper and Row, 1968.

Truitt, Elly Rachel. *Medieval Robots: Mechanism, Magic, Nature, and Art*. Philadelphia: University of Pennsylvania Press, 2015.

Tschumi, Bernard. *Architecture and Disjunction*. Cambridge: MIT Press, 1994.

Tybjerg, Karin. "Wonder-Making and Philosophical Wonder in Hero of Alexandria." *Studies in History and Philosophy of Science* 34 (2003): 443–66.

Uffenbach, Zacharias Konrad von. *Merkwürdige Reisen durch Niedersachsen, Holland und Engelland*. With Johann Georg Schelhorn. 3 vols. Frankfurt: [Christian Ulrich Wagner], 1753–54.

Unger, Richard. *History of Brewing in Holland*. Leiden: Brill, 2001.

Vanhaelen, Angela. "Calvinism and Visual Culture: The Art of Evasion." In *Cultures of Calvinism in Early Modern Europe*, edited by Graeme Murdock and Crawford Gribben, 138–58. Oxford: Oxford University Press, 2019.

———. *Comic Print and Theatre in Early Modern Amsterdam*. Aldershot: Ashgate, 2003.

———. "Local Sites, Foreign Sights: A Sailor's Sketchbook of Human and Animal Curiosities in Early Modern Amsterdam." *RES: Journal of Anthropology and Aesthetics* 45 (Spring 2004): 256–72.

———. "Mapping Angels in Early Modern Amsterdam." *Journal of Early Modern History* 23, nos. 2–3 (2019): 227–56.

———. "Public and Private Intercourse in Dutch Genre Scenes: Soldiers and Enigmatic Women / Painters and Enigmatic Paintings." In *Forms of Association: Making Publics in Early Modern Europe*, edited by Paul Yachnin and Marlene Eberhart, 115–32. Amherst: University of Massachusetts Press, 2015.

———. "Spiritual Culture." In *The Cambridge Companion to the Dutch Golden Age*, edited by Helmer Helmers and Geert Janssen, 225–46. Cambridge: Cambridge University Press, 2018.

———. *The Wake of Iconoclasm: Painting the Church in the Dutch Republic*. University Park: Penn State University Press, 2012.

Vanhaelen, Angela, and Bronwen Wilson. "The Erotics of Looking: Materiality, Solicitation and Netherlandish Visual Culture." *Art History* 35, no. 5 (2012): 875–85.

Vasari, Giorgio. *The Lives of the Artists*. Edited and translated by Julia Conaway Bondanella and Peter Bondanella. Oxford: Oxford University Press, 1991.

———. *Vasari on Technique*. Edited and introduction by G. Baldwin Brown. Translated by Louisa S. Maclehose. New York: Dover, 2012.

Velde, Carl van de. *Classical Mythology in the Netherlands in the Age of Renaissance and Baroque*. Leuven: Peeters, 2009.

Velden, Hugo van der. *The Donor's Image: Gerard Loyet and the Votive Portraits of Charles the Bold*. Turnhout: Brepols, 2000.

———. "Medici Votive Images and the Scope and Limits of Likeness." In *The Image of the Individual: Portraits in the Renaissance*, edited by Nicholas Mann and Luke Syson, 126–37. London: British Museum, 1998.

Veldman, Ilja M. *Crispijn de Passe and His Progeny*. Rotterdam: Sound & Vision Interactive, 2001.

Verbeek, J. "Naschrift bij de Vorstenportretten van de Passes." *Bulletin van het Rijksmuseum* 12, no. 1 (1964): 28–30.

———. "Zeldzame Vorstenportretten door de De Passes in het Rijksprentenkabinet." *Bulletin van het Rijksmuseum* 10, no. 2/3 (1962): 76–84.

Verklaringe van Treffelijcke Konstighe Wercken, en haer Beweginge, Door Oorlogie Werck Gedreven. Amsterdam: Cryspijn van der Pas, 1648.

Verklaringe van Verscheyden Kunst-Rijcke Wercken en haer Beweginghe. Amsterdam: Tymon Houthaak, 1650.

Verklaringe van Verscheyden Kunst-Rijcke Wercken en haer Beweginghe. Amsterdam: Dan. Davidsz. Mansvelt, 1657.

Verklaringe van Verscheyden Kunst-Rijcke Wercken en haer Beweginghe: Door Oorlogie Werck Gedreven. Amsterdam: Gort, 1668.

Verklaringe van Verscheyden Kunst-Rijcke Wercken en haer Beweginghe: Door Oorlogie-Werck Ghedreven. Amsterdam: Tymen Houthaeck, 1648.

Vidler, Anthony. "History of the Folly." In *The Garden of Forking Paths*, edited by Heike Munder, 45–51. Zürich: Migros Museum für Gegenwartskunst, 2011.

Vitruvius. *The Ten Books on Architecture*. Translated by Morris Morgan. New York: Dover, 1960.

Vitry, Paul. "Le sculpteur Michel Bourdin et son Effigie de Henri IV." *La Chronique des Arts et de la Curiosité: Supplément à la "Gazette des beaux-arts"* (1898): 290–92.

Voet, Leon. *The Golden Compasses: The History of the House of Plantin-Moretus*. 2 vols. London: Routledge & Kegan Paul, 1969–72.

Vos, Jan. *D'Oprechte Aenwijser*. Amsterdam: Alexander Iansz. Lintman, 1674.

Voskuhl, Adelheid. *Androids in the Enlightenment: Mechanics, Artisans, and Cultures of the Self*. Chicago: University of Chicago Press, 2015.

Vries, A. D. de. "De Plaatsnijder en Landmeter Cornelis Florissen van Berckenrode en Zijn Familelje." *Oud Holland* 1, no. 1 (1883): 37–43.

Vries, Hans Vredeman de. *Hortorum Viridariorumque Elegantes et Multiplices Formae*. Anvers: Theodor Galle, 1583.

Vriese, J. "Het Nieuwe Doolhof." *Ons Amsterdam* 16, no. 12 (1964): 354–57.

———. "Nog Enkele Amsterdamse Doolhoven." *Ons Amsterdam* 17, no. 3 (1965): 82–86.

———. "Het Oude Doolhof." *Ons Amsterdam* 16, no. 9 (1964): 268–72.

Vrolijk, Arnoud, and Richard Leeuwen. *Arabic Studies in the Netherlands*. Translated by Alistair Hamilton. Leiden: Brill, 2014.

Vrolijk, Arnoud, and Kasper van Ommen, eds. *"All My Books in Foreign Tongues": Scaliger's Oriental Legacy in Leiden, 1609–2009*. Leiden: Leiden University Library, 2009.

Vroom, Wim. *Het Wonderlid van Jan de Witt en Andere Vaderlandse Relieken*. Nijmegen: SUN, 1997.

Waal, J. C. "Oude Doolhoven te Amsterdam." *Buiten* 17, no. 48 (1923): 568–72.

Wachtendorp, Casparus. *Oude Hollandsche Geschiedenissen ofte: Corte Rym-Kronyck; Verdeelt in XIIII Boecken*. The Hague: Tongerloo, Verhoeven and de Putter, 1648.

Wakefield, Colin. "Arabic Manuscripts in the Bodleian Library: The Seventeenth-Century Collections." In *The 'Arabick' Interest of the Natural*

Philosophers, edited by G. A. Russell, 128–46. Leiden: Brill, 1994.

Warburg, Aby. "The Art of Portraiture and the Florentine Bourgeoisie." In *The Renewal of Pagan Antiquity: Contributions to the Cultural History of the European Renaissance*, 185–221. Los Angeles: Getty Research Institute, 1999.

Wassenaer, Nicolaes van. *'T Vyftiende Deel of 't Vervolgh van het Historisch Verhael Aller Gedenckwaerdiger Geschiedeniss die in Europa. . . .* Amsterdam: Jan Jansz, 1629.

Webster, Charles. *The Great Instauration: Science, Medicine, and Reform, 1626–1660*. London: Duckworth, 1975.

Webster, T. B. L. "The Myth of Ariadne from Homer to Catullus." *Greece and Rome* 13, no. 1 (1966): 22–31.

Weemans, Michel. "Herri met de Bles's *Way to Calvary* (1540): A Silenic Landscape." *Art History* 32, no. 2 (2009): 307–31.

Weiss, Allen S. *Mirrors of Infinity: The French Formal Garden and Seventeenth-Century Metaphysics*. New York: Princeton Architectural Press, 1995.

Werrett, Simon. "Wonders Never Cease: Descartes's *Météores* and the Rainbow Fountain." *British Journal for the History of Science* 34, no. 2 (2001): 129–47.

Weststeijn, Thijs. *Art and Antiquity in the Netherlands and Britain: The Vernacular Arcadia of Franciscus Junius (1591–1677)*. Leiden: Brill, 2015.

———. "Between Mind and Body: Painting the Inner Movements According to Samuel van Hoogstraten and Franciscus Junius." *Nederlands Kunsthistorisch Jaarboek* 60, no. 1 (2010): 263–84.

———. "'Painting's Enchanting Poison:' Artistic Efficacy and the Transfer of Spirits." In *Spirits Unseen: The Representation of Subtle Bodies in Early Modern European Culture*, edited by Christine Göttler and Wolfgang Neuber, 141–77. Leiden: Brill, 2007.

———. *The Visible World: Samuel van Hoogstraten's Art Theory and the Legitimation of Painting in the Dutch Golden Age*. Amsterdam: Amsterdam University Press, 2008.

Wiggers, Hinke. "De Stad Amsterdam en haar Vroegste Beelden Collectie." In Jonker, *In Beeld Gebracht*, 60–75.

Wijnman, H. F. "De Hebraïcus Jan Theunisz. Barbarossius alias Johannes Antonides als Lector in het Arabisch aan de Leidse Universiteit (1612–1613): Een Hoofdstuk Amsterdamse Geleerdengeschiedenis." *Studia Rosenthaliania* 2 (1968): 1–41, 149–77.

———. "Jan Theunisz." *Jaarboek Amstelodamum* 25 (1928): 29–123.

Wikam, J. J. "Jacobus Golius (1596–1667) en zijn Handscriften." *Oosters Genootschap in Nederland* 10 (1980): 49–74.

Wilkins, John. *Mathematicall Magick or, the Wonders That May Be Performed by Mechanicall Geometry*. London: Gellibrand, 1648.

Wilson, Bronwen. "The Renaissance Portrait: From Resemblance to Representation." In *The Renaissance World*, edited by John Martin, 452–80. New York: Routledge, 2007.

Wind, Edgar. *Pagan Mysteries in the Renaissance*. New Haven: Yale University Press, 1958.

Winkler, John, and Froma Zeitlin, eds. *Nothing to Do with Dionysos?: Athenian Drama in Its Social Context*. Princeton: Princeton University Press, 1990.

Witkamp, P. H. *Amsterdam in Schetsen*. Amsterdam: G. W. Tielkemeijer, 1869.

Witt, David de. *Jan van Noordt: Painter of History and Portraits in Amsterdam*. Montreal: McGill-Queen's University Press, 2007.

Wood, Gaby. *Edison's Eve: A Magical History of the Quest for Mechanical Life*. New York: Knopf, 2002.

Woodward, Jennifer. *The Theatre of Death: The Ritual Management of Royal Funerals in Renaissance England, 1570–1625*. Woodbridge: Boydell, 1997.

Wright, Craig. *The Maze and Warrior: Symbols in Architecture, Theology, and Music*. Cambridge: Harvard University Press, 2001.

Young, John T. *Faith, Medical Alchemy and Natural Philosophy: Johann Moriaen, Reformed Intelligencer and the Hartlib Circle*. Aldershot: Ashgate, 1998.

Zesen, Philipp von. *Beschreibung der Stadt Amsterdam*. Amsterdam: Nosch, 1664.

Zielinski, Siegfried, and Peter Weibel, eds. *Allah's Automata: Artifacts of Arabic-Islamic Renaissance (800–1200)*. Ostfildern: Hatje/Cantz, 2016.

INDEX

Italicized page references indicate illustrations. Endnotes are referenced with "n" followed by the endnote number.